From Vector Spaces to Function Spaces

Yutaka Yamamoto

Kyoto University
Kyoto, Japan

From Vector Spaces to Function Spaces

Introduction to Functional Analysis with Applications

Society for Industrial and Applied Mathematics
Philadelphia

Library of Congress Cataloging-in-Publication Data

Yamamoto, Yutaka, 1950-
  [Shisutemu to seigyo no sugaku. English]
  From vector spaces to function spaces : introduction to functional analysis with applications / Yutaka Yamamoto.
      p. cm. -- (Other titles in applied mathematics)
  Includes bibliographical references and index.
  ISBN 978-1-611972-30-6 (alk. paper)
1. Functional analysis. 2. Engineering mathematics. I. Title.
  TA347.F86Y3613 2012
  515'.7--dc23
                              2012010732

 is a registered trademark.

Contents

Preface

This book intends to give an accessible account of applied mathematics, mainly of analysis subjects, with emphasis on functional analysis. The intended readers are senior or graduate students who wish to study analytical methods in science and engineering and researchers who are interested in functional analytic methods.

Needless to say, scientists and engineers can benefit from mathematics. This is not confined to a mere means of computational aid, and indeed the benefit can be greater and far-reaching if one becomes more familiar with advanced topics such as function spaces, operators, and generalized functions. This book aims at giving an accessible account of elementary real analysis, from normed spaces to Hilbert and Banach spaces, with some extended treatment of distribution theory, Fourier and Laplace analyses, and Hardy spaces, accompanied by some applications to linear systems and control theory. In short, it is a modernized version of what has been taught as applied analysis in science and engineering schools.

To this end, a more conceptual understanding is required. In fact, conceptual understanding is not only indispensable but also a great advantage even in manipulating computational tools. Unfortunately, it is not always accomplished, and indeed often left aside. Mathematics is often learned by many people as a collection of mere techniques and swallowed as very formal procedures.

This is deplorable, but from my own experience of teaching, its cure seems quite difficult. For students and novices, definitions are often difficult to understand, and mathematical structures are hard to penetrate, let alone the background motivation as to how and why they are formulated and studied.

This book has a dual purpose: one is to provide young students with an accessible account of a conceptual understanding of fundamental tools in applied mathematics. The other is to give those who already have some exposure to applied mathematics, but wish to acquire a more unified and streamlined comprehension of this subject, a deeper understanding through background motivations.

To accomplish this, I have attempted to

- elaborate upon the underlying motivation of the concepts that are being discussed and

- describe how one can get an idea for a proof and how one should formalize the proof.

I emphasized more verbal, often informal, explanations rather than streams of logically complete yet rather formal and detached arguments that are often difficult to follow for nonexperts.

The topics that are dealt with here are quite standard and include fundamental notions of vector spaces, normed, Banach, and Hilbert spaces, and the operators acting on them. They are in one way or another related to linearity, and understanding linear structures forms a core of the treatment of this book. I have tried to give a unified approach to real analysis, such as Fourier analysis and the Laplace transforms. To this end, distribution theory gives an optimal platform. The second half of this book thus starts with distribution theory. With a rigorous treatment of this theory, the reader will see that various fundamental results in Fourier and Laplace analyses can be understood in a unified way. Chapter 9 is devoted to a treatment of Hardy spaces. This is followed by a treatment of a remarkably successful application in modern control theory in Chapter 10.

Let us give a more detailed overview of the contents of the book. As an introduction, we start with some basics in vector spaces in Chapter 1. This chapter intends to give a conceptual overview and review of vector spaces. This topic is often, quite unfortunately, a hidden stumbling point for students lacking an in-depth understanding of what linearity is all about. I have tried to illuminate the conceptual sides of the notion of vector spaces in this chapter. Sometimes, I have attempted to show the idea of a proof first and then show that a complete proof is a "realization" of making such an idea logically more complete. I have also chosen to give detailed treatments of dual and quotient spaces in this chapter. They are often either very lightly touched on or neglected completely in standard courses in applied mathematics. I expect that the reader will become more accustomed to standard mathematical thinking as a result of this chapter.

From Chapter 2 on, we proceed to more advanced treatments of infinite-dimensional spaces. Among them, normed linear spaces are most fundamental and allow rather direct generalizations of finite-dimensional spaces. A new element here is the notion of norms, which introduces the concept of topology. Topology plays a central role in studying infinite-dimensional spaces and linear maps acting on them. Normed spaces give the first step toward such studies.

A problem is that limits cannot, generally, be taken freely for those sequences that may appear to converge (i.e., so-called Cauchy sequences). The crux of analysis lies in limiting processes, and to take full advantage of them, the space in question has to be "closed" under such operations. In other words, the space must be *complete*. Complete normed linear spaces are called Banach spaces, and there are many interesting and powerful theorems derived for them. If, further, the norm is derived from an inner product, the space is called a Hilbert space. Hilbert spaces possess a number of important properties due to the very nature of inner products—for example, the notion of orthogonality. The Riesz representation theorem for continuous linear functionals, as well as its outcome of the orthogonal projection theorem, is a typical consequence of an inner product structure and completeness. Hilbert space appears very frequently in measuring signals due to its affinity with such concepts as energy, and hence in many optimization problems in science and engineering applications. The problem of best approximation is naturally studied through the projection theorem in the framework of Hilbert space. This is also a topic of Chapter 3.

Discussing properties of spaces on their own will give only half the story. What is equally important is their interrelationship, and this exhibits itself through linear operators. In this connection, dual spaces play crucial roles in studying Banach and Hilbert spaces. We give in Chapter 5 a basic treatment of them and prove the spectral resolution theorem for compact self-adjoint operators—what is known as the Hilbert–Schmidt expansion theorem.

We turn our attention to Schwartz distributions in Chapter 6. This theory makes transparent the treatments of many problems in analysis such as differential equations, Fourier analysis (Chapter 7), Laplace transforms (Chapter 8), and Poisson integrals (Chapter 9), and it is highly valuable in many areas of applied mathematics, both technically and conceptually. In spite of this fact, this theory is often very informally treated in introductory books and thereby hardly appreciated by engineers. I have strived to explain why it is important and how some more rigorous treatments are necessary, attempting an easily accessible account for this theory while not sacrificing mathematical rigor too much.

The usefulness of distributions hinges largely on the notion of the delta function (distribution). This is the unity element with respect to convolution, and this is why it appears so frequently in many situations of applied mathematics. Many basic results in applied mathematics are indeed understood from this viewpoint. For example, a Fourier series or Poisson's integral formula is the convolution of an objective function with the Dirichlet or the Poisson kernel, and its convergence to such a target function is a result of the fact that the respective kernel converges to the delta distribution. We take this as a leading principle of the second half of this book, and I attempted to clarify the structure of this line of ideas in the treatments of such convergence results in Chapter 6, and subsequent Chapters 7 and 8 dealing with Fourier and Laplace transforms.

Chapter 9 gives a basic treatment of Hardy spaces, which in turn played a fundamental role in modern control theory. So-called H^∞ control theory is what we are concerned with. Of particular interest is generalized interpolation theory given here, which also plays a fundamental role in this new control theory. We will prove Nehari's theorem as well as Sarason's theorem, along with the applications to the Nevanlinna–Pick interpolation and the Carathéodory–Fejér theorem. We will also discuss the relationship with boundary values and the inner-outer factorization theorem. I have tried to give an easy entry point to this theory.

Chapter 10 is devoted to basic linear control system theory. Starting with an inverted pendulum example, we will see such basic concepts as linear system models, the concept of feedback, controllability, and observability, a realization problem, an input/output framework, and transfer functions, leading to the simplest case of H^∞ control theory. We will see solutions via the Nevanlinna–Pick interpolation, Nehari's theorem, and Sarason's theorem, applying the results of Chapter 9. Fourier analysis (Chapter 7) and the Laplace transforms (Chapter 8) also play key roles here. The reader will no doubt see part of a beautiful application of Hardy space theory to control and systems. It is hoped that this chapter can serve as a concise introduction to those who are not necessarily familiar with the subject.

It is always a difficult question how much preliminary knowledge one should assume and how self-contained the book should be. I have made the following assumptions:

- As prerequisite, I assumed that the reader has taken an elementary course in linear algebra and basic calculus. Roughly speaking, I assumed the reader is at the junior or a higher level in science and engineering schools.

- I did not assume much advanced knowledge beyond the level above. The theory of integration (Lebesgue integral) is desirable, but I chose not to rely on it.

- However, if applied very rigorously, the above principles can lead to a logical difficulty. For example, Fubini's theorem in Lebesgue integration theory, various theorems in general topology, etc., can be an obstacle for self-contained treatments.

I tried to give precise references in such cases and not overload the reader with such concepts.

- Some fundamental notions in sets and topology are explained in Appendix A. I tried to make the exposition as elementary and intuitive as to be beneficial to students who are not well versed in such notions. Some advanced background material, e.g., the Hahn–Banach theorem, is also given here.

This book is based on the Japanese version published by the Asakura Publishing Company, Ltd., in 1998. The present English version differs from the predecessor in many respects. Particularly, it now contains Chapter 10 for application to system and control theory, which was not present in the Japanese version. I have also made several additions, but it took much longer than expected to complete this version. Part of the reason lies in its dual purpose—to make the book accessible to those who first study the above subjects and simultaneously worthwhile for reference purposes on advanced topics. A good balance was not easy to find, but I hope that the reader finds the book helpful in both respects.

It is a pleasure to acknowledge the precious help I have received from many colleagues and friends, to whom I am so much indebted, not only during the course of the preparation of this book but also over the course of a long-range friendship from which I have learned so much. Particularly, Jan Willems read through the whole manuscript and gave extensive and constructive comments. I am also most grateful for his friendship and what I learned from numerous discussions with him. Likewise, I have greatly benefited from the comments and corrections made by Thanos Antoulas, Brian Anderson, Bruce Francis, Tryphon Georgiou, Nobuyuki Higashimori, Pramod Khargonekar, Hitay Özbay, Eduardo Sontag, and Mathukumalli Vidyasagar. I would also like to acknowledge the help of Masaaki Nagahara and Naoki Hayashi for polishing some proofs and also helping me in preparing some figures. I wish to acknowledge the great help I received from the editors at SIAM in publishing the present book.

I would like to conclude this preface by thanking my family, particularly my wife Mamiko for her support in every respect in the past 30 years. Without her help it would not have been possible to complete this work.

Yutaka Yamamoto

Kyoto
February, 2012

Glossary of Notation

A^*	the adjoint operator of a (continuous) linear operator A between Hilbert spaces
A'	the dual operator of a linear operator A
$\mathbb{C}$	the set of complex numbers
$\mathbb{C}_+$	the set of complex numbers with positive real parts
$\mathbb{C}_-$	the set of complex numbers with negative real parts
$C[a,b]$	the space of all continuous functions on the interval $[a,b]$
$C^m[a,b]$	the space of m times continuously differentiable functions on the interval $[a,b]$
$C^\infty[a,b]$	the space of infinitely differentiable functions on the interval $[a,b]$
$C(-\infty,\infty)$	the set of all continuous functions on $(-\infty,\infty)$
$\mathbb{D}$	the open unit disc $\{z \in \mathbb{C} : |z| < 1\}$ in $\mathbb{C}$
δ	Dirac's delta distribution
D_n	the Dirichlet kernel
$\mathcal{D}[a,b]$	the space of infinitely differentiable functions on $(-\infty,\infty)$ with support in $[a,b]$
$\mathcal{D}(\mathbb{R})$	the space of infinitely differentiable functions on $(-\infty,\infty)$ with compact support; may be denoted simply by $\mathcal{D}$
$\mathcal{D}'$	the space of Schwartz distributions on $\mathbb{R}$
$\mathcal{D}'_+(\mathbb{R})$	the space of distributions with support bounded on the left; may also be denoted as $\mathcal{D}'_+$
$\mathcal{F}[f]$	the Fourier transform of f; may also be denoted as $\hat{f}$
$\overline{\mathcal{F}}[f]$	the conjugate Fourier transform of f; also called the inverse Fourier transform
H^p	the Hardy p-space; see Chapter 9
$H(\psi)$	$H^2 \ominus \psi H^2$, where ψ is an inner function
i	the imaginary unit
$\mathbb{K}$	either $\mathbb{R}$ or $\mathbb{C}$, depending on the context
K_n	the Fejér kernel
ℓ^p	the space of sequences whose p-powers are summable $(p < \infty)$
ℓ^∞	the space of bounded sequences
L^p	the space of functions whose p-powers are Lebesgue integrable $(p < \infty)$
L^∞	the space of essentially bounded functions
$\mathcal{L}[f]$	the Laplace transform of f; may also be denoted as $\hat{f}$
$\|f\|_\infty$	H^∞ or L^∞ norm of f

$\|f\|_2$	2-norm of f
$\|f\|_p$	p-norm of f
$\mathcal{L}(X,Y)$	the space of all continuous linear mappings from X to Y; may also be denoted $\mathcal{L}(X)$ when $Y = X$
M^c	the complement of a subset M
$\overset{\circ}{M}$	the interior of a subset M of a topological space
$\overline{M}$	the closure of a subset M of a topological space
$M^\perp$	the orthogonal complement of M
$\langle T, g \rangle$	the value of a linear functional T at g; may also be written as $\langle g, T \rangle$ in the reverse order. This is linear in each variable (i.e., bilinear)
$(f,g)_X$	the inner product of f and g in a Hilbert space X; the subscript X may be omitted when no confusion arises. Note that when the scalar field is the complexes $\mathbb{C}$, this is *not* linear in g $((f,\lambda g) = \overline{\lambda}(f,g))$, and hence is not bilinear
P_r	the Poisson kernel
$\psi^{\sim}(z)$	$\overline{\psi(\overline{z}^{-1})}$
$\mathbb{Q}$	the field of rational numbers
$\mathcal{S}$	the space of rapidly decreasing functions on $\mathbb{R}$
$\mathcal{S}'$	the dual space of $\mathcal{S}$
$T * S$	the convolution of T and S
$\mathrm{Re}\,z, \mathrm{Im}\,z$	the real and imaginary parts of z, respectively
$\mathbb{R}$	the set of real numbers
$\mathbb{R}_+$	the set of nonnegative reals
$\mathbb{R}_-$	the set of nonpositive reals
$\mathbb{T}$	the unit circle in $\mathbb{C}$ $\{z \in \mathbb{C} : |z| = 1\}$
X/M	the quotient space of X modulo M
X'	the topological dual of space X
$X \times Y$	the direct (Cartesian) product of X and Y
$\overline{z}$	the complex conjugate of a complex number z

Chapter 1

Vector Spaces Revisited

The notion of vector spaces is fundamental to any branch of mathematics. We start by reviewing some fundamental concepts needed for discussing more advanced topics. In particular, we will discuss in detail the concepts of quotient spaces and duality that will be fundamental in the subsequent developments.

1.1 Finite-Dimensional Vector Spaces

1.1.1 What Is a Vector Space?

What kind of image does the word "vector" evoke in you? Some may recall a directed arrow, whereas others may think of an ordered tuple of numbers as in Figure 1.1.

Figure 1.1. *Directed arrow and ordered n-tuple*

These are certainly bona fide vectors, with good reason. But some may find it bothersome to further substantiate why they are called vectors.

In high-school or freshman algebra or geometry, have we encountered a very rigorous definition of *why such and such an object is a vector?* One usually does not say that a particular object, for example, a directed arrow, is a vector for such and such reasons. This is quite different from saying that 1 is an integer. The difference between the two may be that the integers refer to a set, while the name *vector* does not. The latter makes sense only after one specifies not only the set itself but also the operational rules it obeys. Hence it may appear natural to call 1 an integer, while the same is not quite true for vectors.[1]

A binary operation, such as addition of vectors, can be defined for two elements. This requires us to specify the universe, i.e., the set, where this occurs. Only after this can we define operations and discuss whether a certain object (e.g., an arrow) is or is not a vector.

Although this may appear to be quite a roundabout way of defining things, we should recognize that a vector is a concept that can be well defined only after we specify where it belongs *and how it behaves* with respect to the defined operations. In short, a seemingly self-contradictory statement that *a vector is an element of a vector space* is the right way of defining it. It is entirely up to you whether you wish to regard an arrow as a vector or not, but only when you consider it as an element of the space consisting of all arrow vectors accompanied by natural vector operations they should obey can it be rightly regarded as a vector.

This may sound confusing. But such a situation is entirely standard in modern mathematics, especially in algebra. Since we deal more with operations and structures induced from them, the peculiarity of an individual element becomes, inevitably, less significant than the totality that constitutes the entire structure.

Let us give a less trivial example. Consider the following second order ordinary differential equation with constant coefficients:

$$\left(\frac{d^2}{dt^2} + 1 \right) x(t) = 0. \tag{1.1}$$

The reader may be familiar with the fact that a general solution can be written as

$$x(t) = C_1 \cos t + C_2 \sin t \tag{1.2}$$

with arbitrary constants C_1, C_2. Let

$$X := \{C_1 \cos t + C_2 \sin t : C_1, C_2 \in \mathbb{R}\}$$

be the set of all solutions that can be obtained by taking all such C_1, C_2. (Here ":=" is the notation meaning that the right-hand side "defines" the left-hand side.) We can then naturally induce the operations of addition and scalar multiplication on the set X, which is a subset of all real-valued functions defined on the real line $(-\infty, \infty)$ as follows:

$$(C_1 \cos t + C_2 \sin t) + (D_1 \cos t + D_2 \sin t) := (C_1 + D_1) \cos t + (C_2 + D_2) \sin t,$$
$$\alpha(C_1 \cos t + C_2 \sin t) := (\alpha C_1) \cos t + (\alpha C_2) \sin t, \ \alpha \in \mathbb{R}.$$

The point is that the sum and scalar multiplication of elements in X again belong to X with respect to such natural function operations. We can then ensure that the space X, along with the operations as defined above, can be called a vector space; that is, it satisfies the axioms of vector spaces. Then, while a solution (1.2) is certainly a function, it can also be referred to as a *vector*, being an element of X. We should keep in mind, however, that it is regarded as a vector in relation to the vector space structure of X defined above. In other words, when we call

$$C_1 \cos t + C_2 \sin t$$

a vector, we should always be conscious of a vector space structure underlying it.

---

[1]This is, however, largely a matter of taste, since, strictly speaking, the notion of integers also refers to the combined notion of a set and the associated rules of basic operations such as addition and multiplication. The underlying psychology here is that the set of integers is familiar to us, while the notion of vectors is more abstract, and we have to be more cautious of such constructs.

1.1.2 Axioms of Vector Spaces

Summarizing the discussions above,

- we may say a vector is an element of a vector space; and

- we may call a set (along with certain structures) a vector space if it satisfies the axioms of vector spaces.

The objective of this section is to clarify the axioms of vector spaces.

First, why in mathematics do we always start with setting up axioms and studying their outcomes? Why can we not content ourselves with individual (and more concrete) arrows and n-tuple vectors? This question may be placed in perspective if we consider more elementary examples.

Mathematics originated from counting: we count the quantities of apples or oranges by 1, 2, 3, etc. However, this leads to an interesting dilemma. How would you answer the following seemingly innocent question:

Suppose we have 3 apples and 5 oranges. How many are there in total?

The "natural" answer to this is to say that there are 8 "objects" or "units of fruit." But this answer makes sense only when we accept the assumption that we count one apple and one orange by the same number 1. Needless to say, an apple and an orange need not be of the same value as "objects." This dilemma would be even clearer if the 5 oranges were replaced by 5 bunches of grapes. Thus, without the *mathematical identification* to count an apple and an orange equally by the number 1, this question *literally* does not make any sense. We are so used to such an identification from elementary school algebra that we are hardly conscious of this identification. If you ask young children the same question, there may be some who would have a hard time absorbing such an identification. Some would be puzzled, speculating on which is more valuable—an apple or an orange.

On the other hand, *if* we *always* make a distinction in counting apples and oranges, or any other objects we encounter in our daily lives, it would be just impossibly inconvenient. Certainly, we do not want to invent "mathematics for apples," "mathematics for oranges," "mathematics for bananas," and so on. They will be entirely equivalent, but distinguishing them with separate characters will make mathematics intolerably redundant. The abstraction of counting objects without really distinguishing the irrelevant individual characters (which may depend on each specific context) will provide us with a universal, although abstract, system that is applicable to any countable objects.[2] Thus even in such an elementary notion of counting, one is required to make the abstraction of identification of different objects.

The universalization and abstraction of counting have yielded the development of the systems of numbers:

$$\text{Natural numbers} \rightarrow \text{Integers} \rightarrow \text{Rational numbers} \rightarrow \text{Reals}.$$

What would then axiomatization of vector spaces bring about?

Very simply stated, it would bring about a universal understanding of linearity. Then what is *"linearity"* to begin with? Let us go over this concept once again.

---

[2] In the Japanese language for example, there are traces of different units for counting different objects. Is this a trace of different mathematics for distinct objects?

Roughly speaking, linearity refers to the preservation property of sums. Consider, for example, an audio amplifier. It amplifies the input signal with a certain amount of magnitude, and as long as the amplitude does not exceed a prespecified limit (called the saturation point), the output of the sum of two input signals is the sum of the respective (independently amplified) outputs. Thus an amplifier "preserves" sums and may be regarded as a linear device.

As another example, consider the differentiation operator d/dt. Since

$$\frac{d}{dt}(x_1(t) + x_2(t)) = \frac{d}{dt}x_1(t) + \frac{d}{dt}x_2(t), \tag{1.3}$$

this may also be regarded as a linear operator.[3]

Operations with such a property are fundamental and can be found everywhere. Examples include integral operators, difference operators, linear electrical circuits, the relationship between a force applied to a spring and its elongation, and the relationship between the opening of a valve and the amount of liquid flow.[4]

It is important to note here that the input and the output may have different dimensions. For example, in the last example of the relationship between a valve opening and the corresponding liquid flow, the two quantities are not measured in the same units. In other words, when we discuss linearity, the meaning of the "sum" can be interpreted differently for inputs and outputs. We can be free from bothersome individual peculiarities by taking advantage of abstraction—without associating a particular units or dimensions to inputs and outputs, as individual units such as apples or oranges occupy a secondary place in counting. Likewise, the concept of vector spaces gives an appropriate basis for discussing the essence of operations such as addition, independently of their interpretations in particular contexts.

Thus, in the axiom of vector spaces, we do not specify the "meaning" of addition but rather list the rules such an operation should obey. This is, on the one hand, convenient in that it is not restricted to a particular context, but, on the other hand, it is inclined to be less intuitive due to the lack of concrete interpretation; it will often lead novices to lose motivation. This is why we gave a rather lengthy discussion of vector spaces above.

Before giving the definition of vector spaces, let us discuss briefly another operation, namely, scalar multiplication.

Yet another aspect of linearity is concerned with proportional relations. If we add an element x to itself, the result would be $x + x$. It is convenient to express this result as $2x$. But, in order for this to make sense, we must be equipped with the "product" of a scalar 2 and a vector x. Scalar multiplication determines this operation. For this to be well defined, we must also specify what class of numbers can be considered scalars. If only the integers are considered, it would be very simple: scalar multiplication is directly defined via addition as above. It will be more convenient, however, to consider such elements as $(1/2)x$ or $(3/4)x$ to ensure more versatility. For the moment, let us understand that the range of our scalars is the set $\mathbb{R}$ of real numbers.

We are now ready to give the definition of vector spaces.

---

[3]The discussion here is informal and ignores the important aspect of scalar multiplication. This will be subsequently discussed.

[4]Some of these, however, may be more nonlinear in practice.

Definition 1.1.1. A vector space is a composite mathematical object consisting of a set X along with two algebraic operations, namely, *addition* and *scalar multiplication*, that satisfy the axioms below. When this is the case, we say that X, along with these operations, is a *vector space* or a *linear space* over the reals $\mathbb{R}$:

1. The first operation, called *addition*, associates to each pair (x, y) of X its *sum* $x + y$ such that the following rules are satisfied:

   (a) For any $x, y, z \in X$, $(x + y) + z = x + (y + z)$ (the *associative law* or *associativity*).

   (b) For any $x, y \in X$, $x + y = y + x$ (the *commutative law* or *commutativity*).

   (c) There exists a unique element $0 \in X$, called zero, such that for every $x \in X$, $0 + x = x$ holds.

   (d) For every x in X, there exists a unique element $x' \in X$ such that $x + x' = 0$. This x' is called the *inverse* of x and is written as $-x$. With this notation, $x + (-y)$ will be written as $x - y$ in what follows.

2. The second operation, called *scalar multiplication*, associates to every $x \in X$ and every $a \in \mathbb{R}$ an element ax such that the following rules are satisfied:

   (a) For every $x \in X$ and $a, b \in \mathbb{R}$, $(a + b)x = ax + bx$.

   (b) For every $x, y \in X$ and $a \in \mathbb{R}$, $a(x + y) = ax + ay$.

   (c) For every $x \in X$ and $a, b \in \mathbb{R}$, $(ab)x = a(bx)$.

   (d) For every $x \in X$, $1x = x$.

Some remarks are in order. First, the meanings of addition are different on the left and right of 2(a). The sum $a + b$ on the left means the sum in the reals, while $ax + bx$ on the right is the sum in the vector space X. In principle, they must be distinguished. But these two additions obey the same rule, so in order not to complicate the notation, we usually use the same symbol, $+$, for both operations. The same remark applies to 2(c). Conditions 2(a) and 2(b) together are referred to as the *distributive law*.

Let us show several consequences of the distributive law.

Lemma 1.1.2. *For every $x \in X$,*

1. $0x = 0$ (*0 on the left-hand side is a real, while 0 on the right-hand side is a vector*).

2. $(-1)x = -x$.

3. *For any positive integer n,* $\underbrace{x + \cdots + x}_{n\text{-times}} = nx$.

Proof.

1. By the distributive law, $0x + 0x = (0 + 0)x = 0x$. Add $-0x$ (its existence is guaranteed by 1(d)) to $0x + 0x$ and use the associative law 1(a) and 1(c) to obtain

$$(0x + 0x) + (-0x) = 0x + (0x + (-0x)) = 0x + 0 = 0x.$$

On the other hand, adding $-0x$ to $0x$ on the right yields

$$0x + (-0x) = 0.$$

Since they are equal, $0x = 0$ follows.

2. By the distributive law 2(a), by 2(d), and by property 1 above, $x + (-1)x = 1x + (-1)x = (1 + (-1))x = 0x = 0$. This implies that $(-1)x$ satisfies the requirement for the inverse of x. Since the inverse must be unique by 1(d), $(-1)x$ must be equal[5] to $-x$.

3. This is clear from

$$\begin{aligned} x + \cdots + x &= 1x + \cdots + 1x \\ &= \underbrace{(1 + \cdots + 1)}_{n}x \\ &= nx \end{aligned}$$

according to conditions 2(d) and 2(a). $\square$

Remark 1.1.3. We add yet another remark. While we have emphasized the importance of the uniqueness of an inverse, this property and the uniqueness of zero can actually be deduced from other axioms. Hence, strictly speaking, such uniqueness need not be included as part of the definition. The proof of uniqueness, albeit quite simple, may however look overly formal and not be of much interest to the majority of the readers, and hence is omitted. We have chosen to list the uniqueness as part of the definition placing more emphasis on the ease of understanding. The reader should keep in mind, when he or she reads other axioms in textbooks, that they are not necessarily written with the least possible set of axioms, and in some instances it is possible that some requirements in the axiom may not always be independent of others. (This, however, is done mainly for convenience or for pedagogical reasons.)

In the definition above, the scalar multiplication is defined between a real number α and a vector x. There is sometimes a need to consider scalar multiplication between complex numbers and vectors. For example, this is so for eigenvalue problems. In such a case, it is more natural to choose the set $\mathbb{C}$ of complex numbers, rather than $\mathbb{R}$, as the set of scalars. In this case, Definition 1.1.1 above works equally well by replacing the reals by the complexes. One may also encounter a situation in which one wants to consider scalars other than the reals or complexes. Although we will not go into detail here, there are indeed many such occasions in mathematics. In such a case, we consider a mathematical object called a *field*. A field is an algebraic object in which we consider addition/subtraction and multiplication/division, as with the reals and complexes. The definition of a vector space is then given precisely in the same way as in Definition 1.1.1 except that the set of scalars is taken to be a field.

---

[5]We should be careful in reading axioms. When an axiom says that an inverse exists *uniquely*, the uniqueness is also part of the requirement expressed by the axiom.

In what follows we consider the case where the set of scalars is either the reals $\mathbb{R}$ or the complexes $\mathbb{C}$. But most of the discussions can be carried out parallelly, so it is convenient to represent the scalars by a symbol $\mathbb{K}$, meaning either $\mathbb{R}$ or $\mathbb{C}$. Thus, when we discuss a vector space over $\mathbb{K}$, we understand that the discussion there applies equally well to the case of real or complex scalars. A vector space over $\mathbb{R}$ is called a *real vector space* and a vector space over $\mathbb{C}$ a *complex vector space*.

A set X can be a vector space over two different sets of scalars, e.g., the reals and complexes. For example, the set $\mathbb{C}$ of complex numbers forms a vector space over itself with respect to the addition and scalar multiplication induced by the ordinary addition and multiplication as complex numbers. On the other hand, it can also be regarded as a vector space over $\mathbb{R}$, by viewing the multiplication of real and complex numbers as scalar multiplication. *These two vector space structures are different*, because they have different sets as scalars.[6] The notion of a vector space is a composite mathematical concept, consisting of not only a set X but also the associated addition and scalar multiplication with required properties; if any one of them is altered, it will become a different vector space, although the underlying set X may remain the same as the example above shows. Thus, calling a set a vector space without reference to the operations defined on it is sloppy shorthand and can be dangerous.

1.1.3   Examples of Vector Spaces

Example 1.1.4. The totality of three-dimensional directed arrows (as a vector space over $\mathbb{R}$). This must be familiar to the readers. The sum of two vectors x and y is defined as the diagonal arrow of the parallelogram formed by them (Figure 1.2). The scalar multiplication is defined by enlarging the size of the vector proportionally to the scalar, taking into account the sign of the scalar.

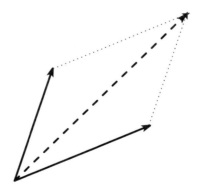

Figure 1.2. *Addition of two arrows*

---

[6]In fact, $\mathbb{C}$ is one-dimensional over $\mathbb{C}$, while it is two-dimensional over $\mathbb{R}$.

Example 1.1.5. The space $\mathbb{R}^n$ of n-tuples of real numbers.

$$\mathbb{R}^n = \left\{ x = \begin{bmatrix} x_1 \\ x_2 \\ \vdots \\ x_n \end{bmatrix} : x_i \in \mathbb{R},\ i = 1,\ldots,n \right\}. \tag{1.4}$$

The addition and scalar multiplication are defined componentwise by

$$\begin{bmatrix} x_1 \\ x_2 \\ \vdots \\ x_n \end{bmatrix} + \begin{bmatrix} y_1 \\ y_2 \\ \vdots \\ y_n \end{bmatrix} := \begin{bmatrix} x_1 + y_1 \\ x_2 + y_2 \\ \vdots \\ x_n + y_n \end{bmatrix}, \quad \alpha \begin{bmatrix} x_1 \\ x_2 \\ \vdots \\ x_n \end{bmatrix} := \begin{bmatrix} \alpha x_1 \\ \alpha x_2 \\ \vdots \\ \alpha x_n \end{bmatrix}. \tag{1.5}$$

Example 1.1.6. The space $\mathbb{C}^n$ of n-tuples of complex numbers. This is given in the same way as above, but $\mathbb{R}$ is replaced by $\mathbb{C}$.

Example 1.1.7. Consider the set of polynomials with real coefficients of degree at most n:

$$P_n := \{p(z) = a_0 + \cdots + a_n z^n : a_i \in \mathbb{R}, \forall i\}.^7 \tag{1.6}$$

Given $p(z) = a_0 + a_1 z + \cdots + a_n z^n$, $q(z) = b_0 + b_1 z + \cdots + b_n z^n$, and a scalar $\alpha \in \mathbb{R}$, we define addition and scalar multiplication coefficientwise as

$$p(z) + q(z) := (a_0 + b_0) + \cdots + (a_n + b_n)z^n, \tag{1.7}$$
$$\alpha p(z) := \alpha a_0 + \cdots + \alpha a_n z^n. \tag{1.8}$$

With these definitions, P_n becomes a vector space over $\mathbb{R}$.

The space P of all polynomials without the degree constraint also becomes a vector space with the same operations. On the other hand, if we imposed the constraint that the degree be *exactly* n, then it would *not* be a vector space any longer (at least with respect to the operations defined above).[8]

Example 1.1.8. The set of real- (or complex-) valued functions defined on the interval $[0,\infty)$ (or on any domain)

$$\mathcal{F}[0,\infty) := \{f : [0,\infty) \to \mathbb{R}\} \tag{1.9}$$

with addition and scalar multiplication defined by

$$(f + g)(t) := f(t) + g(t), \tag{1.10}$$
$$(\alpha f)(t) := \alpha f(t) \tag{1.11}$$

---

[7]The notation $\forall$ means "for every" or "for all."

[8]The sum of two polynomials of degree n is not necessarily of degree n because cancellation may occur at the highest-degree term. Also, the scalar multiplication with 0 makes the result not of degree n.

for $f, g \in \mathcal{F}[0, \infty)$ and scalar $\alpha \in \mathbb{K}$ (with the understanding that $\mathbb{K} = \mathbb{R}$ or $\mathbb{K} = \mathbb{C}$ according as $\mathcal{F}[0, \infty)$ is real- or complex-valued).

In the formulas above, for example, (1.10), the meanings of $+$ are again different on the left and right. The $+$ on the left represents the addition that is to be defined on the function space $\mathcal{F}[0, \infty)$, while the value of the new function $f + g$ at each point t is indeed defined pointwise by the right-hand side value $f(t) + g(t)$. The latter $+$ therefore refers only to the sum of the values of those functions in the space they take their values on (either $\mathbb{R}$ or $\mathbb{C}$ in the present situation). Needless to say, the latter operation is already defined before defining $+$ in $\mathcal{F}[0, \infty)$. The same can be said of the scalar multiplication. The zero element $0 \in \mathcal{F}[0, \infty)$ is defined to be the function that takes the value 0 identically.

Example 1.1.9. The space $C[0, \infty)$ of continuous functions defined on $[0, \infty)$.

We again obtain a vector space by restricting the definitions (1.10) and (1.11) above to those that are continuous. Since the sum and the scalar multiplication definitions (1.10) and (1.11) again give rise to continuous functions when f and g are continuous, they give well-defined operations on $C[0, \infty)$. This example also gives an example of a *vector subspace*, which is to be defined subsequently. Also, the arguments carry over without any change to open/closed intervals (a, b) or $[a, b]$, $a < b$, or other domains, e.g., $[a, b) \cup (c, d]$, $a < b < c < d$, etc.

Example 1.1.10. The space of all solutions of an ordinary linear differential equation with constant coefficients.

Consider the example in subsection 1.1.1, which is the totality X of all solutions of the ordinary differential equation

$$\left(\frac{d^2}{dt^2} + 1 \right) x(t) = 0 \tag{1.12}$$

considered over $[0, \infty)$. Since the solutions are the functions that are expressible in the form

$$x(t) = C_1 \cos t + C_2 \sin t$$

for some C_1 and C_2, it is readily seen that this space constitutes a vector space with respect to the same definitions of addition and scalar multiplication as defined in Example 1.1.8. We need only note that the results of such operations again assume the same expression.

Example 1.1.11. The space $\mathbb{R}^\infty$ of real sequences. Consider the space consisting of real sequences $(x_1, x_2, \ldots, x_n, \ldots)$, $x_i \in \mathbb{R}$:

$$\mathbb{R}^\infty := \{ (x_1, x_2, \ldots, x_n, \ldots) : x_i \in \mathbb{R} \}.$$

Introduce an addition and scalar multiplication as

$$(x_1, x_2, \ldots, x_n, \ldots) + (y_1, y_2, \ldots, y_n, \ldots) := (x_1 + y_1, x_2 + y_2, \ldots, x_n + y_n, \ldots)$$
$$\alpha(x_1, x_2, \ldots, x_n, \ldots) := (\alpha x_1, \alpha x_2, \ldots, \alpha x_n, \ldots)$$

by termwise operations. Needless to say, the 0 element is the sequence where all terms are zero. With these operations, $\mathbb{R}^\infty$ becomes a vector space over $\mathbb{R}$.

Exercise 1.1.12. Verify that the spaces above are indeed vector spaces. Note also that these examples can give vector spaces over $\mathbb{C}$ when $\mathbb{R}$ is replaced by $\mathbb{C}$.

1.1.4 Dimension of a Vector Space

Let us first pose a naive question: what is the *dimension* of a space? Roughly speaking, it is the degree to which an object in a space can move independently. This induces another question: what do we mean by "independently"? More specifically, how should we understand the meaning of dimension so that it gives rise to natural and fruitful outcomes?

We are concerned with vector spaces. We have already seen that the fundamental property here is linearity. We are then naturally led to the following definition of "independence" based on linearity.

Definition 1.1.13. Let X be a vector space over a field $\mathbb{K}$, and let $x_1, \ldots, x_n$ be elements of X. If there exist scalars $a_1, \ldots, a_n \in \mathbb{K}$, not all zero, such that

$$\sum_{j=1}^{n} a_j x_j = 0, \tag{1.13}$$

then we say that $x_1, \ldots, x_n$ are *linearly dependent*. When $x_1, \ldots, x_n$ are not linearly dependent, i.e., when there exist no such scalars, then they are said to be *linearly independent*.

We readily see that, by taking the contrapositive, a necessary and sufficient condition for $x_1, \ldots, x_n$ to be linearly independent is that

$$\sum_{j=1}^{n} a_j x_j = 0 \quad \Rightarrow \quad a_j = 0, \quad j = 1, \ldots, n. \tag{1.14}$$

An element of type

$$\sum_{j=1}^{n} a_j x_j$$

appearing here is called a *linear combination* of $x_1, \ldots, x_n$. In relation to this, the following lemma, although easy, is fundamental.

Lemma 1.1.14. *A set of vectors $x_1, \ldots, x_n$ is linearly dependent if and only if one of them is expressible as a linear combination of others.*

Proof. Suppose, for instance,

$$x_i = \sum_{j \neq i} a_j x_j. \tag{1.15}$$

By moving the left-hand side to the right,[9] we have

$$\sum_{j \neq i} a_j x_j - x_i = 0.$$

Hence $\{x_1, \ldots, x_n\}$ is linearly dependent.

---

[9]To be precise, $-x_i$ is added to both sides. However, such familiar formula chasing procedures on addition and subtraction are guaranteed to hold by the axioms of vector spaces, so we do not have to be overly fussy about this. We will therefore freely use such processes without explicitly mentioning the formal procedures.

Conversely, suppose $\{x_1, \ldots, x_n\}$ is linearly dependent, i.e.,

$$\sum_{j=1}^{n} a_j x_j = 0$$

with a_j, $j = 1, \ldots, n$, not all zero. Suppose a_i is nonzero. Then move the other terms to the right and divide both sides by a_i to obtain

$$x_i = -\sum_{j \neq i} \frac{a_j}{a_i} x_j.$$

Thus x_i is expressible as a linear combination of other x_j's. $\square$

If there exist N linearly independent vectors in a vector space, then we must say that this space has at least N degrees of freedom, since none of them can be expressible as a linear combination of others. Therefore, if we are to define the dimensionality of a vector space as the maximum degree of freedom with respect to linear relations, it is natural to define it as the maximum of all such N's. Before stating it rigorously, however, let us first single out the case where an upper bound for such N exists.

Let X be a vector space. If there exists a number N such that any set of more than N vectors is always linearly dependent, then we say that X is *finite-dimensional*. On the contrary, if such an N does not exist, i.e., if for any n there always exist n linearly independent vectors, we say that X is *infinite-dimensional*.

Example 1.1.15. P in Example 1.1.7 (without restriction on the degree constraint) and Examples 1.1.8, 1.1.9 and 1.1.11 are infinite-dimensional. All in Examples 1.1.4–1.1.6, P_n in Example 1.1.7, and Example 1.1.10 are finite-dimensional.

In what follows, while discussing dimensionality we assume that our vector space is finite-dimensional.

Definition 1.1.16. Let X be a finite-dimensional vector space over $\mathbb{K}$. A family $\{v_1, \ldots, v_n\}$ of vectors in X is a *basis* of X if the following two conditions are satisfied:

1. $\{v_1, \ldots, v_n\}$ is linearly independent.

2. Appending any vector to $\{v_1, \ldots, v_n\}$ makes them linearly dependent.

If this holds, X is said to be *n-dimensional*, and the *dimension* is denoted by $\dim X$.

Two remarks are in order. If appending x to $\{v_1, \ldots, v_n\}$ makes $\{x, v_1, \ldots, v_n\}$ linearly dependent, then this means that x must be expressible as a linear combination of $\{v_1, \ldots, v_n\}$. Otherwise, if

$$ax + \sum a_j v_j = 0,$$

and if $a = 0$, then it would be implied that $\{v_1, \ldots, v_n\}$ is already linearly dependent, which is assumed not to be the case. Hence a cannot be zero, and x can be expressed as a linear combination of $\{v_1, \ldots, v_n\}$.

Another fact to be noted is that the definition of dimension above is not quite complete, for the definition above may appear to depend on a *specific* family $\{v_1,\ldots,v_n\}$ of vectors, but there can be (infinitely many) other linearly independent vectors. If there existed another set $\{w_1,\ldots,w_m\}$ of vectors such that (i) $m > n$, and (ii) $\{w_1,\ldots,w_m\}$ is linearly independent, then we could not sensibly *define* the dimension as a proper attribute of a vector space. An attribute of a space should not depend on a *particular* choice of vectors, and if it does, it does not qualify as a proper quantity of the space.

In fact, such a situation does not occur. If $\{v_1,\ldots,v_n\}$ forms a basis, then any vectors $w_1,\ldots,w_m$ with more than n members are always expressible as linear combinations of $\{v_1,\ldots,v_n\}$, and hence there exist no more than n vectors that are linearly independent.

Also, if we take a set $\{e_1,\ldots,e_r\}$ of linearly independent vectors with less than n members, then it is possible to augment them with $n - r$ vectors to make them a basis (cf. Lemma 1.1.22 below).

These two facts imply that the maximal number of linearly independent vectors is indeed *independent* of particular vectors we started out with. This in turn implies that the definition of dimension is independent of a particular choice of linearly independent vectors.

The facts above may be found in any textbooks on linear algebra, and thus we do not elaborate on them.[10]

We remark only that the fact above is based on the property that a set of n linear equations

$$\begin{bmatrix} a_{11} & a_{12} & \cdots & a_{1m} \\ \vdots & \vdots & \cdots & \vdots \\ a_{n1} & a_{n2} & \cdots & a_{nm} \end{bmatrix} \begin{bmatrix} x_1 \\ x_2 \\ \vdots \\ x_m \end{bmatrix} = \begin{bmatrix} b_1 \\ b_2 \\ \vdots \\ b_n \end{bmatrix}$$

with m unknowns $x_1, x_2, \ldots, x_m$ $(m > n)$ always admits a nontrivial (i.e., nonzero) solution.

Exercise 1.1.17. Verify that P_n in Example 1.1.7 is finite-dimensional and that its dimension is $n + 1$. (Hint: Show that $\{1, t, \ldots, t^n\}$ forms a basis. Note that the zero element here is the polynomial with zero coefficients. If we regard a polynomial as a function in t, this is a function that takes the value identically zero.)

Once a basis is fixed for a finite-dimensional vector space, we can express any element in terms of its coordinates. That is, the following lemma holds.

Lemma 1.1.18. *Let X be an n-dimensional vector space and $\{e_1,\ldots,e_n\}$ a basis for X. Every element x of X can be expressed as a unique linear combination of $\{e_1,\ldots,e_n\}$ as*

$$x = \sum_{i=1}^{n} x_i e_i. \tag{1.16}$$

In this way X becomes isomorphic[11] to $\mathbb{K}^n$ as a vector space.

---

[10]However, the fact on augmentation of linearly independent vectors is important, and it will be proved in Lemma 1.1.22 (p. 13).

[11]That is, there is a one-to-one correspondence between X and $\mathbb{K}^n$, and every vector space structure of X can be translated into one in $\mathbb{K}^n$ and vice versa. See subsection 1.2.1 below for a more detailed discussion.

Proof. By the definition of a basis, $\{x, e_1, \ldots, e_n\}$ must be linearly dependent. Hence from what we have remarked after Definition 1.1.16, x is expressible as a linear combination of $\{e_1, \ldots, e_n\}$.

Suppose now that there exist two such expressions:

$$x = \sum_{i=1}^{n} x_i e_i = \sum_{i=1}^{n} x_i' e_i.$$

It would follow that

$$\sum_{i=1}^{n} (x_i - x_i') e_i = 0.$$

Now since $\{e_1, \ldots, e_n\}$ is linearly independent, the coefficients $x_i - x_i'$ would then be all zero. This implies $x_i = x_i'$; i.e., the expression (1.16) must be unique. $\square$

The numbers $x_1, x_2, \ldots, x_n$ are called the *coordinates* of x with respect to the basis $e_1, e_2, \ldots, e_n$.

Example 1.1.19. The set of vectors

$$e_1 := \begin{bmatrix} 1 \\ 0 \\ 0 \\ \vdots \\ 0 \end{bmatrix}, e_2 := \begin{bmatrix} 0 \\ 1 \\ 0 \\ \vdots \\ 0 \end{bmatrix}, e_n := \begin{bmatrix} 0 \\ 0 \\ 0 \\ \vdots \\ 1 \end{bmatrix}$$

in $\mathbb{R}^n$ forms a basis. This is called the *standard basis* or *natural basis* of $\mathbb{R}^n$.

Example 1.1.20. The two functions $\{\sin t, \cos t\}$ in Example 1.1.10 constitute a basis.

Exercise 1.1.21. Prove this fact.

Concerning bases, the following lemma is fundamental.

Lemma 1.1.22. *Let X be an n-dimensional vector space and $\{e_1, \ldots, e_r\}$ a linearly independent subset. Clearly $n \geq r$; but it is possible to augment $\{e_1, \ldots, e_r\}$ with $n - r$ vectors $e_{r+1}, \ldots, e_n$ to make $e_1, \ldots, e_n$ a basis for X.*

Proof. Since X is n-dimensional, there exists a basis $\{v_1, \ldots, v_n\}$ for it.

If $n = r$, there is nothing to prove because $\{e_1, \ldots, e_r\}$ already constitutes a basis. In this case every $x \in X$ is expressible as a linear combination of $\{e_1, \ldots, e_r\}$.

Now suppose $n > r$. Then there exists an element in X that is not expressible as a linear combination of $\{e_1, \ldots, e_r\}$. This implies that at least one of $\{v_1, \ldots, v_n\}$ is not expressible as a linear combination of $\{e_1, \ldots, e_r\}$. Otherwise, by the very property of bases, every element of X must become expressible as a linear combination of $\{e_1, \ldots, e_r\}$, which is a clear contradiction to $n > r$. Hence let v_1, for example, be one such element. Then define $e_{r+1} := v_1$. Then by Lemma 1.1.14 on p. 10, $\{e_1, \ldots, e_{r+1}\}$ must be linearly independent. If this set does not form a basis, then continue as above to choose one of $\{v_2, \ldots, v_n\}$ that is not linearly dependent over $\{e_1, \ldots, e_{r+1}\}$. Then append it as e_{r+2} to $\{e_1, \ldots, e_{r+1}\}$. Continuing this procedure $n - r$ times will result in a basis. $\Box$

1.2   Linear Mappings and Matrices

1.2.1   Linear Mappings

The concept of linear mappings (maps) is just as important as that of vector spaces. The theory of vector spaces will become complete only after establishing the relationships between the spaces themselves and linear maps defined on them.

We have noted that the fundamental property in vector spaces is linearity. It is represented by addition and scalar multiplication. The most natural mappings with respect to this concept are clearly those that preserve this property. Expressing this idea, we arrive at the following definition.

Definition 1.2.1. Let X, Y be vector spaces over the same field $\mathbb{K}$. A mapping $f : X \to Y$ is a *linear mapping* or a *linear map*[12] if for every $x, x_1, x_2 \in X$, and a scalar $\alpha \in \mathbb{K}$,

$$f(x_1 + x_2) = f(x_1) + f(x_2), \qquad (1.17)$$
$$f(\alpha x) = \alpha f(x) \qquad (1.18)$$

hold. That is, a sum is mapped to the sum of corresponding mapped elements, and a scalar multiplication is also mapped to the scalar multiplication by the same scalar of the mapped element.

Exercise 1.2.2. Conditions (1.17) and (1.18) can be equivalently combined into one condition

$$f(\alpha x_1 + \beta x_2) = \alpha f(x_1) + \beta f(x_2) \quad \forall \alpha, \beta \in \mathbb{K}. \qquad (1.19)$$

Prove this equivalence.

A linear mapping that is also a bijection, that is, an injective (one-to-one) and surjective (onto) mapping, is called an *isomorphism*.

Let $f : X \to Y$ be an isomorphism. Since it is bijective, there exists its inverse mapping $g : Y \to X$. But this does not say anything about the linearity of g. It is, however, automatically guaranteed: Indeed, let y_1, y_2 be any two elements of Y. Since f is bijective, there uniquely exist $x_1, x_2 \in X$ such that $y_1 = f(x_1)$, $y_2 = f(x_2)$. By the very definition of the inverse mapping, $x_1 = g(y_1)$, $x_2 = g(y_2)$. Now

$$y_1 + y_2 = f(x_1) + f(x_2) = f(x_1 + x_2)$$

is clear from the linearity of f. Apply g to both sides to yield $x_1 + x_2 = g(y_1 + y_2)$. That is, g preserves sums. The same can be said of the scalar multiplication, and hence g is a linear mapping.

---

[12] In particular, when $Y = X$, it is often called a *linear transformation* on X. We may also say that f is $\mathbb{K}$-linear when an explicit reference to the scalar field is desired.

Exercise 1.2.3. Verify that g indeed preserves the scalar multiplication.

This inverse mapping g is denoted by f^{-1}. Clearly $f^{-1} \circ f = I_X$, $f \circ f^{-1} = I_Y$ follow. Here I_X is the *identity mapping* that maps an arbitrary element x in X to x itself.[13]

If there exists an isomorphism $f : X \to Y$ between X and Y, the vector space X is said to be *isomorphic* to Y. This is denoted as $X \cong Y$. When this is the case, binary operations $x_1 + x_2$, αx in X are to be mapped by f to $f(x_1) + f(x_2)$, $\alpha f(x)$ in Y. Further, this is a one-to-one correspondence. So X and Y have exactly the same vector space structure. For example, if a fact is proved in X, then it can be equivalently transferred to Y via the isomorphism f.

Let us list some examples of linear mappings.

Example 1.2.4. The mapping

$$A : C[0,1] \to C[0,1] : \phi \mapsto \int_0^t \phi(s)ds$$

is a linear mapping. (The right-hand side is regarded as a function of t.) More generally, for a fixed but arbitrary continuous function $K(t,s)$ in two variables,

$$A : C[0,1] \to C[0,1] : \phi \mapsto \int_0^t K(t,s)\phi(s)ds \tag{1.20}$$

is a linear mapping.

Exercise 1.2.5. In the definition (1.20) above, we must show that the right-hand side indeed gives a continuous function in t. Prove this fact.

Example 1.2.6. The mapping

$$A : C[0,1] \to \mathbb{C} : \phi \mapsto \int_0^1 \phi(s)ds \tag{1.21}$$

is also linear. Note the difference in the upper limit of integration from that in Example 1.2.4 above.

Example 1.2.7. Take any point a in the interval $[0,1]$, and define a mapping A by

$$A : C[0,1] \to \mathbb{C} : \phi \mapsto \phi(a). \tag{1.22}$$

This is a linear mapping.

Example 1.2.8. The mapping

$$A : \mathbb{R}^2 \to \mathbb{R}^2 : \begin{bmatrix} x_1 \\ x_2 \end{bmatrix} \mapsto \begin{bmatrix} -x_2 \\ x_1 \end{bmatrix}$$

(the 90-degree counterclockwise rotation about the origin in the two-dimensional space $\mathbb{R}^2$) is a linear mapping.

---

[13]Needless to say, its matrix representation, to be introduced later, is the identity matrix I.

Example 1.2.9. In the space P in Example 1.1.7, the differentiation

$$\frac{d}{dt} : P \to P : \phi \mapsto \phi'$$

is a linear mapping. If $f, g \in P$, then $(f + g)' = f' + g'$, and likewise for scalar multiplication. This also gives a linear mapping in the space given in Example 1.1.10.

However, we should note that this does not give a linear mapping in function spaces $\mathcal{F}[0, \infty)$ and $C[0, \infty)$ as given in Examples 1.1.8 and 1.1.9. To be more precise, the mapping is not well defined over the whole space in these examples because not every function is differentiable.

Example 1.2.10. The left shift operator in the sequence space $\mathbb{R}^\infty$ in Example 1.1.11

$$\sigma_\ell : (x_1, x_2, \ldots, x_n, \ldots) \mapsto (x_2, x_3, \ldots, x_n, \ldots) \tag{1.23}$$

and the right shift operator

$$\sigma_r : (x_1, x_2, \ldots, x_n, \ldots) \mapsto (0, x_1, x_2, \ldots, x_n, \ldots) \tag{1.24}$$

are both linear mappings.

1.2.2   Matrices

A matrix is a representation of a linear map in a finite-dimensional vector space.

As we will see shortly, to define a linear map, we need only specify where the basis elements are mapped to. Once this is done, we can always extend the mapping to the whole space by using linearity. Furthermore, there is no constraint on how such basis elements should be mapped. That is, a linear map is entirely determined by its action on a basis. We make this precise as follows.

Lemma 1.2.11 (the principle of linear extension). *Let X be a finite-dimensional vector space and $\{e_1, \ldots, e_n\}$ a basis for X. Let Y be another vector space (not necessarily finite-dimensional). For every given set of n vectors, $v_1, \ldots, v_n \in Y$, there exists a linear map $f : X \to Y$ such that*

$$f(e_i) = v_i, \quad i = 1, \ldots, n, \tag{1.25}$$

and this map is unique.

Proof. Take any element x in X, and express it in terms of $\{e_1, \ldots, e_n\}$ as

$$x = \sum_{i=1}^{n} x_i e_i.$$

If f is a linear map, then linearity should imply

$$f(x) = \sum_{i=1}^{n} x_i f(e_i).$$

Thus to satisfy (1.25), we must define $f(x)$ as

$$f(x) = \sum_{i=1}^{n} x_i v_i. \tag{1.26}$$

This clearly yields the uniqueness of f. Also, it is readily ensured that definition (1.26) gives a linear map. □

Now suppose that a linear map $f : X \to Y$ is given between finite-dimensional spaces X, Y and that $\{e_1, \ldots, e_n\}$, $\{v_1, \ldots, v_m\}$ are bases for X and Y, respectively. From what we have seen above, the mapping f is uniquely determined by specifying $f(e_i)$, $i = 1, \ldots, n$. Since $\{v_1, \ldots, v_m\}$ is a basis for Y, each $f(e_i)$ can be expressed as a linear combination of $\{v_1, \ldots, v_m\}$ as follows:

$$f(e_i) = \sum_{j=1}^{m} a_{ji} v_j. \tag{1.27}$$

This can also be written differently in matrix notation as

$$[f(e_1), \ldots, f(e_n)] = [v_1, \ldots, v_m] \begin{bmatrix} a_{11} & \cdots & a_{1n} \\ a_{21} & \cdots & a_{2n} \\ & \cdots & \\ a_{m1} & \cdots & a_{mn} \end{bmatrix}. \tag{1.28}$$

Hence every element $x = \sum_{i=1}^{n} x_i e_i$ is mapped, by the linearity of f, as

$$f\left(\sum_{i=1}^{n} x_i e_i\right) = \sum_{i=1}^{n} x_i f(e_i) = \sum_{i=1}^{n} \sum_{j=1}^{m} a_{ji} x_i v_j = \sum_{i=1}^{m} \left(\sum_{j=1}^{n} a_{ij} x_j\right) v_i.$$

In other words, when we express $f(\sum_{i=1}^{n} x_i e_i)$ as $\sum_{i=1}^{m} y_i v_i$, the coefficients $y_1, \ldots, y_m$ are given by

$$\begin{bmatrix} y_1 \\ y_2 \\ \vdots \\ y_m \end{bmatrix} = \begin{bmatrix} a_{11} & \cdots & a_{1n} \\ a_{21} & \cdots & a_{2n} \\ & \cdots & \\ a_{m1} & \cdots & a_{mn} \end{bmatrix} \begin{bmatrix} x_1 \\ x_2 \\ \vdots \\ x_n \end{bmatrix}. \tag{1.29}$$

This is the *matrix representation* of f (with respect to the bases $\{e_1, \ldots, e_n\}$ and $\{v_1, \ldots, v_m\}$).

Let $f : X \to Y$ and $g : Y \to Z$ be linear maps, and F, G their matrix representations, where we assume that the same basis is taken in Y for these representations. Then it is easy to show that the matrix representation of the composed mapping $g \circ f$ is given by the matrix product GF of G and F. This in particular implies that the associative and distributive laws for matrices automatically follow from the corresponding facts on linear maps.

Exercise 1.2.12. Prove these facts.

1.2.3 Change of Basis

As we have noted, the choice of a basis is not unique. It is therefore important to know how coordinates and matrices change when we alter the basis.

We begin by discussing the change of linear combination expressions in terms of a different basis. Let X be an n-dimensional vector space and $\{e_1,\ldots,e_n\}$, $\{v_1,\ldots,v_n\}$ bases for it. Then there exist scalars t_{ij} and s_{ij}, $i,j = 1,\ldots,n$, such that

$$v_i = \sum_{k=1}^{n} t_{ki} e_k,$$

$$e_j = \sum_{k=1}^{n} s_{kj} v_k.$$

In terms of matrix representation, this can be written as

$$[v_1,\ldots,v_n] = [e_1,\ldots,e_n]T, \quad [e_1,\ldots,e_n] = [v_1,\ldots,v_n]S,$$

where $T = (t_{ij})$, $S = (s_{ij})$. Substituting the first equation into the second, we obtain

$$[e_1,\ldots,e_n] = [e_1,\ldots,e_n]TS.$$

The linear independence of $\{e_1,\ldots,e_n\}$ readily implies $TS = I$. Similarly, $ST = I$. Here I denotes of course the identity matrix, i.e., the matrix with all its diagonal elements equal to 1 and all others being zero.

Thus T and S mutually give the inverse of the other, and hence they are isomorphisms. Now if x is expressed as $\sum_{i=1}^{n} x_i e_i$, then

$$x = [e_1,\ldots,e_n]\begin{bmatrix} x_1 \\ x_2 \\ \vdots \\ x_n \end{bmatrix} = [v_1,\ldots,v_n]S\begin{bmatrix} x_1 \\ x_2 \\ \vdots \\ x_n \end{bmatrix}.$$

This means that the coefficient vector of x when expressed in terms of the basis $\{v_1,\ldots,v_n\}$ is given by

$$S\begin{bmatrix} x_1 \\ x_2 \\ \vdots \\ x_n \end{bmatrix}. \tag{1.30}$$

In other words, the ith coefficient in the new coordinate system (with respect to $\{v_1,\ldots,v_n\}$) is

$$\sum_{k=1}^{n} s_{ik} x_k. \tag{1.31}$$

Let us now see how a change of a basis affects the matrix representation of a linear map f. The following theorem holds.

Theorem 1.2.13. *Let X, Y be n- and m-dimensional vector spaces, respectively, $f : X \to Y$ a linear map, and $\{e_1,\ldots,e_n\}$, $\{e_1',\ldots,e_n'\}$, $\{v_1,\ldots,v_m\}$, $\{v_1',\ldots,v_m'\}$ bases for X and Y, respectively. Further, let $A = (a_{ij})$ be the matrix representation of f with respect to*

$\{e_1,\ldots,e_n\}$, $\{v_1,\ldots,v_m\}$, *and suppose that the basis changes from* $\{e_1,\ldots,e_n\}$ *to* $\{e'_1,\ldots,e'_n\}$ *and from* $\{v_1,\ldots,v_m\}$ *to* $\{v'_1,\ldots,v'_m\}$ *are described by*

$$\left[e'_1,\ldots,e'_n\right] = [e_1,\ldots,e_n]\,Q, \tag{1.32}$$

$$[v_1,\ldots,v_m] = \left[v'_1,\ldots,v'_m\right]P. \tag{1.33}$$

Then the new matrix representation of f with respect to the new bases $\{e'_1,\ldots,e'_n\}$, $\{v'_1,\ldots,v'_m\}$ *is given by* PAQ.

Proof. From (1.28),

$$[f(e_1),\ldots,f(e_n)] = [v_1,\ldots,v_m]\,A. \tag{1.34}$$

We need to change this to a representation in terms of the bases $\{e'_1,\ldots,e'_n\}$ and $\{v'_1,\ldots,v'_m\}$. Substituting (1.32) and (1.33) into (1.34), we obtain

$$\begin{aligned}
\left[f(e'_1),\ldots,f(e'_n)\right] &= f\,[e_1,\ldots,e_n]\,Q \\
&= [v_1,\ldots,v_m]\,AQ \\
&= \left[v'_1,\ldots,v'_m\right]PAQ.
\end{aligned}$$

The first equality above follows from the linearity of f. This shows that the expression of f in terms of $\{e'_1,\ldots,e'_n\}$ and $\{v'_1,\ldots,v'_m\}$ is PAQ. $\qquad\square$

In particular, when $X = Y$, it is of course natural to take $v_i = e_i$ and $v'_i = e'_i$. Then $P = Q^{-1}$ follows from (1.32) and (1.33), and the basis change induces the change in the matrix representation as

$$A \mapsto Q^{-1}AQ.$$

If matrices A and B satisfy $B = Q^{-1}AQ$ for some nonsingular matrix Q, they are said to be *similar* to each other. If $B = PAQ$ for nonsingular matrices P, Q, then the two matrices A, B are said to be *equivalent*.

1.3 Subspaces and Quotient Spaces

Once the concept of vector spaces is specified, the next important concepts in relation to the space itself are those of subspaces and quotient spaces. We now introduce these two fundamental concepts.

1.3.1 Vector Subspaces

Let X be a vector space over $\mathbb{K}$ and V a subset of X. For any x, y in V and scalar $a \in \mathbb{K}$, the sum $x + y$ and the scalar product ax are well defined as elements of X. *If, further, V becomes a vector space with respect to the same binary operations as those given on X, then we say that V is a* (vector) subspace of X.

This will perhaps give the impression that we should check all the axioms in Definition 1.1.1 anew for V. But this is really not so. For example, consider the associative law:

$$(x + y) + z = x + (y + z).$$

The validity of this identity is guaranteed as that for elements of X since X is a vector space and the associative law holds there. The same holds for the other axioms.

Do we have nothing to check then? By no means! If so, then *any subset* would become a subspace. This is absurd. *What is missing here is the requirement for well-defined sums and scalar products to lie in the pertinent subset.* For example, we cannot reasonably define a sum on the subset

$$C := \{(x_1, x_2) \in \mathbb{R}^2 : x_1^2 + x_2^2 = 1\}$$

of the two-dimensional space $\mathbb{R}^2$. While one can well define the sum of (x_1, x_2) and (y_1, y_2) in $\mathbb{R}^2$ as $(x_1 + y_1, x_2 + y_2)$, this goes outside of C, and hence it is not a reasonable candidate as a sum *in C*.

Let us go over Definition 1.1.1 once again. We see that addition and scalar multiplication satisfying the properties following requirement 1 on p. 5 are crucial to this definition. The example of a circle above does not allow the very existence of such sums and scalar multiplications.

Recall further that in the present definition of subspaces we require that V inherit the same addition and scalar multiplication as those for X. In order that they be again legitimate operations in V, they should satisfy

$$x + y \in V, \tag{1.35}$$
$$ax \in V \tag{1.36}$$

for all $x, y \in V$ and $a \in \mathbb{K}$. Otherwise, they cannot qualify as binary operations in V, as observed in the example of a circle. When these two conditions are satisfied, we say that V is *closed* under the addition and scalar multiplication (in X).

Conversely, if V is closed under addition and scalar multiplication as above, then V satisfies the axioms of vector spaces in Definition 1.1.1.

Proposition 1.3.1. *Let X be a vector space over $\mathbb{K}$ and V a nonempty subset that is closed under addition and scalar multiplication. Then V becomes a vector subspace of X.*

Proof. The fundamental laws, such as the associative law, are readily seen to be satisfied from the fact that they are valid in X. Thus what we need to check is the existence of a zero vector 0 and an inverse $-x$ for every $x \in V$. Since $V \neq \emptyset$, take any $x \in V$. By Lemma 1.1.2, $0 = 0x$ and $-x = (-1)x$. Since V is closed under addition and scalar multiplication, they belong to V. □

Example 1.3.2. Let X and Y be vector spaces over $\mathbb{K}$ and $f : X \to Y$ a linear map. Then

$$\ker f := \{x \in X : f(x) = 0\}, \tag{1.37}$$
$$\operatorname{im} f := \{y \in Y : y = f(x) \text{ for some } x \in X\} \tag{1.38}$$

are each subspaces of X and Y, respectively. We call $\ker f$ the *kernel* or the *null space* of f and $\operatorname{im} f$ the *image* or the *range* of f.

More generally, for a given subset M of X, we define

$$f(M) := \{y \in Y : y = f(x) \text{ for some } x \in M\} \tag{1.39}$$

as a subset of Y called the *image of M under f*. In this way, im f may also be denoted as $f(X)$. It is easy to see that when M is a subspace of X, $f(M)$ is also a subspace. Also, for a given subset N of Y,

$$f^{-1}(N) := \{x \in X : f(x) \in N\} \tag{1.40}$$

is also a subset of X and is called the *inverse image, or preimage,* of N under f. It is again easy to verify that $f^{-1}(N)$ is a subspace of X when N is a subspace of Y. Note here that the notation $f^{-1}(N)$ need not refer to the inverse mapping f^{-1}, nor does it assume the existence of such a map.[14] As a special case, ker $f = f^{-1}(\{0\})$.

Exercise 1.3.3. Prove that an image and inverse image of a subspace are subspaces.

Exercise 1.3.4. Let X, Y, and f be as above. Show that

1. if f is injective, then ker $f = \{0\}$;

2. if f is surjective, then im $f = Y$.

Example 1.3.5. Let X be a vector space and $L = \{v_1, \ldots, v_r\}$ be any set of vectors in X. Define V by

$$V = \operatorname{span} L := \left\{ \sum_{j=1}^{r} a_j v_j : a_j \in \mathbb{K} \right\}. \tag{1.41}$$

It is easy to see that V, denoted span L, is the smallest subspace of X that contains L. This space span L is called the *subspace spanned by* $\{v_1, \ldots, v_r\}$, or the *subspace generated by* $\{v_1, \ldots, v_r\}$.

Let us verify that V is indeed the smallest subspace containing L. Let W be a subspace of X containing L. There exists at least one such W because X itself is such a subspace of X. Let us characterize W. Since W contains L, any scalar product $a_j v_j$ must belong to W. Since W is a subspace, any finite sums of all such elements should again belong to W. Now the space V defined by (1.41) consists of all such elements, so $V \subset W$. On the other hand, since

$$\sum_{j=1}^{r} a_j v_j + \sum_{j=1}^{r} b_j v_j = \sum_{j=1}^{r} (a_j + b_j) v_j,$$

$$\alpha \left(\sum_{j=1}^{r} a_j v_j \right) = \sum_{j=1}^{r} (\alpha a_j) v_j,$$

Proposition 1.3.1 implies that V itself is a subspace. Hence V is certainly the smallest among all subspaces containing L.

Exercise 1.3.6. Let X be a vector space and V, W its subspaces. Show that

1. $V \cap W$ is also a subspace;

---

[14]Of course, when the inverse mapping f^{-1} does exist, this coincides with the image of N under f^{-1}.

2. $V \cup W$ is not necessarily a subspace;

3. the smallest subspace containing $V \cup W$ is given by

$$V + W := \{v + w : v \in V, w \in W\}. \tag{1.42}$$

Exercise 1.3.7. Show that for every subset S of a vector space X, there always exists the smallest subspace containing S, which is given by

$$V = \bigcap \{W : W \text{ is a subspace containing } S\}. \tag{1.43}$$

1.3.2   Rank and Nullity of a Linear Mapping

We have now reached the stage of introducing the important concepts of rank and nullity without relying on a particular coordinate system. The rank of a linear mapping is, roughly stated, the extent of freedom of its range (image) space. Unfortunately it is often understood in a more computational or algorithmic way—for example, as the largest size among non-vanishing minors. Such a definition perhaps requires less conceptual effort and is therefore understood to be "easy." However, it is not necessarily always advantageous when discussing more advanced topics, and indeed often requires a somewhat complicated process to actually arrive at the definition, e.g., computation of canonical forms via elementary operations. We here emphasize more conceptual aspects of these notions.

Definition 1.3.8. Let X, Y be vector spaces and $f : X \to Y$ a linear mapping. Then the *rank* of f is defined to be the dimension of its image:

$$\operatorname{rank} f := \dim \operatorname{im} f. \tag{1.44}$$

If $\operatorname{im} f$ is finite-dimensional, $\operatorname{rank} f$ is finite; otherwise, we say that $\operatorname{rank} f$ is infinite and may write $\operatorname{rank} f = \infty$. The *nullity* of f is also defined as the dimension of the kernel of f:

$$\dim \ker f. \tag{1.45}$$

Similarly as above, when $\ker f$ is finite-dimensional, the nullity is finite; otherwise it is infinite.

Let $A : \mathbb{K}^m \to \mathbb{K}^p$ be a $p \times m$ matrix. Represent A as $A = [\boldsymbol{a}_1, \ldots, \boldsymbol{a}_m]$ via an m list of p-dimensional vectors, and let $x = (x_1, \ldots, x_m)^T$ (x^T denotes the transpose of x). Clearly

$$Ax = \sum_{j=1}^{m} x_j \boldsymbol{a}_j. \tag{1.46}$$

Since the image space $\operatorname{im} A$ consists of all such elements, it consists exactly of all linear combinations of the columns of A. Combining this with the example above, we have

$$\operatorname{im} A = \operatorname{span}\{\boldsymbol{a}_1, \ldots, \boldsymbol{a}_m\}.$$

The dimension of this space is clearly the maximum number of linearly independent vectors among $a_1, \ldots, a_m$. This agrees precisely with the usual definition of the rank of matrices.

What is particularly important in the relationship between rank and nullity is the following theorem called the dimension theorem.

Theorem 1.3.9 (dimension theorem). *Let X, Y, and f be as above, and suppose* $\dim X < \infty$. *Then*

$$\dim X = \operatorname{rank} f + \dim \ker f. \tag{1.47}$$

This theorem can be proven more naturally in conjunction with the subsequent treatment of quotient spaces. The proof is thus postponed until Corollary 1.3.22, p. 29.

1.3.3 The Direct Sum of Vector Spaces

Let X be a vector space and V, W subspaces of X. We have seen that the smallest subspace that contains both V and W is $V + W = \{v + w : v \in V, w \in W\}$ (cf. Exercise 1.3.6). In general, the intersection of V and W is nontrivial, so the way of expressing elements of $V + W$ as $v + w$, $v \in V$, $w \in W$ is not necessarily unique. For example, consider

$$V := \left\{ x = \begin{bmatrix} x_1 \\ x_2 \\ 0 \end{bmatrix} : x_1, x_2 \in \mathbb{R} \right\}, \qquad W := \left\{ y = \begin{bmatrix} 0 \\ y_2 \\ y_3 \end{bmatrix} : y_2, y_3 \in \mathbb{R} \right\}$$

in $\mathbb{R}^3$. A vector $(1, 1, 1)^T$ can be expressed as

$$(1, 1, 1)^T = (1, 1, 0)^T + (0, 0, 1)^T = (1, 0, 0)^T + (0, 1, 1)^T$$

in two different ways as sums of vectors in V and W.[15] (There are indeed infinitely many other ways than these.) The reason for this nonuniqueness is due to the redundancy arising from the nontrivial intersection between V and W.

Is there uniqueness of such expressions if V and W have only a trivial intersection, namely, 0? To see this, suppose $V \cap W = \{0\}$, and

$$v_1 + w_1 = v_2 + w_2, \qquad v_i \in V, w_i \in W, i = 1, 2.$$

This yields

$$v_1 - v_2 = w_2 - w_1,$$

where the left-hand side is in V while the right-hand side belongs to W. Hence it belongs to $V \cap W$. By the assumption $V \cap W = \{0\}$, we have

$$v_1 = v_2, \qquad w_1 = w_2;$$

that is, the expression is unique. We have thus seen that when $V \cap W = \{0\}$, the way of expressing elements in $V + W$ is unique. This motivates the following definition.

---

[15]Here A^T and x^T denote, respectively, the transpose of a matrix A and a vector x. In what follows, we often use this notation to represent a column vector in order to save space.

Definition 1.3.10. Let X be a vector space, and V and W its subspaces. If

1. $X = V + W$,

2. $V \cap W = \{0\}$,

then X is said to be the *direct sum* of V and W and is denoted as $X = V \oplus W$. V or W is called a *direct summand* of X.

   In general, the direct sum $V_1 \oplus V_2 \oplus \cdots \oplus V_n$ of n subspaces $V_1, \ldots, V_n$ can be defined similarly. In this case the subspaces must satisfy the condition

$$V_i \cap \left(\sum_{j \neq i} V_j \right) = \{0\} \quad \forall i. \tag{1.48}$$

Exercise 1.3.11. Under condition (1.48), every element w of $V_1 + V_2 + \cdots + V_n$ can uniquely be expressed as

$$w = \sum_{j=1}^{n} v_j, \quad v_j \in V_j.$$

Prove this.

Example 1.3.12. $\mathbb{R}^n = \underbrace{\mathbb{R} \oplus \cdots \oplus \mathbb{R}}_{n}$.

1.3.4   Quotient Spaces

We now introduce another important construction of a vector space. Let X be a vector space and M a subspace of X. Although it may appear somewhat ad hoc, let us first consider the following relation in X associated with M. We will return to its meaning afterward.

Definition 1.3.13. Two elements x and y of X are said to be *congruent modulo M* if

$$x - y \in M$$

holds. This is denoted $x \equiv y \pmod{M}$.[16]

   This relation is an example of what is known as an *equivalence relation*. To be precise, an equivalence relation is a relation in a set that satisfies the three conditions in the following lemma. The congruence satisfies these conditions.

Lemma 1.3.14. *Let M be a subspace of a vector space X. Then the congruence modulo M satisfies the following three properties:*

1. *For every $x \in X$, $x \equiv x \pmod{M}$ (reflexivity).*

---

[16]The term mod M is read "modulo M."

2. *If $x \equiv y \pmod{M}$, then $y \equiv x \pmod{M}$ (symmetry).*

3. *If $x \equiv y \pmod{M}$ and $y \equiv z \pmod{M}$, then $x \equiv z \pmod{M}$ (transversality).*

The proof is easy and left to the reader.

Equivalence relations are important in that they can generate new concepts allowing classification of known objects. In fact, we execute such procedures almost unconsciously in our daily lives, and without them abstract thinking is almost impossible. Consider, for example, the case of polls. In many such cases, a respondent is classified to different categories such as those corresponding to gender, age, and profession. Apparently the main concern here is whether each classified group has a certain specific tendency in opinions on certain questions. The individual opinion of each person is of secondary importance here. Taking the classified groups as above is nothing but considering equivalence classes (i.e., groups). Once we classify a group of people into such subclasses (i.e., equivalence classes), individuality is entirely ignored, and each member is regarded as just the same as others in the same group. See Figure 1.3. We can then talk about *the* opinion shared by all of the members of such a subgroup.

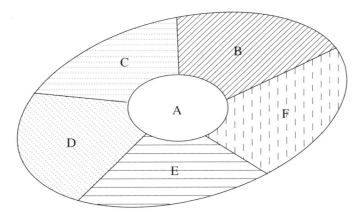

Figure 1.3. *Equivalence classes*

Once an equivalence relation is defined in a set, it is possible to consider a subset of elements that are mutually equivalent via this relation and consider each subset as a single member of a new set that consists of all such subsets of elements. Each such subset of elements is called an *equivalence class* or a *coset*. The equivalence class to which an element x belongs is usually denoted by $[x]$.

In order that the equivalence classes as given above constitute a well-defined classification of the set X, the following two conditions must be fulfilled:

- X is the union of all such equivalence classes.

- Different equivalence classes do not have an intersection.

These properties hold for any equivalence relation, but we here confirm them only for the congruence defined above. The general case is left to the reader as an exercise.

Let us first characterize the equivalence classes with respect to the congruence. Suppose that y is congruent to x modulo M. Then, by definition, $y - x \in M$, so $y - x = m$ for some $m \in M$. Hence $y = x + m$. Conversely, since $x + m - x = m \in M$ for every $m \in M$, we see that the set of all elements that are congruent to x modulo M is given by

$$[x] = x + M := \{x + m : m \in M\}. \tag{1.49}$$

(Note that these subsets are never subspaces of X unless $x \in M$.) Clearly

$$X = \bigcup_{x \in X} (x + M). \tag{1.50}$$

Moreover, if $(x + M) \cap (y + M) \neq \emptyset$, then

$$x + m_1 = y + m_2$$

for some $m_1, m_2 \in M$. (Needless to say, m_1 and m_2 do not have to be equal.) It now follows that for every $m \in M$

$$x + m = (x + m_1) + m - m_1 = y + m_2 + m - m_1 \in y + M. \tag{1.51}$$

This implies $x + M \subset y + M$. By symmetry, we also have $y + M \subset x + M$. Thus, if the intersection is nonempty, then $x + M = y + M$; i.e., the whole equivalence classes coincide. It should also be clear from (1.51) that the condition $(x + M) \cap (y + M) \neq \emptyset$ is equivalent to $x - y \in M$.

The concept of quotient spaces is then introduced as follows: Consider the following question. For a given space X, what is the "space" obtained by ignoring the difference belonging to a subspace M? This is precisely equivalent to identifying those elements that are mutually congruent modulo M. In other words, for every x we take the set $x + M$[17] that is congruent to x, regard it as an *element* of the new set

$$\bigcup (x + M),$$

and study its properties. This set is called the *quotient set* of X (modulo M) and is denoted by

$$X/M \tag{1.52}$$

(pronounced X modulo M).

This quotient set X/M indeed admits a vector space structure that is naturally induced from that of X. This vector space, as will be made precise below, is called the *quotient vector space* of X modulo M.

Before proceeding to a precise definition, let us first give a simple example.

Example 1.3.15. Consider as X the three-dimensional space $\mathbb{R}^3$, and take M as

$$M := \{(x_1, x_2, x_3)^T : x_3 = 0\}.$$

M is the two-dimensional (x_1, x_2)-plane, and two elements $(x_1, x_2, x_3)^T$ and $(x_1', x_2', x_3')^T$ are congruent to each other if and only if $x_3 = x_3'$. In other words, in such equivalence classes,

---

[17]This is called an *affine subspace* parallel to M.

only the difference in the x_3 coordinate matters, and x_1, x_2 coordinates are immaterial. Thus the equivalence class $(x_1, x_2, x_3)^T + M$ is a plane parallel to the (x_1, x_2)-plane, and the quotient set X/M is the totality of all these planes. They are clearly classified by their x_3 coordinates, and hence $X/M \cong \mathbb{R}$.

Let us now introduce a vector space structure to X/M. As the example above suggests, for two cosets $x + M$ and $y + M$, it will be reasonable to define a coset corresponding to their sum as

$$(x + M) + (y + M) := x + y + M. \qquad (1.53)$$

Similarly, for scalar multiplication, we may define

$$a(x + M) := ax + M, \quad a \in \mathbb{K}. \qquad (1.54)$$

The problem here is that there is a freedom in a representation $x + M$, and it is not uniquely determined. That is, for an equivalence class $x + M$, its *representative* x is not unique: indeed, $x + M = x + m + M$ for any m in M. Under such circumstances, do definitions like (1.53) and (1.54) make sense?

Suppose $x + M = x' + M$ and $y + M = y' + M$. In other words,

$$x - x' \in M, \quad y - y' \in M. \qquad (1.55)$$

Naturally $x + y = x' + y'$ does *not* follow. This, however, requires too much and is not necessary for (1.53) to make sense. In fact, in order that it be well defined, it is sufficient that

$$x + y + M = x' + y' + M \qquad (1.56)$$

holds. As we have already seen, this is the case if and only if

$$(x + y) - (x' + y') \in M.$$

But this is obvious from (1.55). The same argument also holds for (1.54), and hence the definitions above for addition and scalar multiplication are well defined.

Once addition and scalar multiplication have been verified to be well defined, it is routine to check the other axioms for vector spaces. They are simply translations of the corresponding properties in X. Let us just note that the zero element in X/M is the equivalence class of 0; that is, $[0] = 0 + M = M$.

The facts established so far are summarized as follows.

Proposition 1.3.16. *Let X be a vector space over $\mathbb{K}$ and M a subspace of X. Then the quotient set X/M can be endowed with a vector space structure over $\mathbb{K}$ with respect to the addition and scalar multiplication defined by (1.53) and (1.54). Furthermore, the correspondence*

$$\pi : X \to X/M : x \mapsto [x] = x + M \qquad (1.57)$$

is a linear map and is surjective. This mapping π is called the canonical (or natural) projection .

Exercise 1.3.17. Show that the above π is a linear map. (Hint: This is immediate from the definitions of addition and scalar multiplication in X/M.)

For the dimensionality of quotient spaces, the following important result holds.

Theorem 1.3.18. *Let X be a finite-dimensional vector space and M a subspace of X. Then*

$$\dim X/M = \dim X - \dim M. \tag{1.58}$$

Proof. Choose a basis $\{e_1,\ldots,e_r\}$ for M, and augment it with $\{e_{r+1},\ldots,e_n\}$ to form a basis for X according to Lemma 1.1.22 (p. 13). It suffices to show that $\{[e_{r+1}],\ldots,[e_n]\}$ forms a basis for X/M.

Take any $x \in X$. Since $\{e_1,\ldots,e_n\}$ is a basis for X, we can express x as

$$x = \sum_{i=1}^{n} x_i e_i, \quad x_i \in \mathbb{K}.$$

Then we have

$$[x] = \sum_{i=1}^{n} x_i [e_i].$$

Since $e_i \in M$ for $i = 1,\ldots,r$, it follows that $[e_i] = M = [0] = 0_{X/M}$[18] for $i = 1,\ldots,r$. Thus

$$[x] = \sum_{i=r+1}^{n} x_i [e_i],$$

and this means that every element $[x]$ in X/M can be expressed as a linear combination of $\{[e_{r+1}],\ldots,[e_n]\}$.

It remains to show that the set $\{[e_{r+1}],\ldots,[e_n]\}$ is linearly independent. Suppose

$$\sum_{i=r+1}^{n} a_i [e_i] = \left[\sum_{i=r+1}^{n} a_i e_i \right] = 0.$$

Since the zero element of X/M is nothing but M, the equality means $\sum_{i=r+1}^{n} a_i e_i \in M$. On the other hand, since $\{e_1,\ldots,e_r\}$ is a basis for M, there exist scalars $b_i, i = 1,\ldots r$, such that

$$\sum_{i=r+1}^{n} a_i e_i = \sum_{i=1}^{r} b_i e_i.$$

This yields

$$\sum_{i=1}^{r} b_i e_i - \sum_{i=r+1}^{n} a_i e_i = 0.$$

Since $\{e_1,\ldots,e_n\}$ is a basis for X, this implies $b_1 = \cdots = b_r = 0$, and $a_{r+1} = \cdots = a_n = 0$. Hence $\{[e_{r+1}],\ldots,[e_n]\}$ is linearly independent. $\square$

The next proposition can be proven in a manner similar to the theorem above.

---

[18]We will write 0_Y when we need to explicitly show that it is the zero element of a space Y.

Proposition 1.3.19. *Let X be a finite-dimensional vector space and M,N subspaces of X. If*

$$X = M \oplus N,$$

then $\dim X = \dim M + \dim N$.

Exercise 1.3.20. Prove the proposition above.

The following theorem is called the *homomorphism theorem*.

Theorem 1.3.21 (homomorphism theorem). *Let X, Y be vector spaces over $\mathbb{K}$ and $f : X \to Y$ a linear map. Then*

$$X/\ker f \cong \operatorname{im} f;$$

i.e., $X/\ker f$ is isomorphic to $\operatorname{im} f$ (p. 15).

Proof. Consider the following mapping $\overline{f} : X/\ker f \to Y$:

$$\overline{f} : [x] = x + \ker f \mapsto f(x).$$

A commutative diagram[19] would express this as

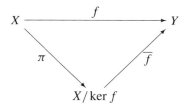

We first show that the mapping $\overline{f}$ is well defined. It suffices to show that $[x_1] = [x_2]$ implies $f(x_1) = f(x_2)$. By the definition of congruence, $x_1 - x_2 \in \ker f$. This means $f(x_1 - x_2) = 0$, and hence $f(x_1) = f(x_2)$, and hence $\overline{f}$ is well defined. The linearity of mapping $\overline{f}$ is easy to verify.

Now suppose $\overline{f}([x]) = 0$. Then $f(x) = 0$, which implies $x \in \ker f$. This means $[x] = 0_{X/\ker f}$. Then by Exercise 1.3.4, $\overline{f}$ is injective (a one-to-one mapping). Hence if we confine the codomain[20] of $\overline{f}$ to $\operatorname{im} f$, then $\overline{f}$ becomes a bijection and hence gives an isomorphism between $X/\ker f$ and $\operatorname{im} f$. □

This now yields the following *dimension theorem* (cf. Theorem 1.3.9, p. 23).

Corollary 1.3.22 (dimension theorem). *Let X, Y be vector spaces and $f : X \to Y$ a linear map. Suppose that X and $\operatorname{im} f$ are finite-dimensional. Then*

$$\dim X = \operatorname{rank} f + \dim \ker f. \tag{1.59}$$

---

[19]A commutative diagram consists of several sets and arrows where arrows represent mappings, and every composition of several mappings in which the results of compositions with the same domain and codomain (see the next footnote) should coincide. In the present diagram, this means $f = \overline{f} \circ \pi$.

[20]For a mapping $f : X \to Y$, X is called the *domain* and Y the *codomain*. See Appendix A.

Proof. The homomorphism theorem, Theorem 1.3.21, implies $X/\ker f \cong \operatorname{im} f$. Hence $\dim(X/\ker f) = \dim(\operatorname{im} f)$. On the other hand, Theorem 1.3.18 yields $\dim(X/\ker f) = \dim X - \dim \ker f$. Since $\dim(\operatorname{im} f) = \operatorname{rank} f$ (Definition 1.44), (1.59) readily follows. $\square$

1.3.5 Linear Equations

We now study some fundamental properties of the linear equation

$$Ax = b \tag{1.60}$$

as an application. Here $b \in \mathbb{R}^m$, and $A = (a_{ij}) = [a_1, \ldots, a_m]$ is an $m \times n$ matrix over $\mathbb{R}$. Regarding A as a linear map $A : \mathbb{R}^n \to \mathbb{R}^m$, we see that (1.60) admits at least one solution if and only if

$$b \in \operatorname{im} A.$$

Rewriting this, we have

$$\operatorname{span} A = \operatorname{span}[a_1, \ldots, a_n] = \operatorname{span}[a_1, \ldots, a_n, b]. \tag{1.61}$$

That is to say, if we append the vector b to the columns of A as $\begin{bmatrix} A & b \end{bmatrix}$, the space spanned by these column vectors does not increase. Since $\operatorname{span} A \subset \operatorname{span} \begin{bmatrix} A & b \end{bmatrix}$ always holds, we obtain a solvability condition equivalent to (1.61) as follows:

$$\operatorname{rank} A = \operatorname{rank} \begin{bmatrix} A & b \end{bmatrix}. \tag{1.62}$$

Let us examine (1.60) a little further. How much freedom is there for the solution space $\{x : Ax = b\}$?

Suppose there exist two (distinct) solutions x_1, x_2. Since $A(x_1 - x_2) = Ax_1 - Ax_2 = b - b = 0$, the difference of two arbitrary solutions should belong to $\ker A$, that is, the solution space of $Ax = 0$. Conversely, if x_0 is an arbitrary solution to (1.60), and x is any element of $\ker A$, then $A(x_0 + x) = Ax_0 + Ax = b + 0 = b$, and hence $x_0 + x$ is also a solution to (1.60). Hence the solution space of $Ax = b$ can be expressed as

$$x_0 + \ker A$$

with a special solution x_0 to $Ax_0 = b$. This is precisely the equivalence class $[x_0]$ represented by x_0 in the quotient space $\mathbb{R}^n/\ker A$. Its degree of freedom is $\dim \ker A$, and, according to the dimension theorem, this is $n - \operatorname{rank} A$.

As a special case, when $m = n$ and $\operatorname{rank} A = n$, the space of solutions has the zero-dimensional degree of freedom; i.e., in this case the solution exists and is unique. This can also be seen from the fact that A is invertible (nonsingular) by $\operatorname{rank} A = n$.

1.4 Duality and Dual Spaces

1.4.1 Dual Spaces

Duality plays a crucial role in various situations in mathematics. It is, however, difficult to elucidate what it precisely is in a way applicable to all cases. Roughly speaking, the word duality refers not to a single, isolated mathematical entity, but rather to a situation in which two objects appear in pairs and one is somehow derived from the other, or the two may be understood together in some generalized context. Examples include Fleming's right-hand and left-hand rules in electromagnetics, graphs and dual graphs in graph theory, and primal and dual problems in linear programming. One may be regarded as the mirror image of the other.

In the context of linearity, however, duality can be defined much more clearly. The duality here refers to various relationships between vector spaces and their dual spaces, and those between linear maps and their dual (adjoint) mappings.

Let us examine this in more detail. Suppose we are given a vector space X. This is, after all, a collection of points that as a whole satisfy certain properties (namely, the axioms for vector spaces). Of course, each "point" may actually be a function or something that may not look like a "point" at first glance, but such a difference in concrete terms is rather immaterial to the essence of the argument. This is the standpoint of abstract vector spaces. For such a collection X of "points," one can associate with X a family of functions defined on X. What kind of functions? Since we are now in the world of linearity, with a given vector space X, it is quite natural to take linear mappings as the objects to be considered on it. Furthermore, we take the codomain to be the scalar field $\mathbb{K}$ as the simplest choice. If X is a real vector space, then we take $\mathbb{R}$ for $\mathbb{K}$, and if it is a complex vector space, then we take $\mathbb{C}$ for $\mathbb{K}$. A scalar-valued linear mapping

$$\varphi : X \to \mathbb{K}$$

is called a *linear functional* or a *linear form*[21] of X.

Definition 1.4.1. Let X be a vector space. The space

$$X^* = \{\varphi : X \to \mathbb{K} : \varphi \text{ is linear}\}$$

is called the *dual (space)* of X.

Let now $\alpha, \beta \in \mathbb{K}$ be scalars, and let φ, ψ be linear functionals. Then a new functional $\alpha\varphi + \beta\psi$ can be defined as follows:

$$(\alpha\varphi + \beta\psi)(x) := \alpha\varphi(x) + \beta\psi(x). \tag{1.63}$$

Exercise 1.4.2. Verify that $f := \alpha\varphi + \beta\psi$ defined by (1.63) indeed gives a linear functional (i.e., an element of X^*). (Hint: Compute $f(ax + by)$ according to the definition.)

As a special element, there is a linear functional 0 that maps every element $x \in X$ to $0 \in \mathbb{K}$.[22] We have thus seen that the sum and scalar multiplication of elements in X^* again belong to X^*. Moreover, the following theorem holds.

---

[21]This is a very important concept; in fact, Schwartz distributions, which we will encounter in Chapter 6, will be defined using this concept.

[22]Unless there is a specific reason for doing otherwise, we will denote both the zero element in X and the zero functional by the same symbol 0. The distinction will be clear from the context; the notation would be quite awkward were we to distinguish them by different symbols.

Theorem 1.4.3. *Let X be a vector space over $\mathbb{K}$ and X^* its dual space. Then X^* is also a vector space over $\mathbb{K}$ with the functional 0 as the zero element, and with addition and scalar multiplication defined by* (1.63).

Exercise 1.4.4. Prove Theorem 1.4.3. (One must check the axioms for vector spaces; definition (1.63) is essential.)

Now let x be an element of a vector space X and X^* the dual space of X. To this x, let us associate the mapping

$$T_x : X^* \to \mathbb{K} : f \mapsto f(x). \tag{1.64}$$

This mapping T_x is clearly a functional[23] on X^*. Moreover, for each $f, g \in X^*, \alpha, \beta \in \mathbb{K}$,

$$\begin{aligned} T_x(\alpha f + \beta g) &= (\alpha f + \beta g)(x) \quad \text{(by definition (1.64))} \\ &= \alpha f(x) + \beta g(x) \quad \text{(by (1.63))} \\ &= \alpha T_x(f) + \beta T_x(g) \quad \text{(again by (1.64))} \end{aligned} \tag{1.65}$$

holds. In other words, T_x is a linear functional on X^*, namely, an element of $(X^*)^*$ (usually denoted as X^{**}). Furthermore, the following theorem holds.

Lemma 1.4.5. *$T_x = T_y$ holds if and only if $x = y$.*

Proof. If $x = y$, then $T_x = T_y$ trivially follows from its definition (1.64). Conversely, if $T_x = T_y$, then $T_x(f) = T_y(f)$ for every $f \in X^*$. The definition (1.64) then yields $f(x) = f(y)$ for every $f \in X^*$. The following theorem, Theorem 1.4.6, then implies $x = y$.   □

Theorem 1.4.6 (uniqueness principle). *Let x be an element of a vector space X, and suppose that*

$$f(x) = 0$$

for every $f \in X^*$. Then $x = 0$.

Proof. See Corollary A.7.3 (p. 244) to the Hahn–Banach theorem.   □

When X is finite-dimensional, we can directly prove this by characterizing X^* with a dual basis (see the next section) without recourse to such big machinery as the Hahn–Banach theorem. However, this theorem is valid even for infinite-dimensional spaces, so we stated in its general form. It indeed plays a key role in many places. Actually, what is to be proven in Corollary A.7.3 is a special case when f is confined to be a continuous linear functional.[24]

According to Lemma 1.4.5, we see that every x may be identified with a linear functional T_x on X^* defined by (1.64). Furthermore, in view of

$$\begin{aligned} T_{x+y}(f) &= f(x+y) = f(x) + f(y) = T_x(f) + T_y(f) \\ &= (T_x + T_y)(f) \quad \text{(by definition (1.63) of the addition in } X^*) \end{aligned}$$

---

[23]A function on X taking values in $\mathbb{K}$ is called a functional.

[24]This, of course, covers the result in the finite-dimensional case.

and (1.65), the action of this correspondence $x \mapsto T_x$ is entirely symmetrical both in $x \in X$ and in $f \in X^*$. It is thus more logical and convenient not to employ an asymmetrical notation such as $f(x)$ or $T_x(f)$ to denote the value of f at x in dealing with such a duality. Hence we will hereafter denote $f(x)$ or $T_x(f)$ by

$$\langle x, f \rangle = \langle f, x \rangle. \tag{1.66}$$

In such a case, we will not be so particular about the order of f and x.[25]

Exercise 1.4.7. Express equalities (1.63) and (1.65) in the notation of (1.66).

1.4.2 Dual Bases

The dual space X^* of a vector space X has, aside from the fundamental property of being a vector space itself, various properties. In particular, when X is finite-dimensional, it is also of finite dimension, and indeed of the same dimension. We will show this in this subsection.

Let X be an n-dimensional vector space, and let $\{e_1, \ldots, e_n\}$ be a basis. Define $\{f_1, \ldots, f_n\}$ in X^* by

$$\langle e_i, f_j \rangle = \delta_{ij}, \quad i, j = 1, \ldots, n. \tag{1.67}$$

Here δ_{ij} denotes the so-called Kronecker delta:

$$\delta_{ij} = \begin{cases} 1 & \text{(when } i = j\text{),} \\ 0 & \text{(when } i \neq j\text{).} \end{cases} \tag{1.68}$$

Since $\{e_1, \ldots, e_n\}$ is a basis for X, (1.67) certainly defines n linear functionals $f_1, \ldots, f_n \in X^*$, according to the linear extension principle (Lemma 1.2.11, p. 16). For $f_1, \ldots, f_n$ thus defined, the following fact holds.

Theorem 1.4.8. *The elements $f_1, \ldots, f_n$ defined by (1.67) constitute a basis for the dual space X^*. Hence if X is finite-dimensional, then X^* is also finite-dimensional, and their dimensions coincide.*

Proof. Let us try to give a proof in a rather informal but more illustrative way.

We must first show that every $f \in X^*$ can be expressed as a linear combination of $f_1, \ldots, f_n$. A usual trick in such a case is to check first "what would occur if this were the case." So suppose that there exist scalars $\alpha_1, \ldots, \alpha_n$ such that

$$f = \sum_{i=1}^{n} \alpha_i f_i.$$

---

[25]The notation $\langle \cdot, \cdot \rangle$ looks like that for inner products. In fact, some authors use this notation for inner products also. However, while an inner product is defined on $X \times X$, the above quantity is defined on $X \times X^*$. Moreover, an inner product is only *conjugate linear* in the second argument, i.e., multiplication by α in the second argument results in the value multiplied by $\bar{\alpha}$, as opposed to (1.66), which is also linear in the second argument. To avoid confusion, we will make the distinction of using $\langle \cdot, \cdot \rangle$ for a linear form (linear functionals) and $(\cdot, \cdot)$ for an inner product.

Apply this representation to each of the elements $e_1, \ldots, e_n$ of X. Then we must have

$$
\begin{aligned}
\langle f, e_j \rangle &= \sum_{i=1}^{n} \alpha_i \langle f_i, e_j \rangle \\
&= \sum_{i=1}^{n} \alpha_i \delta_{ij} = \alpha_j, \quad j = 1, \ldots, n.
\end{aligned}
\tag{1.69}
$$

This, conversely, naturally induces candidates for α_j as

$$
\alpha_j := \langle f, e_j \rangle, \quad j = 1, \ldots, n.
$$

Then it has become nearly routine to introduce a candidate g as

$$
g := \sum_{i=1}^{n} \alpha_i f_i.
$$

Noticing that f and g agree on the basis elements $\{e_1, \ldots, e_n\}$, we see that f and g agree on the whole space X by the linear extension principle, Lemma 1.2.11. In other words, $\{f_1, \ldots, f_n\}$ spans X^*.

It remains only to show that the set $\{f_1, \ldots, f_n\}$ is linearly independent. Suppose

$$
\sum_{i=1}^{n} \alpha_i f_i = 0.
$$

Applying this to e_j, we see $\alpha_j = 0$ similarly to (1.69). Hence the set $\{f_1, \ldots, f_n\}$ is linearly independent.[26]   □

Definition 1.4.9. The basis $\{f_1, \ldots, f_n\}$ for X^* thus defined is called the *dual basis* corresponding to $\{e_1, \ldots, e_n\}$.

Example 1.4.10. We give one important example. One may say that every instance of duality in finite-dimensional spaces virtually reduces to this case.

Take as X the n-dimensional vector space $\mathbb{K}^n$, and let $\{e_1, \ldots, e_n\}$ be the standard basis of X, e_i being the column vector with only one nonzero entry at the ith component (Example 1.1.19 on p. 13). From what we have seen above, its dual basis $\{f_1, \ldots, f_n\}$ is determined by

$$
\langle e_i, f_j \rangle = \delta_{ij}.
$$

---

[26]It is often observed that some people tend to have an overly formal idea about mathematical logic. In the present context, one might try to give α_i before computing what it should be; this is not a good idea. Of course, many textbooks often give such a proof by first giving α_i as if it were natural, but this is possible only when one already knows the result or has tried to find α_i by "inspection"; a streamlined formal proof is given only after such α_i's are found. Giving too formal a proof may thus be misleading, and here we give a rather roundabout (but more instructive) proof for the benefit of showing the idea.

In view of the definition of matrix multiplication, it is clear that f_j should be given by

$$f_j \;=\; [0 \ldots 0 \quad \underset{\hat{j}}{1} \quad 0 \ldots 0]$$

and X^* consists precisely of all n-dimensional *row* vectors.

   In this sense, we can say that the dual concept of column vectors is that of row vectors, and, similarly, the dual of rows is given by columns.

1.4.3 Dual Mappings

The theory of vector spaces consists of the study of not merely the spaces themselves but also the linear mappings intertwining these spaces.

   We may then ask what position linear mappings occupy in the theory of duality. As we see below, duality theory exhibits the richest results in relation to the interconnections between linear mappings and their adjoints (dual mappings).

   Suppose we are given a linear mapping

$$\mathcal{A} : X \to Y \tag{1.70}$$

from X to Y.[27] Let us ask the following question: Does $\mathcal{A}$ induce some kind of relationship between dual spaces X^* and Y^*?

   Let $y^* \in Y^*$ be given. This is of course a linear functional

$$y^* : Y \to \mathbb{K}.$$

In view of (1.70), we are naturally led to the composed mapping

$$y^* \circ \mathcal{A} : X \to \mathbb{K} : x \mapsto y^*(\mathcal{A}x). \tag{1.71}$$

Exercise 1.4.11. Make sure that this mapping $y^* \circ \mathcal{A}$ is an element of X^*.

   We now view this linear functional $y^* \circ \mathcal{A}$ as the image of y^* under $\mathcal{A}$ and denote it as $\mathcal{A}^* y^*$.

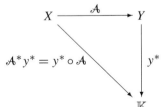

Exercise 1.4.12. Show that this mapping

$$\mathcal{A}^* : Y^* \to X^* : y^* \mapsto \mathcal{A}^* y^* = y^* \circ \mathcal{A} \tag{1.72}$$

is linear.

---

[27]We here use the script symbol $\mathcal{A}$ to distinguish it from its matrix representation, which is to be denoted by A.

Definition 1.4.13. The mapping $\mathcal{A}^*$ defined above is called the *dual mapping*, or the *adjoint mapping*, of $\mathcal{A}$.

The definition (1.71) can be expressed as

$$\langle x, \mathcal{A}^* y^* \rangle = \langle \mathcal{A}x, y^* \rangle \tag{1.73}$$

with the notation introduced in (1.66). This is perhaps easier to remember because of its similarity to the notions of inner products and adjoints.

The following proposition is fundamental to dual mappings and dual bases.

Proposition 1.4.14. *Let X and Y be vector spaces of dimensions n and m with bases $\{e_1, \ldots, e_n\}$ and $f_1, \ldots, f_m$, respectively. Let $\{e_1', \ldots, e_n'\}$ and $\{f_1', \ldots, f_m'\}$ be their respective dual bases. Let $\mathcal{A} : X \to Y$ be a linear mapping and $A = (a_{ij})$ its matrix representation with respect to bases $\{e_1, \ldots, e_n\}$ and $\{f_1, \ldots, f_m\}$. Then the matrix representation of the dual mapping $\mathcal{A}^*$ of $\mathcal{A}$ in terms of the dual bases $\{e_1', \ldots, e_n'\}$ and $\{f_1', \ldots, f_m'\}$ is the transposed matrix A^T of A.*

Proof. Recall that the matrix representation of a linear mapping $\mathcal{A}$ amounts to expressing the image $\mathcal{A}e_i, i = 1, \ldots, n$, of $\{e_1, \ldots, e_n\}$ under $\mathcal{A}$ in terms of a basis $\{f_1, \ldots, f_m\}$ of Y as follows:

$$[\mathcal{A}e_1, \ldots, \mathcal{A}e_n] = [f_1, \ldots, f_m] \begin{bmatrix} a_{11} & \cdots & a_{1n} \\ a_{21} & \cdots & a_{2n} \\ & \cdots & \\ a_{m1} & \cdots & a_{mn} \end{bmatrix} =: [f_1, \ldots, f_m]A. \tag{1.74}$$

To find the representation of $\mathcal{A}^*$ in terms of the dual bases, let us compute $[\mathcal{A}^* f_1', \ldots, \mathcal{A}^* f_n']$. To this end, assume the representation of form

$$[\mathcal{A}^* f_1', \ldots, \mathcal{A}^* f_m'] = [e_1', \ldots, e_n'] \begin{bmatrix} b_{11} & \cdots & b_{1m} \\ b_{21} & \cdots & b_{2m} \\ & \cdots & \\ b_{n1} & \cdots & b_{nm} \end{bmatrix} =: [e_1', \ldots, e_m']B. \tag{1.75}$$

This is always possible since $\{e_1', \ldots, e_n'\}$ is a basis for X^*. To find the coefficients b_{ij}, let us compute $\langle e_i, \mathcal{A}^* f_j' \rangle$:

$$\langle e_i, \mathcal{A}^* f_j' \rangle = \left\langle e_i, \sum_{k=1}^{n} b_{kj} e_k' \right\rangle$$

$$= \sum_{k=1}^{n} b_{kj} \delta_{ik}$$

$$= b_{ij}.$$

We here used the definition of the dual bases (1.67). On the other hand, using the definition (1.73) of $\mathcal{A}^*$, we have

$$\langle e_i, \mathcal{A}^* f_j' \rangle = \langle \mathcal{A} e_i, f_j' \rangle$$

$$= \left\langle \sum_{k=1}^{m} a_{ki} f_k, f_j' \right\rangle \quad \text{(by (1.74))}$$

$$= \sum_{k=1}^{m} a_{ki} \langle f_k, f_j' \rangle$$

$$= \sum_{k=1}^{m} a_{ki} \delta_{kj}$$

$$= a_{ji}.$$

This yields $b_{ij} = a_{ji}$. That is, B is the transpose A^T of A. $\square$

Remark 1.4.15. Note that even when $\mathbb{K} = \mathbb{C}$ the representation of the adjoint $\mathcal{A}^*$ is given by A^T and *not* the conjugate transpose of A. This situation is different from the Hilbert space adjoint, which will be treated in Chapter 3.

Exercise 1.4.16. Suppose that $A : X \to Y$ is a linear map and A^* its dual mapping. Prove the following:

- A is surjective (onto) if and only if A^* is injective (one-to-one).

- A is injective if and only if A^* is surjective.

Problems

1. Let $A : X \to Y$ be a linear map between vector spaces X and Y. Show that $A(0) = 0$, $A(-x) = -Ax$.

2. Let X be a finite-dimensional vector space and M a subspace of X. Show that there exists a subspace N of X such that

$$X = M \oplus N.$$

3. Let M_1 and M_2 be subspaces of a finite-dimensional vector space X. Show that $\dim(M_1 + M_2) = \dim M_1 + \dim M_2 - \dim(M_1 \cap M_2)$.

Chapter 2

Normed Linear Spaces and Banach Spaces

This chapter introduces normed linear spaces and Banach spaces—two of the key concepts in this book. In particular, we deal with such fundamental concepts as continuity of mappings and convergence of sequences which can be rigorously treated with the concept of norms. This leads to the concept of completeness that characterizes Banach spaces.

2.1   Normed Linear Spaces

2.1.1   Norms and Topology

Topology is one of the most important concepts in mathematics. It is used to define, for example, a concept such as how distant two points are in a space. As an example, in spaces such as the reals $\mathbb{R}$ and complexes $\mathbb{C}$, for a pair of points x, y, the notion of the absolute value (modulus) $|x - y|$ is defined, and this can be used to measure the distance between the two points. In n-dimensional spaces $\mathbb{R}^n$ or $\mathbb{C}^n$, there is, for example, the notion of the Euclidean norm $\|x\|$ defined by

$$\|x\|_2 := \left\{ \sum_{k=1}^{n} |x_k|^2 \right\}^{1/2}, \tag{2.1}$$

as a generalization of the notion of the absolute value in $\mathbb{R}$ or $\mathbb{C}$. This again serves the purpose of measuring the length of a vector or the distance between two points.

Let us first examine basic properties of norms in $\mathbb{R}^n$. A *norm* $\|\cdot\| : X \to \mathbb{R}$ is a nonnegative function that satisfies the following three properties:

$$\|x\| = 0 \quad \text{if and only if } x = 0, \tag{2.2}$$

$$\|\alpha x\| = |\alpha| \, \|x\| \quad \forall \alpha \in \mathbb{K}, \tag{2.3}$$

$$\|x + y\| \leq \|x\| + \|y\|. \tag{2.4}$$

The last inequality is called the *triangle inequality*. It refers to the fact that the length of $x + y$ does not exceed the sum of the lengths of the respective sides.

Example 2.1.1. Each of the following gives a norm in $\mathbb{R}^n$ (or $\mathbb{C}^n$). See Example 2.1.12 below.

1. $\|x\|_2 := \{\sum_{k=1}^{n} |x_k|^2\}^{1/2}$.

2. $\|x\|_1 := \sum_{k=1}^{n} |x_k|$.

3. $\|x\| := \max_{1 \le k \le n} |x_k|$.

4. $\|x\|_p := \{\sum_{k=1}^{n} |x_k|^p\}^{1/p}, \quad p \ge 1$.

When a norm is defined, we can naturally introduce a *distance* by

$$d(x,y) := \|x - y\|. \tag{2.5}$$

We can use this distance to further introduce various topological notions, for example, as follows[28]:

- A sequence $\{x_n\}$ *converges* to x if and only if $d(x_n, x) = \|x_n - x\| \to 0$ (we denote this as $x_n \to x$); x is called the *limit*[29] of $\{x_n\}$.

- A mapping f is *continuous* at x if and only if $x_n \to x$ implies $f(x_n) \to f(x)$.

- An *open ϵ-neighborhood* of a point x is defined by $B(x,\epsilon) := \{y : \|y - x\| < \epsilon\}$.

- A subset F is a *closed set* if and only if for every convergent sequence x_n, $x_n \in F$, its limit x belongs to F.

- A subset V is a *neighborhood* of x if and only if there exists $\epsilon > 0$ such that $B(x,\epsilon) \subset V$.

- A subset O is an *open set* if and only if for every point x in O there exists a neighborhood V of x such that $V \subset O$.

The reason we introduce a topology in a vector space is that we wish to develop a theory based on continuity. This is absolutely indispensable for infinite-dimensional spaces, such as Banach or Hilbert spaces.[30] Actually, we may even have to go one step further to consider infinite-dimensional spaces such as those called topological vector spaces in dealing with spaces of distributions or spaces with the weak topology, which we will encounter later (and is indeed just as useful as the notions of Banach or Hilbert spaces). But for the moment we postpone the general treatment of those spaces (see Appendix A) and start with topological notions as indicated above for more elementary normed linear spaces and their fundamental properties.

---

[28] A rigorous treatment will be given in a more general setting in the subsequent sections.

[29] At this point we do not yet know that a limit of a sequence is unique. It will be shown, however, that if a limit exists at all, then it must be unique.

[30] Of course, there are finite-dimensional Banach or Hilbert spaces, but we typically need them in an infinite-dimensional context.

2.1.2 Definition of Normed Linear Spaces

A normed linear space is a vector space equipped with a special type of nonnegative-valued function called a *norm* satisfying (2.2)–(2.4). Let us start by giving its definition.

Definition 2.1.2. Let X be a vector space over a field $\mathbb{K}$. A function defined on X

$$\|\cdot\| : X \to [0,\infty) : x \mapsto \|x\| \tag{2.6}$$

satisfying the conditions (2.2)–(2.4) (p. 39) is called a *norm*. The composite concept of X with a norm is called a *normed linear space*.[31]

A normed linear space can be naturally equipped with a topology[32] in the following way. This is precisely parallel to the way of defining convergence, etc., of sequences in $\mathbb{R}^n$.

Definition 2.1.3. Let X be a normed linear space with norm $\|\cdot\|$. A sequence $\{x_n\}$ in X *converges* to $x \in X$ as $n \to \infty$ if

$$\|x_n - x\| \to 0.$$

The point x is called the *limit* of $\{x_n\}$. The (open) ϵ-*neighborhood* of a point x is the subset

$$B(x,\epsilon) := \{y \in X : \|y - x\| < \epsilon\}. \tag{2.7}$$

We may also consider $\|y - x\| \le \epsilon$ in place of $\|y - x\| < \epsilon$ above. In such a case, $B(x,\epsilon)$ would become a closed set, while (2.7) is an open set. (See subsection 2.1.4 for open and closed sets.) This closed neighborhood will be called a *closed ϵ-neighborhood*.

Remark 2.1.4. Topology is a mathematical concept that is introduced to a space X by specifying the notion of "closeness" there. For a point x in the space X, this is done by giving a family of neighborhoods of x. For a normed space, a subset V is called a *neighborhood* of x if it contains an open ϵ-neighborhood $B(x,\epsilon)$. The totality of such neighborhoods as a whole determines a *topology* for X. We can then discuss such notions as continuity of mappings, etc., based solely on this notion of topology, but we will not delve into this general treatment further here. For a further discussion, see Appendix A.

Rewriting $y_n := x_n - x$, we see that $x_n \to x$ if and only if $y_n \to 0$. The ϵ-neighborhood (2.7) of x is also the parallel shift by x of the ϵ-neighborhood

$$B(0,\epsilon) := \{y \in X : \|y\| < \epsilon\}$$

of the origin.[33] Thus the topology of a normed linear space is determined by suitably shifting the whole family of neighborhoods of the origin 0. In other words, it is uniform

---

[31]In short, often abbreviated as *normed space*. Strictly speaking, we may have to write $(X, \|\ \|)$. But it is rather bothersome to write as precisely as this, so we may often write X. Of course, it is quite possible that a different norm is introduced on the same vector space X. In such a case, the two spaces should be regarded as different normed linear spaces. Also, when we wish to explicitly show where the norm is given, one may write $\|\cdot\|_X$, indicating the space X. For example, for an ℓ^2 space to be defined later (Example 2.1.17, p. 44), its norm will be denoted by $\|x\|_{\ell^2}$, or simply by $\|x\|_2$.

[32]For a precise definition in full generality, see Appendix A. For the moment, it is enough to understand that it refers to a concept related to the convergence of sequences, etc., and a notion of closeness of points.

[33]That is, $B(x,\epsilon) = x + B(0,\epsilon)$.

over the whole space, and hence many topological arguments (such as a proof of continuity) can be reduced to those at the origin.

Definition 2.1.5. A mapping $f : X \to Y$ (not necessarily linear) between normed linear spaces X and Y is said to be *continuous at* $x \in X$ if $x_n \to x$ implies $f(x_n) \to f(x)$. It is said to be *continuous* if it is continuous at every point $x \in X$.

Remark 2.1.6. If a mapping A is linear, then the above definition is equivalent to requiring that $x_n \to 0$ imply $Ax_n \to 0$. In fact, if $x_n \to x$, putting $y_n := x_n - x$ reduces the argument to that at the origin because $y_n \to 0$ implies $Ay_n \to 0$, which means $Ax_n \to Ax$.

The norm $\|\cdot\|$ itself is a real-valued function defined on the associated normed linear space. This function is indeed a continuous function in the sense defined above.

Lemma 2.1.7. *The mapping*

$$\|\cdot\| : X \to \mathbb{R} : x \mapsto \|x\|$$

is always continuous in a normed linear space X (equipped with norm $\|\cdot\|$).

Proof. Observe the inequality

$$\big|\|x\| - \|y\|\big| \le \|x - y\|, \quad x, y \in X. \tag{2.8}$$

This follows immediately from applying the triangle inequality (2.4) to $x = (x - y) + y$. Then $x_n \to x$ implies $\|x_n - x\| \to 0$, and hence by (2.8) $\big|\|x_n\| - \|x\|\big| \to 0$ follows. □

We now give a characterization of continuous linear maps. To this end, let us first introduce the notion of boundedness.

Definition 2.1.8. Let X be a normed linear space. A subset E of X is said to be *bounded* if

$$\sup\{\|x\| : x \in E\} < \infty. \tag{2.9}$$

A linear mapping $A : X \to Y$ between normed linear spaces X, Y is said to be *bounded* if for every bounded subset $E \subset X$, $AE = \{Ax : x \in E\}$ is bounded in Y.

Remark 2.1.9. A linear map (or mapping) between normed linear spaces is often called a linear *operator*.

Lemma 2.1.10. *For a linear map $A : X \to Y$ between normed linear spaces X and Y to be bounded, it is necessary and sufficient that there exist $M > 0$ such that*

$$\|Ax\| \le M \|x\| \quad \forall x \in X. \tag{2.10}$$

Proof. Suppose (2.10) holds. Let E be any bounded subset of X. Then

$$\sup\{\|x\| : x \in E\} \le C$$

for some constant $C > 0$. Then for every $x \in E$, $\|Ax\| \le MC$.

Conversely, suppose A is bounded. For $E := \{x : \|x\| \leq 1\}$, AE must be bounded, so that $\|Ax\| \leq M$ for all $x \in E$. For every nonzero $x \in X$, $x/\|x\| \in E$, and hence $\|A(x/\|x\|)\| \leq M$. By linearity of A and (2.3), $\|Ax\| \leq M\|x\|$. □

We then have the following proposition.

Proposition 2.1.11. *A linear mapping $A : X \to Y$ between normed linear spaces X and Y is continuous if and only if it is bounded.*

Proof. Let us first show the sufficiency. By Lemma 2.1.10, (2.10) holds. Suppose $x_n \to 0$. It readily follows that

$$\|Ax_n\| \leq M\|x_n\| \to 0.$$

Conversely, suppose that (2.10) does not hold for any positive M. Then for every positive integer n, there exists x_n that does not satisfy (2.10) for this n, namely,

$$\|Ax_n\| > n\|x_n\|, \quad n = 1, 2, \ldots.$$

Since x_n is clearly nonzero, put $w_n := x_n/(\sqrt{n}\|x_n\|)$. Then (2.3) yields

$$\|Aw_n\| = \frac{1}{\sqrt{n}\|x_n\|}\|Ax_n\| > \frac{1}{\sqrt{n}\|x_n\|}n\|x_n\| = \sqrt{n}.$$

Again by (2.3), $\|w_n\| = 1/\sqrt{n} \to 0$, but $\|Aw_n\| = \sqrt{n} \to \infty$. This contradicts the continuity of A. □

2.1.3 Examples of Normed Linear Spaces

Example 2.1.12. *Euclidean space $\mathbb{R}^n$.* Its norm is defined by

$$\|x\|_2 := \left\{\sum_{k=1}^{n}|x_k|^2\right\}^{1/2}, \tag{2.11}$$

where $x = [x_1, \ldots, x_n]^T$.

If we introduce a norm by

$$\|x\|_1 := \sum_{k=1}^{n}|x_k|, \tag{2.12}$$

or by

$$\|x\| := \max_{1 \leq k \leq n}|x_k|, \tag{2.13}$$

or still differently by

$$\|x\|_p := \left\{\sum_{k=1}^{n}|x_k|^p\right\}^{1/p}, \quad p \geq 1, \tag{2.14}$$

each space becomes a (different) normed linear space.[34]

---

[34]They are not called Euclidean spaces.

Let us make sure that they are indeed normed linear spaces. The only property that should be checked is the triangle inequality (2.4). The other two properties are obvious. The cases for (2.12) and (2.13) follow from the respective inequality for each component. Those for (2.11) and (2.14) are so-called Minkowski's inequality:

$$\left\{\sum_{k=1}^{n}|x_k+y_k|^p\right\}^{1/p} \leq \left\{\sum_{k=1}^{n}|x_k|^p\right\}^{1/p} + \left\{\sum_{k=1}^{n}|y_k|^p\right\}^{1/p}.$$

(For the latter, see Theorem A.8.2 (p. 245).)

Example 2.1.13. $C[a,b]$ denotes the set of all continuous functions on the closed interval $[a,b]$. Introduce the following norm:

$$\|f\| := \sup_{a \leq t \leq b} |f(t)|. \tag{2.15}$$

Then $C[a,b]$ becomes a normed linear space.

Exercise 2.1.14. Show the above.

Example 2.1.15. It is possible to introduce a different norm to the same set $C[a,b]$:

$$\|f\|_1 := \int_a^b |f(t)|dt \tag{2.16}$$

or

$$\|f\|_2 := \left\{\int_a^b |f(t)|^2 dt\right\}^{1/2}. \tag{2.17}$$

$C[a,b]$ becomes a normed linear space with respect to either of these norms.

Exercise 2.1.16. Verify the above for $\|\cdot\|_1$. (The case for $\|\cdot\|_2$ will become clear in Chapter 3 on Hilbert spaces.)

Example 2.1.17. Sequence spaces ℓ^p, $p \geq 1$. Consider the space ℓ^p consisting of infinite sequences $\{x_n\}$ as follows:

- $\ell^1 := \{\{x_n\} : \sum_{n=1}^{\infty}|x_n| < \infty\}$.

$$\|\{x_n\}\|_1 := \sum_{n=1}^{\infty}|x_n|. \tag{2.18}$$

- $\ell^p := \{\{x_n\} : \sum_{n=1}^{\infty}|x_n|^p < \infty\}, \quad 1 < p < \infty$.

$$\|\{x_n\}\|_p := \left\{\sum_{n=1}^{\infty}|x_n|^p\right\}^{1/p}. \tag{2.19}$$

- $\ell^\infty := \{\{x_n\} : \sup_{1 \leq n < \infty} |x_n| < \infty\}.$

$$\|\{x_n\}\|_\infty := \sup_{1 \leq n < \infty} |x_n|. \tag{2.20}$$

Each of the spaces is a normed linear space with respect to the respective norm.

To see this, we must verify the triangle inequality as in Example 2.1.12. But this is also an immediate consequence of Minkowski's inequality (see Appendix A, Theorem A.8.2 (p. 245)).

Example 2.1.18. The space $L^p(a,b)$, $a < b$, $p \geq 1$, of all functions whose p-powered modulus is Lebesgue integrable[35] on interval (a,b) (we may also take a closed interval $[a,b]$; for (half) open intervals $a = -\infty$ or $b = \infty$ is also allowed). This may be regarded as a variant of the above Example 2.1.17, in which the sums are replaced by integrals, and is defined as follows:

- $L^1(a,b) := \{f : (a,b) \to \mathbb{K} : \int_a^b |f(t)| dt < \infty\}.$

$$\|f\|_1 := \int_a^b |f(t)| dt. \tag{2.21}$$

- $L^p(a,b) := \{f : (a,b) \to \mathbb{K} : \int_a^b |f(t)|^p dt < \infty\}.$

$$\|f\|_p := \left\{\int_a^b |f(t)|^p dt\right\}^{1/p}, \quad 1 < p < \infty. \tag{2.22}$$

- $L^\infty(a,b) := \{f : (a,b) \to \mathbb{K} : \sup_{a < t < b} |f(t)| < \infty\}.$

$$\|f\|_\infty := \sup_{a < t < b} |f(t)|. \tag{2.23}$$

The triangle inequality, and hence the axioms of norm, is likewise seen to be satisfied by the integral form of Minkowski's inequality (Appendix A, Theorem A.8.4, p. 246).

It is known that $L^q \subset L^p$ when a, b are finite and $0 < p < q \leq \infty$ (Lemma A.8.5, p. 246). In particular, $L^2(a,b) \subset L^1(a,b)$ and $L^\infty(a,b) \subset L^p(a,b)$ (but only when a, b are finite).

Exercise 2.1.19. Give an example of a function in $L^1(0,1)$ that does not belong to $L^2(0,1)$. Also, give an example of a function in $L^2(0,\infty)$ that does not belong to $L^1(0,\infty)$.

The spaces in Example 2.1.18 are defined in terms of integrals, in contrast to those in Example 2.1.17. Hence we should take such integrals in the sense of Lebesgue [22, 37], and this induces some subtlety in their treatment. For example, the value of a function at one point does not alter any of the values of the integrals above, so it can well be changed without really altering the essential properties of a function in such spaces. To be more precise, we are allowed to change the values of functions on a set of measure zero (see [20]

---

[35]This is for precision, and if the reader is not familiar with Lebesgue integration theory, he/she should not worry too much about this. See also the discussions after Exercise 2.1.19.

for details). The notion of sup above should also obey this convention. We do not discuss Lebesgue integration theory in this book; for concrete examples of functions, it is usually not an obstacle to take the integral as a Riemann integral. However, the key point here is that those L^p spaces above are closed under limits (i.e., *complete* in a more precise terminology; see section 2.2 below) only by taking the integration in the sense of Lebesgue. This has greatly advanced the theory of function spaces. Refinement of the notion of integration has been very important historically.

Example 2.1.20. $C^m[a,b]$: The space of all functions on $[a,b]$ whose derivatives up to order m exist and are continuous—often referred to as *the space of m times continuously differentiable functions*. This space is a normed linear space with respect to the norm

$$\|\psi\|_{(m)} := \sup_{a \leq t \leq b, 0 \leq k \leq m} \left| \frac{d^k}{dt^k} \psi(t) \right|. \tag{2.24}$$

Example 2.1.21. $C(a,b)$: The space of all real-valued continuous functions on an open interval (a,b). It is not possible to introduce a suitable norm to this space. For if $f \in C(a,b)$, it is possible that $\lim_{t \to a} f(t)$ or $\lim_{t \to b} f(t)$ may be divergent; standard examples of norms do not work for this property. The same is true for $C(-\infty, \infty)$ or $C^m(a,b)$. However, while these spaces are not normed spaces, we can nevertheless define an appropriate topology in the sense indicated in Remark 2.1.4; i.e., we can define the notion of convergence, etc. This motivates a more elaborate treatment of the general structure of topological vector spaces for such spaces, which plays a crucial role in the theory of distributions. See Chapter 6 and Appendix A.

Example 2.1.22. $C^\infty[a,b]$: The space of all infinitely continuously differentiable functions on a closed interval $[a,b]$. Note first that

$$C^\infty[a,b] = \bigcap_{m=0}^\infty C^m[a,b].$$

For any function $f \in C^\infty[a,b]$, it is possible to associate the norm $\|f\|_{(m)}$ (2.24) to it. But none of these are quite appropriate for this space $C^\infty[a,b]$, because there exists a sequence that may converge with respect to such a norm, but the limit no longer belongs to $C^\infty[a,b]$. In other words, the space is not closed under the limiting procedure if we take such a norm— similar to the case of rationals that are not closed with respect to limits. This is the issue of *completeness* which we will discuss in section 2.2. Of course, strictly speaking, there is nothing logically wrong in introducing a topology that does not make the space complete. However, it is far more advantageous to have this property.

The trouble here is that the norm $\|f\|_{(m)}$ governs the behavior of f only up to the derivative of mth order, whereas one wants to control it to derivatives of arbitrary high order to make it compatible with C^∞ functions.

It is thus more appropriate (or advantageous) to equip a topology on this space using *all* of this family of norms. That is, a sequence f_n of functions is defined to converge to f if $\|f_n - f\|_{(m)}$ converges to 0 for each m.

This space with the topology thus introduced is no longer a normed linear space, much less a Banach or a Hilbert space. It, however, belongs to the class called Fréchet spaces and plays an important role in the formulation of distributions we discuss in later chapters.

Similarly as above, we can also introduce a topology on $C^m(-\infty,\infty)$ with infinitely many real-valued functions

$$\|f\|_{n,m} := \sup_{-n\le t\le n; 0\le k\le m}\left|\frac{d^k f}{dt^k}\right|, \quad n=1,2,3,\ldots,$$

called *seminorms*. The reason for calling each of them a *seminorm* and not a norm is that $\|f\|_{n,m}=0$ does not imply $f=0$. But if $\|f\|_{n,m}=0$ for all n, then $f=0$ follows, of course.

The same can be said of $C^\infty(-\infty,\infty)$ with infinitely many seminorms:

$$\|f\|_{n,m} := \sup_{-n\le t\le n; 0\le k\le m}\left|\frac{d^k f}{dt^k}\right|, \quad n,m=1,2,3,\ldots.$$

Here $f_p\to 0$ is defined to mean that $\|f_p\|_{n,m}\to 0$ for every (fixed) m,n. The difference from $C^m(-\infty,\infty)$ lies in the fact that we also take infinitely many seminorms with respect to the order m of differentiation. Again, this is not a normed linear space.

2.1.4 Open and Closed Sets

The notions of open and closed sets are central to discussing topological properties. Indeed, a topology can be introduced using these notions. However, here we give an absolute minimum on these concepts by confining ourselves to normed linear spaces.

Definition 2.1.23. Let M be a subset of a normed linear space X. A point x in M is called an *interior point* if there exists $\epsilon > 0$ such that $B(x,\epsilon) \subset M$. The set of all interior points (can be empty $\emptyset$) of M is called the *interior* of M and is denoted by $\overset{\circ}{M}$. A subset consisting only of interior points, i.e., $M = \overset{\circ}{M}$, is called an *open set*.

A point x of X is a *closure point* or an *adherent point* of M if $M \cap B(x,\epsilon) \neq \emptyset$ for every $\epsilon > 0$. This means that every ϵ-neighborhood of x always has some intersection with M (Figure 2.1). The set of all closure points of M is called the *closure* of M and is denoted by $\overline{M}$. (Note that $M \subset \overline{M}$ always.) A subset M such that $M = \overline{M}$ is called a *closed set*.

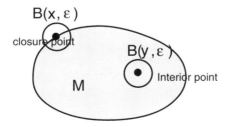

Figure 2.1. *Interior point and closure point*

Exercise 2.1.24. The whole space X and the empty set $\emptyset$ are both open and, simultaneously, closed sets. A singleton $\{x\}$ and closed unit ball

$$\{x \in X : \|x\| \leq 1\}$$

are closed sets. Show these facts.

Proposition 2.1.25. *The complement M^c of an open set M is closed and, conversely, the complement N^c of a closed set N is open.*

Proof. Recall that the complement M^c of M is

$$M^c = \{x \in X : x \notin M\}.$$

Suppose $m \notin M^c$, i.e., $m \in M$. Since M is open, there exists $\epsilon > 0$ such that $B(m, \epsilon) \subset M$. Hence $B(m, \epsilon) \cap M^c = \emptyset$. Thus m cannot be a closure point of M^c. This means that M^c contains all its closure points, so that it is a closed set.

Suppose now that N is a closed set. If $x \notin N$, x cannot be a closure point of N because N is closed. Then by definition there exists $\epsilon > 0$ such that $B(x, \epsilon) \cap N = \emptyset$. In other words, $B(x, \epsilon) \subset N^c$. That is, x is an interior point of N^c. Since $x \in N^c$ is arbitrary, N^c is an open set. □

Exercise 2.1.26. Prove the following statements:

1. The union of arbitrary (can be infinitely many) open sets is again an open set.

2. The intersection of finitely many open sets is open.

3. The intersection of arbitrary (can be infinitely many) closed sets is again closed.

4. The union of finitely many closed sets is closed.

The next lemma should also be obvious. We leave its proof as an exercise.

Lemma 2.1.27. *Let M be a subset of a normed linear space X. A necessary and sufficient condition for M to be closed is that if $x_n \in M$ converges to $x \in X$, then x belongs to M, i.e., $x \in M$.*

The following proposition is an important consequence of continuity.

Proposition 2.1.28. *Let X, Y be normed linear spaces and $A : X \rightarrow Y$ a continuous linear map. Let N be a closed subset of Y. Then $A^{-1}(N)$ is a closed subset of X. In particular, $\ker A$ is a closed set. Similarly, for every open subset $O \subset Y$, $A^{-1}(O)$ is an open set.*

Proof. Take any sequence x_n in $A^{-1}(N)$ that converges to $x \in X$. It suffices to show that $x \in A^{-1}(N)$. By definition, $Ax_n \in N$, and the continuity of A implies $Ax_n \rightarrow Ax$. On the other hand, since N is closed, Lemma 2.1.27 yields $Ax \in N$, that is, $x \in A^{-1}(N)$.

The assertion for ker A follows from the fact that the singleton $\{0\}$ is a closed set.

Finally, if O is open, then O^c is closed, and hence $A^{-1}(O^c)$ is closed. Since clearly $A^{-1}(O^c) = (A^{-1}(O))^c$, $(A^{-1}(O))$ is open. □

2.2   Banach Spaces

Normed linear spaces are not enough to develop a rich theory of analysis. The drawback is that we cannot freely take limits there. For this we need the concept of completeness which we will discuss subsequently.

2.2.1   Completeness

The length of the diagonal of the square with side length 1 is $\sqrt{2}$. It is well known that the Greeks were puzzled by the fact that it cannot be expressed as the ratio of two integers.

Admitting that $\sqrt{2}$ cannot be expressed as such, we still encounter an annoying question: is it really a number? Is it really an entity we should deal with in mathematics? One may be quite surprised to see that this is not at all an easy question with a ready-made answer. For, recalling the historical development of numbers,

$$\text{Natural numbers} \rightarrow \text{Integers} \rightarrow \text{Rationals},$$

we see that they developed according to the need for required operations (integers for addition and subtraction, and rationals for multiplication and division). These procedures have been constructive. The requirement of justifying the length of the diagonal of a square is, to the contrary, a marked contrast. Imagine a "number" that becomes 2 when squared; this sounds quite tautological if we do not yet know what we mean by "numbers." The situation looks different from the case of rationals—for example, calling $1/5$ the quantity obtained by equally dividing a unit five times.[36] On the other hand, no matter how we may choose a rational x, we cannot make x^2 exactly equal to 2, although we may be able to make it close to 2 (in fact, arbitrarily close). This arises from the fact that we cannot take limits freely in the set of rationals. This is the source of the bewilderment of the Greeks.

We had to wait until the construction of real numbers by Dedekind, Cantor, Weierstrass, and others for the resolution of this problem. We will not delve into this question further here; for details, see, for example, the treatment by Dedekind [6].[37] We cannot, however, emphasize strongly enough the importance of the fact that the greatest benefit of the construction of the reals is that limits can freely be taken there; that is, this has provided a safe ground where analysis can be flawlessly developed. The set of real numbers is "closed" under the operation of taking limits. Such a property is called *completeness* in analysis.

The importance of completeness is not necessarily restricted to the reals. Another example is the set of complex numbers. Moreover, since we are attempting to develop

---

[36]One could argue that even rationals require a more conceptual and abstract set-up if we want to introduce them rigorously. This is indeed correct in that we need the concept of equivalence classes to justify them. But we will not delve into this question further. What we emphasize is the level of abstraction needed for real numbers and the difference from the impression when dealing with these concepts. The difference perhaps stems from the difficulty of dealing with infinity (or a procedure involving infinity such as taking limits).

[37]See also Appendix A.

analysis over normed linear spaces, their completeness is, needless to say, of central importance. We thus start with the following definition.

Definition 2.2.1. A sequence $\{x_n\}$ in a normed linear space X is called a *Cauchy sequence* if $\|x_n - x_m\| \to 0$ as $m, n \to \infty$; in other words, for every $\epsilon > 0$, there exists an integer N such that for all $n, m > N$

$$\|x_n - x_m\| < \epsilon.$$

The space X is said to be *complete* if every Cauchy sequence is convergent, i.e., always has a limit in X. A complete normed linear space is called a *Banach space*.

The definition above may appear to lack a motivation, so let us add some further descriptions. First note that a convergent sequence is always a Cauchy sequence. This is almost trivial, but we here state it in the form of a lemma.

Lemma 2.2.2. *Let X be a normed linear space, and let $x_n \to x$. Then $\{x_n\}$ is a Cauchy sequence.*

Proof. Apply the triangle inequality (2.4) to $x_n - x_m = (x_n - x) + (x - x_m)$ to obtain

$$\|x_n - x_m\| \leq \|x_n - x\| + \|x_m - x\|.$$

This readily implies that $\|x_n - x_m\| \to 0$ whenever $n, m \to \infty$.   □

Remark 2.2.3. Note that when we say a "convergent sequence," the existence of a limit is part of its definition.

As seen above, a convergent sequence is always a Cauchy sequence; i.e., the distance between x_n and x_m approaches 0 as $n, m \to \infty$. The point of accumulation is the limit of this sequence. On the other hand, a sequence with its mutual distance between two elements approaching to zero, i.e., a Cauchy sequence, does not necessarily have a limit. That is, the converse of the above lemma is not always true; the limit (so to speak) does not necessarily exist in the space. Consider, for example, a positive sequence of rationals whose squares converge to 2. When considered over the space of rational numbers, we readily see that this sequence does not admit a limit (in the rationals). We thus say that a space is complete (i.e., a Banach space) if such a phenomenon does not occur, i.e., every Cauchy sequence admits a limit.

2.2.2   Examples of Banach Spaces

Example 2.2.4. The set $\mathbb{R}$ of real numbers. Let $\{x_n\}_{n=1}^{\infty}$ be a Cauchy sequence in $\mathbb{R}$. Then there exists $M > 0$ such that $|x_n| \leq M$ for all n. In fact, there exists N such that for $n, m \geq N$, $|x_n - x_m| < 1$. Then, let $M := \max\{|x_1|, |x_2|, \ldots, |x_N|\} + 1$ to satisfy this requirement. By the Bolzano–Weierstrass theorem, there is an accumulation point x_0 of this sequence $\{x_n\}_{n=1}^{\infty}$. This is clearly a unique accumulation point, and also $\lim_{n \to \infty} x_n = x_0$.

Likewise, n-dimensional spaces $\mathbb{R}^n$ and $\mathbb{C}^n$ are complete.

Example 2.2.5. The space $C[a,b]$ of all real-valued continuous functions on the interval $[a,b]$ considered in Example 2.1.13. Let us show that this space is complete. Let $\{f_n\}_{n=1}^{\infty}$ be a Cauchy sequence. That is,

$$\|f_n - f_m\| \to 0, \quad n,m \to \infty.$$

Clearly, for each fixed t,

$$|f_n(t) - f_m(t)| \leq \|f_n - f_m\| \tag{2.25}$$

holds. Hence for each t, $\{f_n(t)\}_{n=1}^{\infty}$ is a Cauchy sequence as a sequence of real numbers. The completeness of $\mathbb{R}$ given above now yields its (unique) limit, and we denote it by $f(t)$. Since this holds for every t, we obtain a function f on $[a,b]$. We will now prove that f belongs to $C[a,b]$ and is a limit of f_n.

Now take any $\epsilon > 0$. There exists $N(\epsilon) > 0$ such that $n,m \geq N(\epsilon)$ implies $\|f_n - f_m\| < \epsilon$. Now fix $n > N(\epsilon)$ and let $m \to \infty$ in (2.25) to obtain

$$|f_n(t) - f(t)| \leq \limsup_{m \to \infty} \|f_n - f_m\| \leq \epsilon.$$

Observe here that the right-hand side does not depend on t. Thus the sequence $\{f_n\}$ converges uniformly to f. Hence f is continuous as a uniform limit of continuous functions; that is, $f \in C[a,b]$. This means that $C[a,b]$ is complete and is a Banach space.

Example 2.2.6. $C^m[a,b]$ given in Example 2.1.20. We again need to prove the completeness. To this end, we can make use of Example 2.2.5 above as follows.

Take a Cauchy sequence $\{f_n\}_{n=1}^{\infty}$. The definition of the norm (2.24) readily implies that the derivatives $d^k f_n / dt^k$ of each order k form a Cauchy sequence in $C[a,b]$. As in Example 2.2.5 above, each of them converges uniformly to a continuous function $f^{(k)}$. Furthermore, the relation

$$\frac{d}{dt}\frac{d^k f_n}{dt^k} = \frac{d^{k+1} f_n}{dt^{k+1}}$$

also holds in the limit $n \to \infty$. This implies that $f^{(k)}$ must be the kth derivative of f (:= limit of f_n). Clearly it also follows that $\|f_n - f\|_{(m)} \to 0$ in the sense of the norm (2.24) (see p. 46), and this completes the proof.

2.3 Closed Linear Subspaces and Quotient Spaces

Let X be a normed linear space and M a subspace of X. If M is a closed subset of X in the sense of Definition 2.1.23, M is a *closed (linear) subspace* of X. M then becomes a normed linear space with respect to the same norm of X. If, further, X is a Banach space, M is also a Banach space.

Exercise 2.3.1. Verify the facts above.

As a particular example, if a subspace is finite-dimensional as a vector space, then it is always a closed subspace. This fact is important, but its proof is not so trivial, despite its innocent-looking finite-dimensionality hypothesis. Let us here give a proof.

Theorem 2.3.2. *Let X be a normed linear space and M a finite-dimensional subspace of X. Then M is a closed subspace.*

Proof. We prove this by induction on the dimension N of M. First let $N = 1$. There exists $e \neq 0$ such that

$$M = \{\alpha e \,:\, \alpha \in \mathbb{K}\}.$$

Suppose that $\alpha_n e \to x$. Lemma 2.2.2 implies that $\{\alpha_n e\}$ is a Cauchy sequence. On the other hand, since $\|\alpha_n e - \alpha_m e\| = |\alpha_n - \alpha_m| \, \|e\|$, $\{\alpha_n\}$ itself is a Cauchy sequence. By the completeness of $\mathbb{K}$ (Example 2.2.4 above), there exists $\alpha \in \mathbb{K}$ such that $\alpha_n \to \alpha$. Then $\alpha_n e \to \alpha e$, and hence M is a closed subset. (Moreover, $x = \alpha e$ because $\alpha_n e \to x$.)

   Now suppose that the conclusion holds for the subspaces with dimension less than or equal to $N - 1$, and also that $\dim M = N$. Take a basis $\{e_1, \ldots, e_N\}$ for M, and let M_k be the subspace spanned by $[e_1, \ldots, e_{k-1}, e_{k+1}, \ldots, e_N]$. Define

$$\delta_k := d(e_k, M_k) := \inf_{x \in M_k} \{\|e_k - x\|\}.$$

Observe $\delta_k > 0$. On one hand, if $\delta_k = 0$, then there would exist, by the definition of inf (see Appendix A), a sequence $\{x_n\}$ of M_k such that $x_n \to e_k$. On the other hand, M_k is $N - 1$-dimensional, and is closed by the induction hypothesis. This would imply $e_k \in M_k$, but this contradicts the linear independence of $\{e_1, \ldots, e_N\}$. Hence $\delta_k > 0$. Thus let $\delta := \min_{1 \leq k \leq N} \delta_k$, and then $\delta > 0$. Now suppose

$$x_n = \sum_{j=1}^{N} \alpha_j^n e_j \to x, \quad n \to \infty.$$

Again by Lemma 2.2.2, $\{x_n\}$ is a Cauchy sequence. For each fixed k, we have

$$\|x_n - x_m\| = \left\| (\alpha_k^n - \alpha_k^m) e_k + \sum_{j \neq k} (\alpha_j^n - \alpha_j^m) e_j \right\| \geq |\alpha_k^n - \alpha_k^m| \delta$$

by the definition of δ and δ_k. On the other hand, since $\|x_n - x_m\| \to 0$, $|\alpha_k^n - \alpha_k^m| \to 0$ follows, so that each $\{\alpha_k^n\}$ is a Cauchy sequence. By the completeness of $\mathbb{K}$, it converges to $\alpha_k \in \mathbb{K}$. We then have

$$\left\| \sum_{k=1}^{N} \alpha_k^n e_k - \sum_{k=1}^{N} \alpha_k e_k \right\| \leq \sum_{k=1}^{N} |\alpha_k^n - \alpha_k| \, \|e_k\| \leq N \cdot \max_{1 \leq k \leq N} |\alpha_k^n - \alpha_k| \, \|e_k\|.$$

Hence x_n converges to $\sum_{k=1}^{N} \alpha_k e_k$. (Thus we have $x = \sum_{k=1}^{N} \alpha_k e_k$.) Since $\sum_{k=1}^{N} \alpha_k e_k \in M$, M is a closed subspace. $\qquad\qquad\qquad\square$

   We now turn our attention to the quotient space X/M introduced in Chapter 1. We will also show that X/M can naturally be regarded as a normed linear space by suitably introducing a norm to this quotient space.

Definition 2.3.3. Let $(X, \|\cdot\|)$ be a normed linear space, M a closed subspace of X, and X/M the quotient space of X modulo M. Define the norm of the equivalence class $[x] = x + M$ by

$$\|[x]\| := \inf_{m \in M} \|x + m\|_X. \tag{2.26}$$

This is the least value among all norms $\|x + m\|$ in the equivalence class $x + M$.

Let us show that this definition indeed satisfies the axioms for norms and hence that X/M is a normed linear space with respect to this norm.

Proposition 2.3.4. *The norm defined by* (2.26) *satisfies the axioms* (2.2)–(2.4) *for a norm. Hence X/M becomes a normed linear space with respect to this norm. Moreover, if X is a Banach space, then X/M is also a Banach space.*

Proof. First, if $[x] = 0$, then $[x] = M$ as a set. Since M is a subspace, $0 \in M$. Then $\|[0]\| = 0$ clearly follows from (2.26). Conversely, if $\|[x]\| = 0$, then definition (2.26) and the property of the infimum (p. 237, Lemma A.2.1) imply that for each n there exists $u_n \in M$ such that

$$\|x + u_n\| < \|[x]\| + \frac{1}{n} = \frac{1}{n}.$$

Then $u_n \to -x$, and since M is a closed subspace, $-x \in M$ follows. Hence $0 = x - x \in M$, which implies $[x] = [0]$.

For scalar multiplication, when $\alpha \neq 0$, we have

$$\begin{aligned}
\|\alpha[x]\| = \|[\alpha x]\| \quad &(\text{note } \alpha[x] = [\alpha x]) \\
&= \inf_{m \in M} \|\alpha x + m\|_X \\
&= \inf_{m \in M} \|\alpha(x + m)\|_X = \inf_{m \in M} |\alpha| \|x + m\|_X \\
&= |\alpha| \|[x]\|.
\end{aligned}$$

Since this equality clearly holds for $\alpha = 0$, (2.3) also follows.

Let us now prove the triangle inequality (2.4). In view of the property of the infimum (Lemma A.2.1), for every $\epsilon > 0$, there exist $u, v \in M$ such that

$$\|x + u\| - \frac{\epsilon}{2} < \|[x]\|, \quad \|y + v\| - \frac{\epsilon}{2} < \|[y]\|.$$

We then have

$$\begin{aligned}
\|[x] + [y]\| = \|[x + y]\| \\
= \inf_{m \in M} \|x + y + m\|_X \\
\leq \|x + y + u + v\|_X \\
\leq \|x + u\|_X + \|y + v\|_X \\
< \|[x]\| + \|[y]\| + \epsilon.
\end{aligned}$$

Since $\|[x] + [y]\|$, $\|[x]\| + \|[y]\|$ are both constants and $\epsilon > 0$ is arbitrary, it follows that

$$\|[x] + [y]\| \leq \|[x]\| + \|[y]\|.$$

This is the triangle inequality.

Finally, let us prove that when X is a Banach space, so is X/M, i.e., it is complete. Let $[x_n]$ be a Cauchy sequence, i.e.,

$$\|[x_n] - [x_m]\| \to 0, \quad n, m \to \infty.$$

By taking a suitable subsequence $[x_{n_i}]$, we can make

$$\|[x_{n_i}] - [x_{n_j}]\| < \frac{1}{2^i}, \quad i \le j.$$

Hence we can assume from the outset that

$$\|[x_n] - [x_m]\| < \frac{1}{2^n}, \quad n \le m.$$

Take any element x_n in $[x_n]$. Then by the property of the infimum,

$$\|x_{n+1} - x_n + z_n\| < \frac{1}{2^n}$$

for some $z_n \in M$. Now set $w_1 := x_1$, $w_n := x_n - x_{n-1} + z_{n-1}$, and $y_n := w_1 + w_2 + \cdots + w_n$. Then we have $y_n = x_n + (z_{n-1} + \cdots + z_1) \in x_n + M = [x_n]$, so that $[y_n] = [x_n]$. Moreover,

$$\|y_{n+1} - y_n\| = \|x_{n+1} - x_n + z_n\| < \frac{1}{2^n}.$$

This yields

$$\|y_{n+p} - y_n\| \le \sum_{k=1}^{p} \|y_{n+k} - y_{n+k-1}\| < \sum_{k=1}^{p} \frac{1}{2^{n+k-1}} < \frac{1}{2^{n-1}}.$$

Hence $\{y_n\}$ is a Cauchy sequence in X. Thus there exists a limit y, and $y_n \to y$ holds. Since $[y_n] = [x_n]$, $[x_n] \to [y]$; that is, the Cauchy sequence $\{[x_n]\}$ is convergent. □

2.4   Banach's Open Mapping and Closed Graph Theorems

Banach's Open Mapping and Closed Graph Theorems play important roles in applications. We first state these theorems. The proofs will be given in section 2.5 using Baire's category argument.

Theorem 2.4.1 (Banach's Open Mapping Theorem). *Let X, Y be Banach spaces and $f : X \to Y$ a continuous linear mapping. If $f(X) = Y$, then f is an open mapping; that is, for every open set $O \subset X$, $f(O)$ is an open subset of Y.*

This theorem then yields the following closed graph theorem.

Theorem 2.4.2 (Banach's Closed Graph Theorem). *Let X, Y be Banach spaces and $f : X \to Y$ a linear mapping. Let $G(f)$ denote the* graph *of f, i.e., the subset of $X \times Y$ defined by*

$$G(f) := \{(x, f(x)) \in X \times Y : x \in X\}.$$

If $G(f)$ is a closed subset of $X \times Y$, then f is a continuous mapping.

2.5 Baire's Category Theorem

Baire's category argument plays a crucial role in various aspects of analysis. We start with the following definition.

Definition 2.5.1. Let X be a topological space and M a subset of X. M is said to be *nowhere dense* if the closure $\overline{M}$ does not contain a nonempty open subset of X. M is said to be of *the first category* if M is expressible as the union of a countable number of sets each of which is nowhere dense in X. M is said to be of *the second category* if it is not of the first category.

The central theorem is the following. (For metric spaces, see Appendix A.)

Theorem 2.5.2 (Baire's Category Theorem). *A nonempty complete metric space (X,d) is of the second category, where d denotes the distance function.*

Proof. Suppose

$$X = \bigcup_{n=1}^{\infty} M_n, \tag{2.27}$$

where each M_n is nowhere dense. Since M_1 is nowhere dense, there exists an element $x_1 \in \overline{M_1}^c$. Since $(\overline{M_1})^c$ is an open set, there exists a closed ball $K_1 := \{x : d(x,x_1) \leq r_1\}$ such that $K_1 \cap M_1 = \emptyset$. Now consider the ball $L_2 := \{x : d(x,x_1) \leq r_1/2\}$. Since M_2 is nowhere dense, $L_2 \cap M_2 \neq L_2$. Hence there exists $x_2 \in L_2 \cap M_2^c$. Then there exists a sufficiently small $r_2 > 0$ such that $K_2 := \{x : d(x,x_2) \leq r_2\} \subset K_1$ such that $K_2 \cap M_2 = \emptyset$. We may assume $r_2 < r_1/2$ without loss of generality. Proceeding inductively, we obtain a sequence of closed balls $K_1 \supset K_2 \supset K_3 \supset \cdots$ such that

1. the radii r_n of K_n converge to 0 as $n \to \infty$, and

2. $K_n \cap M_n = \emptyset$ for every n.

The centers x_n of K_n clearly form a Cauchy sequence, and hence they converge to a limit x_0 by the completeness of X. Clearly x_0 belongs to all K_n. On the other hand, by the condition $K_n \cap M_n = \emptyset$, x_0 belongs to no M_n, and this contradicts (2.27). $\qquad\square$

Remark 2.5.3. The same result holds if we replace X by any nonempty open subset of X.

2.5.1 Proof of The Open Mapping Theorem 2.4.1

We now give a proof of the Open Mapping Theorem, Theorem 2.4.1. While the theorem holds for more general spaces (e.g., Fréchet spaces), we here content ourselves with the case for Banach spaces. The same is true for the Closed Graph Theorem. For a more general treatment, see, for example, [29, 51, 66]. We start with the following lemma.

Lemma 2.5.4. *Let X, Y be normed linear spaces, and let $f : X \to Y$ be a continuous linear mapping. Suppose that the image $\operatorname{im} f = f(X)$ is a set of the second category in Y. Then the closure $\overline{f(B_1)}$ of the image of the unit ball $B_1 := \{x : \|x\| < 1\}$ contains a neighborhood of the origin of Y.*

Proof. First observe that $X = \cup_{n=1}^{\infty}(nB_1)$, where $nB_1 = \{nx : x \in B_1\}$. Hence $f(X) = \cup_{n=1}^{\infty} f(nB_1)$. Since $f(X)$ is of the second category, $\overline{f(nB_1)}$ contains a neighborhood for some n. Since $f(nB_1) = nf(B_1)$, this clearly means that $\overline{f(B_1)}$ itself contains a neighborhood. By linearity of f, this also clearly means that $\overline{f(B_1)}$ contains a neighborhood of the origin.   □

Proof of Theorem 2.4.1. Since Y is a complete metric space, it is of the second category. Hence by the previous lemma $\overline{f(B_1)}$ contains a neighborhood of the origin.

Let us first prove that the image $f(B_1)$ of the unit ball B_1 in X contains a ball in Y. To this end, denote, for the moment, by $B_{X,\epsilon}$ and $B_{Y,\epsilon}$, the ϵ balls of the origin of X and Y, respectively.

By the previous lemma, there exists a positive sequence $\delta_i \to 0$ such that

$$B_{Y,\delta_i} \subset \overline{f(B_{X,1/2^i})}, \quad i = 1, 2, \ldots. \tag{2.28}$$

Our objective is to show that $B_{Y,\delta_1} \subset f(B_{X,1})$.

Take any $y \in B_{Y,\delta_1}$. We want to show that $y = f(x)$ for some $x \in B_{X,1}$. By (2.28), this y can be approximated arbitrarily closely by elements in $\overline{f(B_{X,1/2})}$. Hence for δ_2 there exists $x_1 \in B_{X,1/2}$ such that

$$\|y - f(x_1)\| < \delta_2.$$

Since $y - f(x_1) \in B_{Y,\delta_2}$, (2.28) again implies that there exists $x_2 \in B_{X,1/2^2}$ such that

$$\|y - f(x_1) - f(x_2)\| < \delta_3.$$

Proceeding inductively, we obtain a sequence $\{x_i\}_{i=1}^{\infty}$, $x_i \in B_{X,1/2^i}$, such that

$$\left\| y - f\left(\sum_{i=1}^{n} x_i\right) \right\| < \delta_{n+1}, \quad n = 1, 2, \ldots. \tag{2.29}$$

Since

$$\left\| \sum_{i=m+1}^{n} x_i \right\| \leq \sum_{i=m+1}^{n} \|x_i\| \leq \sum_{i=m+1}^{n} 1/2^i = 1/2^m,$$

$\{\sum_{i=1}^{n} x_i\}$ forms a Cauchy sequence. The completeness of X readily implies that there exists a limit $x = \sum_{i=1}^{\infty} x_i$. It clearly follows from (2.29) and the continuity of f that $y = f(x)$. Also,

$$\|x\| = \lim_{n \to \infty} \left\| \sum_{i=1}^{n} x_i \right\| \leq \lim_{n \to \infty} \sum_{i=1}^{n} \|x_i\| \leq \left(\sum_{i=1}^{\infty} 2^{-i}\right) = 1. \tag{2.30}$$

That is, $x \in B_{X,1}$. Hence $y = f(x)$ belongs to $f(B_{X,1})$. Since $y \in B_{Y,\delta_1}$ was arbitrary, this implies $B_{Y,\delta_1} \subset f(B_{X,1})$.

Note that by the linearity of f, any ball $B_{X,\epsilon}$ is mapped surjectively to a set that contains a ball $B_{Y,\delta}$ for some $\delta > 0$.

Now let O be a nonempty open set in X, and let $x \in O$. Take sufficiently small ϵ such that $x + B_{X,\epsilon} \subset O$. Then $f(B_{X,\epsilon})$ contains a neighborhood $B_{Y,\delta}$ for some δ. It then follows that $f(O) \supset f(x + B_{X,\epsilon}) = f(x) + f(B_{X,\epsilon}) \supset f(x) + B_{Y,\delta}$. This means that $f(O)$

contains a neighborhood of each of its points. This means that f maps each open set of X onto an open set of Y. □

The following corollary is an easy consequence of Theorem 2.4.1.

Corollary 2.5.5. *Let f be a continuous linear mapping from a Banach space X onto another Banach space Y, which is also injective. Then the inverse mapping f^{-1} is also a continuous linear mapping from Y onto X.*

2.5.2   Proof of the Closed Graph Theorem, Theorem 2.4.2

We are now ready to give a proof of the Closed Graph Theorem.

Let $f : X \to Y$ be a linear mapping with closed graph; i.e., $G(f)$ is closed. First note that the product space $X \times Y$ is a Banach space with respect to the product norm:

$$\|(x, y)\|_{X \times Y} := \|x\|_X + \|y\|_Y.$$

Hence $G(f)$ is also a Banach space, as a closed subspace of $X \times Y$. The linear mapping $U : G(f) \to Y$ defined by

$$U : G(f) \ni (x, f(x)) \mapsto x \in X \tag{2.31}$$

is clearly a continuous linear mapping from $G(f)$ to X. Furthermore, this mapping is bijective. Hence its inverse U^{-1} is a continuous mapping (Corollary 2.5.5). The linear mapping $V : G(f) \to f(X) : (x, f(x)) \mapsto f(x)$ is clearly continuous since it is a projection. Hence $f = V U^{-1}$ is a continuous mapping.

2.5.3   The Uniform Boundedness Principle

The following theorem is called the uniform boundedness principle.

Theorem 2.5.6 (uniform boundedness principle). *Let $\{T_\lambda : \lambda \in \Lambda\}$ be a nonempty family of bounded linear mappings from a Banach space X to a normed linear space Y. Suppose that for each $x \in X$*

$$\sup_{\lambda \in \Lambda} \|T_\lambda x\| < \infty. \tag{2.32}$$

Then $\|T_\lambda\|$ is bounded.

Proof. Define X_n, $n = 1, 2, \ldots$, by

$$X_n := \left\{ x \in X : \sup_{\lambda \in \Lambda} \|T_\lambda x\| \leq n \right\}.$$

By the continuity of T_λ, each X_n is clearly a closed subset of X, and by (2.32)

$$X = \bigcup_{n=1}^{\infty} X_n.$$

Then by Baire's Category Theorem, Theorem 2.5.2, one of X_n contains an open set. By the linearity of each T_λ, we may assume without loss of generality that X_1 indeed contains an open set. Hence there exist $x \in X_1$ and $\delta > 0$ such that whenever $\|x - y\| < \delta$, y belongs to X_1.

Now take any z with $\|z\| < \delta$. It follows that $x + z \in X_1$. Then

$$\|T_\lambda z\| = \|T_\lambda(x + z) - T_\lambda x\| \leq \|T_\lambda(x + z)\| + \|T_\lambda x\| \leq 1 + 1 \leq 2$$

because both $x + z$ and x belong to X_1.

Then for every $w \in X$, $\|\delta w/(2\|w\|)\| < \delta$, and $\|T_\lambda \delta w/(2\|w\|)\| \leq 2$. This readily implies $T_\lambda w \leq \frac{4}{\delta}\|w\|$, irrespective of $\lambda \in \Lambda$. This implies $\sup \|T_\lambda\| \leq \frac{4}{\delta}$. ☐

The following is a direct consequence of this theorem.

Corollary 2.5.7. *Let T_n, $n = 1, 2, \ldots$, be a sequence of bounded linear operators from a Banach space X to a normed linear space Y. Suppose that for each $x \in X$, $\lim_{n \to \infty} T_n x$ exists. Define $T : X \to Y$ by*

$$Tx := \lim_{n \to \infty} T_n x, \quad x \in X.$$

Then T is a bounded linear operator from X to Y.

Proof. The linearity of T follows readily from the linearity of each T_n. Since $\|T_n x\|$ is bounded for each x, $\|T_n\|$ is bounded by the previous theorem, say, $\|T_n\| \leq M$. It clearly follows that $\|T\| \leq M$. ☐

Remark 2.5.8. Both Theorem 2.5.6 and Corollary 2.5.7 are also referred to as the *Banach–Steinhaus theorem*. We also note that they are also valid in a much more general context. For example, when X is a space known as a barreled space, the conclusion remains intact. See [29, 48, 51, 58, 64, 66] for details and related facts.

Problems

1. Show that in a normed linear space, a limit of a sequence is always unique. That is, if $x_n \to x$ and $x_n \to y$, then $x = y$.

2. Show that in a normed linear space, a convergent sequence $\{x_n\}$ is always bounded; i.e., there exists a constant $M > 0$ such that $\|x_n\| \leq M$ for all n.

3. Give an example of a sequence in $C^1[0, 1]$ that is uniformly convergent in itself, but its derivatives do not converge uniformly, so that the limit function does not belong to $C^1[0, 1]$. (Hint: Let $f_n(t) = (2t)^n$ ($0 \leq t \leq 1/2$), and 1 ($1/2 \leq t \leq 1$), and consider their indefinite integrals.)

4. The inclusion relation $L^2(a, b) \subset L^1(a, b)$ given in Example 2.1.18 does not hold if a, b are not finite. Give a counterexample for the case $(a, b) = (0, \infty)$.

Chapter 3

Inner Product and Hilbert Spaces

Most of the readers may already be familiar with the concept of inner products. An inner product space is a vector space equipped with a special kind of metric (a concept measuring length, etc.) called an inner product. This naturally induces a norm, so an inner product space also becomes a normed linear space. Generally speaking, however, norms measure only the size of a vector, whereas inner products introduce more elaborate concepts related to angles and directions such as orthogonality and thus allow for much more detailed study.

In this chapter we start by introducing inner product spaces and Hilbert spaces and then discuss their fundamental properties, particularly the projection theorem, the existence of the orthogonal complement, and orthogonal expansion, and their application to best (optimal) approximation. The orthogonal expansion is later seen to give a basis for Fourier expansion, and the best approximation in this context is important for various engineering applications.

3.1 Inner Product Spaces

We start with the following definition without much intuitive justification.

Definition 3.1.1. Let X be a vector space over $\mathbb{K}$. An *inner product* is a mapping from the product space[38] $X \times X$ into $\mathbb{K}$

$$(\cdot, \cdot) : X \times X \to \mathbb{K}$$

that satisfy the following properties for all $x, y, z \in X$, and every scalar $\alpha \in \mathbb{K}$:

1. $(x, x) \geq 0$, and equality holds only when $x = 0$.

2. $(y, x) = \overline{(x, y)}$.

3. $(\alpha x, y) = \alpha (x, y)$.

4. $(x + y, z) = (x, z) + (y, z)$.

---

[38]A product of sets X, Y is the set of all ordered pairs $\{(x, y) : x \in X\}$. See Appendix A for more details (p. 234).

The reader may already be familiar with this definition, but let us add some explanation for its background. In a more elementary situation, an inner product may be introduced as

$$(x, y) := \|x\| \|y\| \cos \theta.$$

Here $\|x\|$ is supposed to represent the "length" of the vector x, and θ is the angle between x, y. A drawback of this more intuitive definition is that concepts such as length and angle are generally not defined in general vector spaces, so it is not a feasible idea to attempt to define an inner product based on these "undefined" quantities. The right procedure is rather to derive these concepts from that of inner products.

A popular example is the following *standard inner product* defined on $\mathbb{R}^n$:

$$(x, y) := \sum_{i=1}^{n} x_i y_i. \tag{3.1}$$

The space $\mathbb{R}^n$ equipped with this inner product is called the n-dimensional *Euclidean space*.

While this gives a perfect example of an inner product in $\mathbb{R}^n$, this is not the only one. In fact, let $A = (a_{ij})$ be a positive definite matrix, i.e., a symmetric matrix with positive eigenvalues only. Then

$$(x, y)_A := (Ax, y) = \sum_{i,j=1}^{n} a_{ji} x_i y_j \tag{3.2}$$

also induces an inner product. Here (Ax, y) on the right-hand side denotes the standard inner product. Unless A is the identity matrix, this defines an inner product different from the standard one.

A space X equipped with inner product $(\cdot, \cdot)$ is called an *inner product space*.[39] Let $(X, (\cdot, \cdot))$ be an inner product space. Then we can define the *length* of the vector x as

$$\|x\| := \sqrt{(x, x)}. \tag{3.3}$$

As we see below in Proposition 3.1.8, this definition satisfies the axioms of norms (2.2)–(2.4). Hence an inner product space can also be regarded as a normed space.

When two vectors x, y satisfy $(x, y) = 0$, we say that x, y are mutually *orthogonal*, and we write $x \perp y$. If $x \perp m$ for every m in a subset M, we also write $x \perp M$.

If we introduce a different inner product in a vector space, then the notion of the length of the vector will naturally change, so the resulting inner product space should be regarded as a different inner product space.

Example 3.1.2. The n-dimensional complex vector space $\mathbb{C}^n$ endowed with the inner product

$$(x, y) := \sum_{i=1}^{n} x_i \overline{y_i} = y^* x, \quad \text{note the order of } x \text{ and } y,$$

where $x = [x_1, \ldots, x_n]^T$ and $y = [y_1, \ldots, y_n]^T$. This is called the n-dimensional (*complex Euclidean space* or the *unitary space*. The inner product here is an extension of the standard inner product in $\mathbb{R}^n$ and is also referred to under the same terms.

---

[39] If this space is complete, then it is called a Hilbert space. An inner product space (not necessarily complete) is sometimes called a pre-Hilbert space.

Example 3.1.3. Let us give a less trivial example of an inner product space. Let $C[0,1]$ be the vector space consisting of all $\mathbb{K}$-valued ($\mathbb{K} = \mathbb{C}$ or $\mathbb{R}$) continuous functions on the interval $[0,1]$. Then

$$(f,g) := \int_0^1 f(t)\overline{g(t)}dt \tag{3.4}$$

defines an inner product in $C[0,1]$.

Let us first show some basic properties of inner products. We then show that an inner product space naturally becomes a normed linear space.

Proposition 3.1.4. *Let X be an inner product space.*[40] *For every x,y,z in X and every scalar $\alpha \in \mathbb{K}$, the following properties hold:*

1. $(x,y+z) = (x,y)+(x,z)$.

2. $(x,\alpha y) = \overline{\alpha}(x,y)$.

3. $(x,0) = 0 = (0,x)$.

4. *If $(x,z) = 0$ for every $z \in X$, then $x = 0$.*

5. *If $(x,z) = (y,z)$ for every $z \in X$, then $x = y$.*

Proof. The proofs for 1 and 2 are easy and left as an exercise. One needs only use property 2 in Definition 3.1.1 to switch the order of elements and then invoke properties 3 and 4 in the same definition.

Property 3 above follows directly from $(0x,x) = 0 \cdot (x,x) = 0$. Property 4 reduces to property 1 of Definition 3.1.1 by putting $z = x$.

Finally, property 5 follows by applying property 4 to $x - y$. $\square$

Theorem 3.1.5 (Cauchy–Schwarz).[41] *Let $(X,(\cdot,\cdot))$ be an inner product space. For every $x,y \in X$,*

$$|(x,y)| \leq \|x\|\,\|y\|, \tag{3.5}$$

where $\|x\| = \sqrt{(x,x)}$ as defined by (3.3). The equality holds only when x,y are linearly dependent.

Proof. If $y = 0$, then the equality clearly holds. So let us assume $y \neq 0$. For every scalar α,

$$0 \leq (x - \alpha y, x - \alpha y) = (x,x) - \alpha(y,x) - \overline{\alpha}(x,y) + |\alpha|^2(y,y). \tag{3.6}$$

To see the idea, let us temporarily assume that (x,y) is real-valued and also that α is real. Then the right-hand side of (3.6) becomes

$$\|y\|^2 \alpha^2 - 2(x,y)\alpha + \|x\|^2 \geq 0.$$

---

[40]It is often bothersome to write $(X,(\cdot,\cdot))$ to explicitly indicate the dependence on the equipped inner product, so we often write X. When we want to show the dependence on the inner product explicitly, we write $(X,(\cdot,\cdot))$.

[41]Often called Schwarz's inequality.

Completing the square, this is equal to

$$\left(\|y\|\alpha - \frac{(x,y)}{\|y\|}\right)^2 - \left(\frac{(x,y)}{\|y\|}\right)^2 + \|x\|^2.$$

This attains its minimum

$$-\left(\frac{(x,y)}{\|y\|}\right)^2 + \|x\|^2$$

when $\alpha = (x,y)/\|y\|^2$, and hence $|(x,y)|^2 \le \|x\|^2\|y\|^2$.

Considering this in the general case (i.e., α and (x,y) are not necessarily real-valued), let us substitute $\alpha = (x,y)/\|y\|^2$ into (3.6). This yields

$$0 \le \|x\|^2 - \frac{|(x,y)|^2}{\|y\|^2},$$

and hence

$$|(x,y)| \le \|x\|\,\|y\|$$

follows.

Conversely, suppose that x, y are linearly dependent and that $y \ne 0$. Then $x = \lambda y$ for some scalar λ. Then $|(\lambda y, y)| = |\lambda||(y,y)| = |\lambda|\|y\|^2 = \|\lambda y\|\,\|y\|$, and hence the equality holds. $\qquad\square$

Schwarz's inequality readily yields the following *triangle inequality*.

Proposition 3.1.6. *Let X be an inner product space. Then*

$$\|x+y\| \le \|x\| + \|y\|$$

for every x, y in X.

Proof. Apply Schwarz's inequality to

$$\|x+y\|^2 = \|x\|^2 + 2\operatorname{Re}(x,y) + \|y\|^2 \le \|x\|^2 + 2|(x,y)| + \|y\|^2. \qquad\square$$

Proposition 3.1.7 (parallelogram law). *Let X be an inner product space. For every x, y in X,*

$$\|x+y\|^2 + \|x-y\|^2 = 2\|x\|^2 + 2\|y\|^2.$$

Proof. The proof is immediate from

$$\|x+y\|^2 = (x+y, x+y) = \|x\|^2 + (x,y) + (y,x) + \|y\|^2,$$
$$\|x-y\|^2 = \|x\|^2 - (x,y) - (y,x) + \|y\|^2. \qquad\square$$

Now let $(X, (\cdot,\cdot))$ be an inner product space. As we have already seen in (3.3), we can introduce a norm into this space by

$$\|x\| := \sqrt{(x,x)}. \tag{3.7}$$

When we wish to explicitly indicate that this is a norm in space X, it may be denoted as $\|x\|_X$ with subscript X.

Let us ensure that this indeed satisfies the axioms for norms (Definition 2.1.2).

Proposition 3.1.8. *The norm* (3.7) *defined on the inner product space* $(X, (\cdot, \cdot))$ *satisfies the axioms of norms* (2.2)–(2.4). *Thus an inner product space can always be regarded as a normed linear space.*

Proof. That $\|x\| \geq 0$ and $\|x\| = 0$ only when $x = 0$ readily follow from the axioms of inner products (Definition 3.1.1).

The triangle inequality $\|x + y\| \leq \|x\| + \|y\|$ has already been proven in Proposition 3.1.6. For scalar products, the axioms of inner products and Proposition 3.1.4 yield

$$\|\alpha x\|^2 = (\alpha x, \alpha x) = \alpha \overline{\alpha}(x, x) = |\alpha|^2 (x, x) = |\alpha|^2 \|x\|^2. \qquad \square$$

Notice that the inner product is continuous with respect to the topology induced by the norm as defined above. That is, the following proposition holds.

Proposition 3.1.9 (continuity of inner products). *Let X be an inner product space. Suppose that sequences $\{x_n\}$ and $\{y_n\}$ satisfy $\|x_n - x\| \to 0$, $\|y_n - y\| \to 0$ as $n \to \infty$. Then $(x_n, y_n) \to (x, y)$.*

Proof. Since $x_n \to x$, there exists M such that $\|x_n\| \leq M$ for all n (Chapter 2, Problem 2). Then the Cauchy–Schwarz inequality (Theorem 3.1.5, p. 61) implies

$$
\begin{aligned}
|(x_n, y_n) - (x, y)| &= |(x_n, y_n - y) + (x_n - x, y)| \\
&\leq |(x_n, y_n - y)| + |(x_n - x, y)| \\
&\leq M \|y_n - y\| + \|x_n - x\| \|y\|.
\end{aligned}
$$

The right-hand side converges to 0. $\qquad \square$

As we have noted, an inner product space can always be regarded as a normed space by introducing a norm by (3.7). This normed space is, however, not necessarily complete, i.e., not necessarily a Banach space. For example, the space $C[0, 1]$ in Example 3.1.3 with inner product (3.4) is not complete[42] (cf. Problem 4). When this becomes complete, it is called a Hilbert space. This is the topic of the next section.

3.2 Hilbert Space

The following definition arises naturally from what we have seen so far.

Definition 3.2.1. An inner product space $(X, (\cdot, \cdot))$ that becomes a Banach space with respect to the norm naturally induced by (3.7) is called a *Hilbert space*.

We will see in due course that a Hilbert space exhibits various properties that resemble those in finite-dimensional spaces. But before going into detail, let us give two typical examples of a Hilbert space.

---

[42]To be precise, with respect to the norm induced by it.

Example 3.2.2. The space $L^2[a,b]$ of all square integrable functions on the interval $[a,b]$ in the sense of Lebesgue, given in Example 2.1.18 (p. 45) by setting $p = 2$. Define an inner product by

$$(f,g) := \int_a^b f(t)\overline{g(t)}dt. \tag{3.8}$$

Clearly

$$|f(t)\overline{g(t)}| \leq |f(t)|^2 + |g(t)|^2,$$

so the integral on the right-hand side of (3.8) exists and is finite. For $f,g \in L^2[a,b]$,

$$|f(t) + g(t)|^2 \leq |f(t)|^2 + 2|f(t)g(t)| + |g(t)|^2$$

(at each t), so that $f + g$ also belongs to $L^2[a,b]$.

The crucial step is in guaranteeing that (3.8) is well defined for $L^2[a,b]$ and that $L^2[a,b]$ constitutes a vector space; verifying that (3.8) indeed satisfies the axioms of norms is a rather easy exercise and hence is left to the reader.

Thus $L^2[a,b]$ is an inner product space. Further, it is clear that the norm

$$\|\phi\|_2 := \left\{ \int_a^b |\phi(t)|^2 dt \right\}^{1/2}$$

defined by setting $p = 2$ in (2.22) is indeed induced from this inner product via (3.7).

It remains only to show the completeness. But, as with $L^1[a,b]$, this requires knowledge of the Lebesgue integral, so we omit its proof here.

Example 3.2.3. The sequence space ℓ^2 defined in Example 2.1.17. Introduce an inner product by

$$(\{x_n\},\{y_n\}) := \sum_{n=1}^{\infty} x_n \overline{y_n}. \tag{3.9}$$

Let us show that this is indeed an inner product. Note that $|ab| \leq a^2 + b^2$ for every real a,b. Then

$$\sum_{n=1}^{\infty} |x_n \overline{y_n}| \leq \sum_{n=1}^{\infty} \left\{ |x_n|^2 + |y_n|^2 \right\} < \infty,$$

so that (3.9) always converges. It is almost trivial to see that the axioms for inner products (Definition 3.1.1, p. 59) are satisfied. Further, as with Example 3.2.2 above, for $\{x_n\}$, $\{y_n\} \in \ell^2$,

$$|x_n + y_n|^2 \leq |x_n|^2 + 2|x_n y_n| + |y_n|^2,$$

and hence $\{x_n + y_n\} \in \ell^2$. This shows that ℓ^2 constitutes a vector space.

It remains to show the completeness. Let $x^{(k)} := \{x_n^{(k)}\}_{n=1}^{\infty}$, $k = 1,2,\ldots$, be a Cauchy sequence of elements in ℓ^2. Note that

$$|x_n^{(k)} - x_n^{(\ell)}| \leq \left\| x^{(k)} - x^{(\ell)} \right\|_2$$

for all n, k, ℓ. Then for each n, $x_n^{(k)}$, $k = 1, 2, \ldots$, constitute a Cauchy sequence in $\mathbb{K}$. Hence there exists $x_n \in \mathbb{K}$ such that $x_n^{(k)} - x_n \to 0$ as $k \to \infty$. Define $x := \{x_n\}_{n=1}^{\infty}$. Showing $x \in \ell^2$ and $\|x^{(k)} - x\|_2 \to 0$ $(k \to \infty)$ would complete the proof.

Take any $\epsilon > 0$. There exists N such that whenever $k, \ell \geq N$, $\|x^{(k)} - x^{(\ell)}\|_2 \leq \epsilon$. Take any positive integer L. Then

$$\left\{ \sum_{n=1}^{L} |x_n^{(k)} - x_n^{(\ell)}|^2 \right\}^{1/2} \leq \left\| x^{(k)} - x^{(\ell)} \right\|_2 \leq \epsilon$$

whenever $k, \ell \geq N$. Letting ℓ go to infinity shows that

$$\left\{ \sum_{n=1}^{L} |x_n^{(k)} - x_n|^2 \right\}^{1/2} \leq \epsilon$$

for each fixed L and for $k \geq N$. Since this estimate is independent of L, we can let L tend to infinity to obtain

$$\left\| x^{(k)} - x \right\|_2 = \left\{ \sum_{n=1}^{\infty} |x_n^{(k)} - x_n|^2 \right\}^{1/2} \leq \epsilon \tag{3.10}$$

whenever $k \geq N$. Then $x = x - x^{(k)} + x^{(k)}$ $(k \geq N)$ implies $x \in \ell^2$ because $x - x^{(k)}$ and $x^{(k)}$ both belong to ℓ^2. It also follows from (3.10) that $\|x^{(k)} - x\|_2 \to 0$.

We will subsequently see that in a Hilbert space there always exists a perpendicular, as well as its foot, from an arbitrary element to a given, arbitrary, closed subspace. This in turn implies the existence of a direct sum decomposition for an arbitrary closed subspace. These properties are certainly valid in finite-dimensional Euclidean spaces but not necessarily so for Banach spaces. It is these properties that enable us to solve the minimum distance (and hence the minimum-norm approximation) problem, and this in turn (see section 3.3, p. 73) gives rise to a fairly complete treatment of Fourier expansion in the space L^2 (see subsection 3.2.2 (p. 69) and section 7.1 (p. 141)).

The following theorem gives the foundation for this fact.

Theorem 3.2.4. *Let X be a Hilbert space, and M a nonempty closed subset of X. For every $x \in X$, there exists m_0 in M that minimizes $\|x - m_0\|$ among all m in M. Furthermore, such an m_0 is unique.*

A set M is said to be *convex* if for every pair $x, y \in M$ and $0 < t < 1$, $tx + (1 - t)y \in M$ holds.

Proof of Theorem 3.2.4. Suppose

$$d := \inf_{m \in M} \|x - m\|. \tag{3.11}$$

Since M is nonempty, $d < \infty$. By the definition of infimum (see Lemma A.2.1 in Appendix A), for every $\epsilon > 0$ there always exists $m \in M$ such that

$$d \leq \|x - m\| < d + \epsilon.$$

Hence for each positive integer n there exists $m_n \in M$ such that

$$\|x - m_n\|^2 < d^2 + \frac{1}{n}. \tag{3.12}$$

Then by the parallelogram law, Proposition 3.1.7 (p. 62), we have

$$\|(x - m_k) - (x - m_\ell)\|^2 + \|x - m_k + x - m_\ell\|^2 = 2\|x - m_k\|^2 + 2\|x - m_\ell\|^2$$
$$< 4d^2 + 2\left(\frac{1}{k} + \frac{1}{\ell}\right).$$

This yields

$$\|m_k - m_\ell\|^2 < 4d^2 + 2\left(\frac{1}{k} + \frac{1}{\ell}\right) - 4\left\|x - \frac{m_k + m_\ell}{2}\right\|^2.$$

Since M is convex, $(m_k + m_\ell)/2$ is also an element of M, so that by (3.11)

$$\left\|x - \frac{m_k + m_\ell}{2}\right\|^2 \geq d^2$$

holds. Hence

$$\|m_k - m_\ell\|^2 < 4d^2 + 2\left(\frac{1}{k} + \frac{1}{\ell}\right) - 4d^2$$
$$= 2\left(\frac{1}{k} + \frac{1}{\ell}\right).$$

Since the last expression approaches 0 as $k, \ell \to \infty$, the sequence $\{m_n\}$ is a Cauchy sequence. The completeness of the Hilbert space then implies that there exists a limit $m_0 \in X$ such that $m_n \to m_0$ as $n \to \infty$. Furthermore, since M is closed, $m_n \to m_0$ implies $m_0 \in M$. Hence by (3.11) we have

$$\|x - m_0\| \geq d.$$

On the other hand, letting $n \to \infty$ in (3.12), we obtain by the continuity of a norm (Lemma 2.1.7) that

$$\|x - m_0\| \leq d.$$

Hence $\|x - m_0\| = d$. Thus there exists a point m_0 that attains the infimum of the distance.

Let us show the uniqueness. Suppose that there exists another $m_0' \in M$ such that $\|x - m_0'\| = d$. Since M is convex, $(m_0 + m_0')/2 \in M$, and hence

$$\left\|x - \frac{m_0 + m_0'}{2}\right\| \geq d.$$

Applying the parallelogram law (Proposition 3.1.7) to $x - m_0$ and $x - m_0'$, we have

$$\|m_0 - m_0'\|^2 = 2\|x - m_0\|^2 + 2\|x - m_0'\|^2 - 4\left\|x - \frac{m_0 + m_0'}{2}\right\|^2$$
$$\leq 2d^2 + 2d^2 - 4d^2 = 0.$$

Hence $m_0 = m_0'$.   □

3.2.1 Orthogonal Complements

Definition 3.2.5. Let E be a subset of a Hilbert space X. A subset of X given by

$$\{x \in X : (x, y) = 0 \text{ for all } y \in E\} \tag{3.13}$$

is called the *orthogonal complement* of E and is denoted by $E^\perp$ or by $X \ominus E$.

Proposition 3.2.6. *The orthogonal complement $E^\perp$ of any subset E of a Hilbert space X is always a closed subspace.*

Proof. This is easy and left as an exercise for the reader. $\square$

As a consequence of the completeness, there always exists the perpendicular foot to a closed subspace. Let us prove the following theorem.

Theorem 3.2.7 (projection theorem). *Let X be a Hilbert space and M a closed subspace. For any $x \in X$ there exists a unique $x_0 \in M$ such that*

$$(x - x_0, m) = 0 \quad \forall m \in M. \tag{3.14}$$

The element x_0 is called the foot of the perpendicular from x to M. Conversely, when (3.14) *holds,*

$$\|x - x_0\| \le \|x - m\| \quad \forall m \in M$$

follows. That is, the perpendicular from x to M gives the shortest distance from x to M.

Proof. Let us first prove the uniqueness (if it exists) of such x_0 satisfying (3.14). Suppose that there exists another $x_1 \in M$ such that

$$(x - x_1, m) = 0 \quad \forall m \in M.$$

This readily implies $(x_0 - x_1, m) = 0$ for all $m \in M$. Since $x_0 - x_1$ is itself an element of M, we can substitute $x_0 - x_1$ into m to obtain $(x_0 - x_1, x_0 - x_1) = 0$. Then by the first axiom of inner products (p. 59) $x_0 - x_1 = 0$ follows.

Let us now prove the existence. Since every subspace is clearly convex, Theorem 3.2.4 implies that there exists $x_0 \in M$ such that

$$\|x - x_0\| \le \|x - y\| \quad \forall y \in M.$$

For every $m \in M$ and a complex number λ, we have

$$\|x - x_0 - \lambda m\|^2 = \|x - x_0\|^2 - \bar{\lambda}(x - x_0, m) - \lambda(m, x - x_0) + |\lambda|^2 \|m\|^2. \tag{3.15}$$

Since $(x - x_0, m) = e^{i\theta}|(x - x_0, m)|$ for some θ, we may set $\lambda = te^{i\theta}$ (t is a real parameter here) in the above and obtain

$$0 \le \|x - x_0 - \lambda m\|^2 - \|x - x_0\|^2 = -2t|(x - x_0, m)| + t^2 \|m\|^2.$$

Hence, for $t \geq 0$, we have

$$|(x - x_0, m)| \leq \frac{t}{2} \|m\|^2.$$

Letting $t \to 0$, we obtain $(x - x_0, m) = 0$.

Conversely, suppose (3.14) holds. Then setting $\lambda = 1$ in (3.15) and using (3.14), we obtain

$$\|x - x_0 - m\|^2 = \|x - x_0\|^2 + \|m\|^2 \geq \|x - x_0\|^2.$$

Hence $\|x - x_0\|$ gives the shortest distance from x to M.   □

This theorem yields the following theorem, which states that for every closed subspace M of X, X can be decomposed into the direct sum of M and its orthogonal complement $M^\perp$.

Theorem 3.2.8. *Let X be a Hilbert space and M a closed subspace. For any $x \in X$, take $x_0 \in M$ uniquely determined by Theorem 3.2.7, and then define*

$$p_M(x) := x_0,$$
$$p_{M^\perp}(x) := x - x_0.$$

Then

$$x = p_M(x) + p_{M^\perp}(x) \tag{3.16}$$

gives a direct sum decomposition of X into M and $M^\perp$. (See Figure 3.1.)

Proof. That every x is expressible as (3.16) and that $x_0 \in M$, $x - x_0 \in M^\perp$ are already shown in Theorem 3.2.7. Also, $M \cap M^\perp = \{0\}$ follows because $x \in M \cap M^\perp$ readily implies $(x, x) = 0$ by definition.   □

The projection theorem may often be used in the following shifted form.

Corollary 3.2.9. *Let X be a Hilbert space and M a closed subspace. Take any x in X and set*

$$V := x + M = \{x + m : m \in M\}.$$

Then there uniquely exists an element x_0 in V with minimum norm that is characterized by $x_0 \perp M$.

The proof reduces to the projection theorem, Theorem 3.2.7, by shifting V by $-x$. This is left as an exercise for the reader.

Exercise 3.2.10. Prove Corollary 3.2.9.

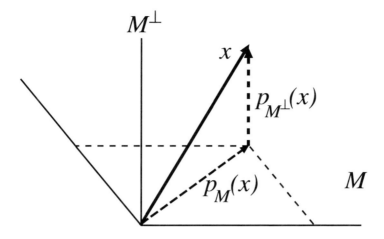

Figure 3.1. *Orthogonal complement*

3.2.2 Orthogonal Expansion

Definition 3.2.11. A family $\{e_n\}_{n=1}^\infty$ of nonzero elements in a Hilbert space X is said to be an *orthogonal system* if

$$(e_i, e_j) = 0 \quad \forall i \neq j.$$

If, further, $\|e_i\| = 1$ for all i, then $\{e_n\}_{n=1}^\infty$ is called an *orthonormal system*.

If the family $\{e_1, \ldots e_n\}$ is an orthonormal system, then it is a linearly independent family. Indeed, if

$$\sum_{j=1}^n \alpha_j e_j = 0,$$

then by taking the inner product of both sides with e_k, we obtain that

$$0 = \left(\sum_{j=1}^n \alpha_j e_j, e_k \right) = \alpha_k$$

holds for each k, and this shows the linear independence. Likewise, we can easily modify the proof to show that an orthogonal system is linearly independent.

Example 3.2.12. The family

$$e_n(t) := \frac{1}{\sqrt{2\pi}} e^{int}, \quad n = 0, \pm 1, \pm 2, \ldots,$$

in $L^2[-\pi, \pi]$ forms an orthonormal system. One can easily check by direct calculation that $(e_n, e_m) = \delta_{nm}$ (δ_{nm} denotes the Kronecker delta, i.e., the function that takes value 1 when $n = m$ and 0 otherwise).

As we have just seen in the above example, an orthonormal system typically appears in the theory of Fourier series. Let us adopt the following definition by generalizing this.

Definition 3.2.13. Let $\{e_n\}$ be a given orthonormal system in a Hilbert space X. For every $x \in X$, (x, e_n) is called the *nth Fourier coefficient* of x and the series $\sum(x, e_n)e_n$ the *Fourier series* of x.

We should note, however, that at this stage this series is only a formal object, and we say nothing about its convergence. However, the following fact can easily be verified.

Theorem 3.2.14. *Let X be a Hilbert space and $\{e_k\}_{k=1}^n$ an orthonormal system in it. Take any $x \in X$. The Fourier series $\sum(x, e_k)e_k$ gives the best approximant among all elements of form $\sum c_k e_k$ in the sense of norm, i.e., making $\|x - \sum c_k e_k\|$ minimal. The square of the error $\|x - \sum(x, e_k)e_k\|$ is given by*

$$d^2 = \|x\|^2 - \sum_{k=1}^n |(x, e_k)|^2. \tag{3.17}$$

Proof. Let $M := \operatorname{span}\{e_k\}_{k=1}^n$, i.e., the finite-dimensional subspace consisting of all linear combinations of $\{e_k\}_{k=1}^n$. By Theorem 2.3.2 (p. 52), M is a closed subspace. By Theorem 3.2.7, the minimum of $\|x - m\|$, $m \in M$, is attained when $(x - m) \perp M$. Set $m = \sum c_k e_k$, and write down this condition to obtain

$$\left(x - \sum c_k e_k, e_i\right) = 0, \quad i = 1, \dots, n. \tag{3.18}$$

According to the orthonormality $(e_k, e_i) = \delta_{ki}$, we have

$$(x, e_i) = c_i(e_i, e_i) = c_i.$$

This gives precisely the Fourier coefficients.

Finally, the orthonormality of $\{e_k\}$ readily yields

$$\left\|x - \sum(x, e_k)e_k\right\|^2 = \left(x - \sum(x, e_k)e_k, x - \sum(x, e_k)e_k\right)$$

$$= \|x\|^2 - \sum_{k=1}^n (x, e_k)\overline{(x, e_k)}$$

$$= \|x\|^2 - \sum_{k=1}^n |(x, e_k)|^2,$$

which is (3.17). $\square$

Theorem 3.2.15 (Bessel's inequality). *Let $\{e_k\}_{k=1}^\infty$ be an orthonormal system in a Hilbert space X. For every $x \in X$, the following inequality, called Bessel's inequality, holds:*

$$\sum_{k=1}^\infty |(x, e_k)|^2 \le \|x\|^2. \tag{3.19}$$

Proof. Take any n and set $y_n := \sum_{k=1}^{n}(x,e_k)e_k$. Apply (3.17) to y_n to obtain

$$\sum_{k=1}^{n}|(x,e_k)|^2 \leq \|x\|^2.$$

Letting $n \to \infty$, we obtain (3.19). $\qquad\square$

The convergence of a series is of course defined by the convergence of the sequence of its partial sums, as with real series. But unlike the case of the reals, testing its convergence is in general not easy. However, for the case of series of linear sums with respect to an orthonormal system as with Fourier series, the following simple criterion is available.

Proposition 3.2.16. *Let X be a Hilbert space and $\{e_n\}_{n=1}^{\infty}$ an orthonormal system. A necessary and sufficient condition for the series $\sum_{k=1}^{\infty}\alpha_k e_k$ with complex coefficients α_k to be convergent in X is that*

$$\sum_{k=1}^{\infty}|\alpha_k|^2 < \infty.$$

Proof. Define the nth partial sum s_n by

$$s_n := \sum_{k=1}^{n}\alpha_k e_k,$$

and suppose that $s = \lim_{n\to\infty}s_n$ is convergent. Since

$$\left(\sum_{k=1}^{n}\alpha_k e_k, e_j\right) = \alpha_j,$$

the continuity of inner products implies $(s,e_k) = \alpha_k$. Then by Bessel's inequality (3.19)

$$\sum_{k=1}^{\infty}|\alpha_k|^2 = \sum_{k=1}^{\infty}|(s,e_k)|^2 \leq \|s\|^2 < \infty.$$

Conversely, if $\sum_{k=1}^{\infty}|\alpha_k|^2 < \infty$, then for $m \geq n$ we have

$$\|s_m - s_n\|^2 = \left\|\sum_{k=n+1}^{m}\alpha_k e_k\right\|^2$$

$$= \sum_{k=n+1}^{m}|\alpha_k|^2 \to 0 \quad (n \to \infty).$$

That is, the sequence $\{s_n\}$ is a Cauchy sequence and is convergent by the completeness of X. $\qquad\square$

Definition 3.2.17. An orthonormal system $\{e_n\}_{n=1}^{\infty}$ in a Hilbert space X is said to be *complete* if it has the property that the only element orthogonal to all e_n is 0.

Some remarks may be in order. First, the term "complete" is confusing, since it has no relevance to the completeness in the sense of convergence of Cauchy sequences. Second, the condition above is equivalent to the property that every element of X is expressible as an (infinite) linear combination of $\{e_n\}_{n=1}^{\infty}$, and hence we have the term "completeness"[43] for $\{e_n\}_{n=1}^{\infty}$.

Theorem 3.2.18. *Let $\{e_n\}_{n=1}^{\infty}$ be a complete orthonormal system in a Hilbert space X. Then for every $x \in X$, the following equalities hold:*

1. $$x = \sum_{n=1}^{\infty} (x, e_n) e_n, \tag{3.20}$$

2. $$\|x\|^2 = \sum_{n=1}^{\infty} |(x, e_n)|^2. \tag{3.21}$$

The second equality is called Parseval's identity.

Proof. Let us first note that Bessel's inequality (3.19) along with Proposition 3.2.16 imply that $\sum_{n=1}^{\infty} (x, e_n) e_n$ converges. Observe that

$$\left(x - \sum_{n=1}^{\infty} (x, e_n) e_n, e_j \right) = (x, e_j) - \sum_{n=1}^{\infty} (x, e_n)(e_n, e_j) = (x, e_j) - (x, e_j) = 0.$$

That is, $x - \sum_{n=1}^{\infty} (x, e_n) e_n$ is orthogonal to all e_j. Hence by the completeness of $\{e_n\}_{n=1}^{\infty}$, we have $x - \sum_{n=1}^{\infty} (x, e_n) e_n = 0$, i.e., (3.20) follows.

Clearly

$$\left\| \sum_{n=1}^{N} (x, e_n) e_n \right\|^2 = \sum_{n=1}^{N} |(x, e_n)|^2.$$

Letting $N \to \infty$, we have by the continuity of a norm that the left-hand side converges to $\|x\|^2$, while the right-hand side converges to $\sum_{n=1}^{\infty} |(x, e_n)|^2$. This proves (3.21). $\qquad\square$

Examining the proof above, we see that the completeness is the key to showing that the error $x - \sum_{n=1}^{\infty} (x, e_n) e_n$ is indeed 0. Conversely, if the family is not complete, i.e., there exists a nonzero element x that is orthogonal to all e_n, then this expansion theorem is no longer valid. In fact, for such an element x, its Fourier expansion $\sum_{n=1}^{\infty} (x, e_n) e_n$ is clearly 0, but $x \neq 0$.

This is not an exceptional case at all. Indeed, for a given orthonormal system $\{e_n\}_{n=1}^{\infty}$, extracting one element, say, e_1, out of this system makes it noncomplete, and the condition above will not be satisfied. For example, in space ℓ^2, the family

$$e_2 = (0, 1, 0, \ldots), e_3 = (0, 0, 1, 0, \ldots), \ldots, e_n = (0, \ldots, 1, 0, \ldots)$$

is not complete. The element $e_1 = (1, 0, 0, \ldots)$ is missing here.

---

[43]in the sense that nothing is missing.

The next proposition gives a characterization of completeness.

Proposition 3.2.19. *Each of the following statements gives a necessary and sufficient condition for an orthonormal system $\{e_n\}_{n=1}^{\infty}$ in a Hilbert space X to be complete:*

1. $\mathrm{span}\{e_n : n = 1, 2, \ldots\}$ *is dense[44] in X.*

2. *Parseval's identity, i.e.,*

$$\|x\|^2 = \sum_{n=1}^{\infty} |(x, e_n)|^2,$$

holds for every $x \in X$.

Proof. The necessity is already shown in Theorem 3.2.18.

The sufficiency of condition 1 can be seen as follows: Suppose that x is orthogonal to all e_n. Then it is orthogonal to $\mathrm{span}\{e_n : n = 1, 2, \ldots\}$. Since this space is dense in X, this implies that x is orthogonal to every element of X by the continuity of the inner product. Then by Proposition 3.1.4 (p. 61), $x = 0$ follows; that is, $\{e_n\}_{n=1}^{\infty}$ is complete.

To show that Parseval's identity yields completeness, assume that $\{e_n\}_{n=1}^{\infty}$ is not complete. Then there exists a nonzero x that is orthogonal to all e_n. Clearly we have $\sum_{n=1}^{\infty} |(x, e_n)|^2 = 0$. On the other hand, $\|x\|^2 \neq 0$, and this contradicts Parseval's identity. $\square$

3.3 Projection Theorem and Best Approximation

To state it simply, the notion of inner product is a generalization of bilinear forms. The projection theorem, Theorem 3.2.7, asserts that the minimization of such bilinear forms is given by the orthogonal projection onto a given subspace.

We here give a simple application related to such an approximation problem. Where there is a performance index (also called a penalty function or a utility function depending on the context and field) of second order such as energy, its optimization (minimization) is mostly reduced to the projection theorem. Least square estimation, Kalman filtering, and the Yule–Walker equation are typical examples. For details, the reader is referred to, for example, Luenberger [31].

Let X be a Hilbert space and $M = \mathrm{span}\{v_1, \ldots, v_n\}$ its finite-dimensional subspace. We may assume, without loss of generality, that $\{v_1, \ldots, v_n\}$ is linearly independent by deleting redundant vectors if necessary. The problem here is to find $\hat{x} \in M$ such that

$$\|x - \hat{x}\| = \inf_{m \in M} \|x - m\|$$

for any given $x \in X$.

First note that M is a closed subspace as shown in Theorem 2.3.2. Hence by the projection theorem, Theorem 3.2.7, $\hat{x}$ is given by the foot of the perpendicular to M, and it is characterized by

$$(x - \hat{x}, m) = 0, \quad m \in M.$$

---

[44]A subset in a topological space X is *dense* if its closure (Definition 2.1.23, p. 47) agrees with the whole space X; see Appendix A, section A.3 (p. 238).

Since $M = \text{span}\{v_1,\ldots,v_n\}$, this is clearly equivalent to

$$(x - \hat{x}, v_i) = 0, \quad i = 1,\ldots,n.$$

Setting $\hat{x} = \sum_{j=1}^{n} \alpha_j v_j$, this can be rewritten as

$$
\begin{array}{ccccccc}
(v_1,v_1)\alpha_1 & + & (v_2,v_1)\alpha_2 & + & \cdots & (v_n,v_1)\alpha_n & = & (x,v_1), \\
(v_1,v_2)\alpha_1 & + & (v_2,v_2)\alpha_2 & + & \cdots & (v_n,v_2)\alpha_n & = & (x,v_2) \\
\vdots & & \vdots & & & \vdots & & \vdots \\
(v_1,v_n)\alpha_1 & + & (v_2,v_n)\alpha_2 & + & \cdots & (v_n,v_n)\alpha_n & = & (x,v_n),
\end{array}
\tag{3.22}
$$

which in turn is written in matrix form as

$$
\begin{bmatrix}
(v_1,v_1) & (v_2,v_1) & \cdots & (v_n,v_1) \\
(v_1,v_2) & (v_2,v_2) & \cdots & (v_n,v_2) \\
\vdots & & & \vdots \\
(v_1,v_n) & (v_2,v_n) & \cdots & (v_n,v_n)
\end{bmatrix}
\begin{bmatrix}
\alpha_1 \\ \alpha_2 \\ \vdots \\ \alpha_n
\end{bmatrix}
=
\begin{bmatrix}
(x,v_1) \\ (x,v_2) \\ \vdots \\ (x,v_n)
\end{bmatrix}.
\tag{3.23}
$$

This equation is called the *normal equation*, and the transpose of the left-hand side matrix of (3.23) is called the *Gram matrix* and is denoted by $G(v_1, v_2, \ldots, v_n)$.

One can also solve its dual problem, namely, that of finding a vector of minimum norm satisfying finitely many linear constraints, according to Corollary 3.2.9 as follows.

Set $M := \text{span}\{v_1,\ldots,v_n\}$ as above, and suppose that $\{v_1,\ldots,v_n\}$ is linearly independent. Consider the constraints

$$
\begin{array}{ccc}
(x,v_1) & = & c_1, \\
(x,v_2) & = & c_2 \\
\vdots & & \vdots \\
(x,v_n) & = & c_n,
\end{array}
\tag{3.24}
$$

and find an optimal element with $\|x\|$ minimal. Denote such an element by $\hat{x}$.

Let x_0 be a solution of (3.24). Then every solution of (3.24) is expressible as (an element of) $x_0 + M^{\perp}$.

Hence, by Corollary 3.2.9, the solution to the minimum norm problem $\hat{x}$ belongs to $M^{\perp\perp}$, and $M = M^{\perp\perp}$ because M is a closed subspace (Problem 3). Hence $\hat{x} \in M$, and we can set

$$\hat{x} = \sum_{j=1}^{n} \beta_j v_j.$$

Writing down the conditions for $\hat{x} \in M$ to satisfy (3.24), we obtain

$$
\begin{bmatrix}
(v_1,v_1) & (v_2,v_1) & \cdots & (v_n,v_1) \\
(v_1,v_2) & (v_2,v_2) & \cdots & (v_n,v_2) \\
\vdots & & & \vdots \\
(v_1,v_n) & (v_2,v_n) & \cdots & (v_n,v_n)
\end{bmatrix}
\begin{bmatrix}
\beta_1 \\ \beta_2 \\ \vdots \\ \beta_n
\end{bmatrix}
=
\begin{bmatrix}
c_1 \\ c_2 \\ \vdots \\ c_n
\end{bmatrix}.
\tag{3.25}
$$

As a slight modification, consider the case where M is given as the kernel of a linear mapping $L : X \to Y$. To give a solution to this problem, we need the concept of adjoint mappings in Hilbert space. This is to be introduced later in Chapter 5, section 5.4 (p. 94), and we here give a solution to the minimum norm problem assuming the knowledge of this concept. The reader not familiar with this concept can skip the material in the rest of this section without much loss.

Let $L : X \to Y$ be a linear map from a Hilbert space X to another Hilbert space Y; actually we can assume X and Y are both just inner product spaces. The adjoint mapping L^* of L is the mapping $L^* : Y \to X$ defined by

$$(Lx, y)_Y = (x, L^*y)_X, \quad x \in X, \ y \in Y \tag{3.26}$$

(see section 5.4 (p. 94), Theorem 5.4.1). Then the following fact holds.

Theorem 3.3.1. *Let X, Y, L be as above, and suppose that Y is finite-dimensional and that $LL^* : Y \to Y$ admits an inverse. Then for any $y_0 \in Y$, $Lx = y_0$ admits a solution, and the minimum-norm solution is given by*

$$x_0 = L^*(LL^*)^{-1}y_0. \tag{3.27}$$

Proof. Since $Lx_0 = LL^*(LL^*)^{-1}y_0 = y_0$, x_0 is clearly a solution to $Lx = y_0$.

Define $M := \ker L$, and take any $x \in M$. Clearly

$$(x, x_0) = \left(x, L^*(LL^*)^{-1}y_0\right) = \left(Lx, (LL^*)^{-1}y_0\right) = \left(0, (LL^*)^{-1}y_0\right) = 0,$$

so that x_0 is orthogonal to M. Hence, by Corollary 3.2.9 (p. 68), x_0 is the solution with minimum norm. $\qquad\square$

Needless to say, a typical example of orthogonal expansion is the Fourier expansion. To actually prove its completeness, we will further need some analytic arguments. Furthermore, its essence is perhaps better appreciated for its connection with Schwartz distributions. We will thus discuss this topic in the later chapter on Fourier analysis.

Problems

1. (**Gram–Schmidt orthogonalization**). Let $(X, (\cdot, \cdot))$ be an inner product space and $\{x_1, \ldots, x_n, \ldots\}$ a sequence of linearly independent vectors (i.e., any finite subset is linearly independent). Show that one can choose an orthonormal family $\{e_1, \ldots, e_n, \ldots\}$ such that for every n, $\mathrm{span}\{e_1, \ldots, e_n\} = \mathrm{span}\{x_1, \ldots, x_n\}$ holds. (Hint: Set $e_1 := x_1 / \|x_1\|$, and suppose that $\{e_1, \ldots, e_{n-1}\}$ have been chosen. Then define $z_n := x_n - \sum_{i=1}^{n-1}(x_n, e_i)e_i$ and $e_n := z_n / \|z_n\|$. This procedure is called the *Gram–Schmidt orthogonalization*.)

2. Let Q be an $n \times n$ real positive definite matrix, let A be an $m \times n$ real matrix such that $\mathrm{rank}\, A = m$, and let $b \in \mathbb{R}^m$. Find the minimum of $x^T Q x$ under the constraint $Ax = b$. (Hint: Define an inner product in $\mathbb{R}^n$ by $(x, y)_Q = y^T Q x$, and then reduce this problem to Theorem 3.3.1.)

3. Let X be a Hilbert space and M a closed subspace. Show that $(M^\perp)^\perp = M$.

4. Show that $C[0,1]$ is *not* complete with respect to the inner product (3.4). (Hint: Consider the sequence f_n of functions given in Problem 3 of Chapter 2.)

Chapter 4

Dual Spaces

The dual space of a vector space has already been introduced in Chapter 1. It is simply the set of linear functionals (i.e., linear forms) defined on the original vector space. In the case of normed linear spaces, however, we require these linear functionals to be continuous. The reason for introducing a norm is to study properties that depend on topology, for example, continuity. Hence it is not of much interest to consider discontinuous functionals in this context. To state it differently, if we were to consider discontinuous functionals, then the space need not be normed. In what follows we therefore consider continuous linear functionals only, and we will distinguish this dual space from the dual space X^* defined in Chapter 1 without considering continuity, calling it an *algebraic dual space*.[45]

This chapter introduces normed dual spaces and discusses their completeness, the Riesz–Fréchet theorem, which is most fundamental to the duality in Hilbert space, weak and weak-star (weak*) topologies, and duality between subspaces and quotient spaces.

4.1 Dual Spaces and Their Norms

Definition 4.1.1. Let $(X, \|\cdot\|)$ be a normed linear space. The set X' of all continuous linear functionals on $(X, \|\cdot\|)$ defined by

$$X' := \{\varphi : X \to \mathbb{K} : \varphi \text{ is continuous and linear on } X\} \tag{4.1}$$

is called the *(normed) dual space* of $(X, \|\cdot\|)$.

Recall that the value of the linear functional $f \in X'$ at point $x \in X$ is denoted by $\langle f, x \rangle$ ((1.66) in Chapter 1, p. 33). For linear functionals $f, g \in X'$, their sum and the scalar product are defined by

$$\langle f + g, x \rangle := \langle f, x \rangle + \langle g, x \rangle, \tag{4.2}$$

$$\langle \alpha f, x \rangle := \alpha \langle f, x \rangle, \quad x \in X, \quad \alpha \in \mathbb{K}, \tag{4.3}$$

---

[45]However, when X is finite-dimensional, any linear functional becomes automatically continuous; hence this distinction is not necessary in that case.

as in (1.63) (p. 31). That they define linear functionals has been ensured already in Chapter 1, but we here have to further ensure that they belong to X', i.e., that they are continuous. The following proposition guarantees this.

Proposition 4.1.2. *Let X be a normed linear space and X' its dual space. Let $f, g \in X'$, $\alpha \in \mathbb{K}$. Then $f + g, \alpha f \in X'$.*

Proof. Since f, g are continuous linear functionals, Proposition 2.1.11 implies that there exist $C_1, C_2 > 0$ such that

$$|\langle f, x \rangle| \leq C_1 \|x\|,$$
$$|\langle g, x \rangle| \leq C_2 \|x\|, \quad x \in X.$$

Hence

$$\begin{aligned} |\langle f + g, x \rangle| &= |\langle f, x \rangle + \langle g, x \rangle| \\ &\leq |\langle f, x \rangle| + |\langle g, x \rangle| \\ &\leq C_1 \|x\| + C_2 \|x\| = (C_1 + C_2) \|x\|. \end{aligned}$$

Therefore, again by Proposition 2.1.11, $f + g$ is continuous. Similarly for αf. □

Hence X' is closed under addition and scalar multiplication in the algebraic dual X^*. So X' is a vector subspace of X^*, and hence it is in itself a vector space.

Remark 4.1.3. As a special case, when X is finite-dimensional, a linear functional on it can be, as seen in subsection 1.4.2, expressed by a matrix (in fact, a row vector). Hence in this case an element of X^* is always continuous, and hence $X' = X^*$. On the other hand, when X is infinite-dimensional, there always exists, although we will not prove it here, a discontinuous linear functional, and hence the distinction between X' and X^* becomes crucial.

X' is in fact not only a vector space, but can also be naturally endowed with a norm, and becomes a normed linear space. Moreover, it can be shown to be a Banach space (even when X is not).

Definition 4.1.4. Let $(X, \|\cdot\|)$ be a normed linear space, and X' its dual space. Define

$$\|\varphi\|_{X'} := \sup_{x \in X, x \neq 0} \frac{|\langle \varphi, x \rangle|}{\|x\|}. \tag{4.4}$$

This defines a norm on X' (see Theorem 4.1.5 below). This is called the *dual norm* induced by the norm of X.

The norm $\|\varphi\|_{X'}$ defined here amounts to giving the largest magnification of the functional φ among all $x \in X$. In particular,

$$|\langle \varphi, x \rangle| \leq \|\varphi\|_{X'} \|x\|. \tag{4.5}$$

Theorem 4.1.5. *The function $\|\cdot\|_{X'}$ defined above indeed gives a norm on X'. With respect to this norm, $(X', \|\cdot\|_{X'})$ becomes a Banach space.*

Proof. First, if $\|\varphi\|_{X'} = 0$, then $\langle \varphi, x \rangle = 0$ for all $x \in X$. This means that φ is 0 as a function on X, that is, it is the zero element of X'.

We next show

$$\|\varphi + \psi\|_{X'} \leq \|\varphi\|_{X'} + \|\psi\|_{X'}. \tag{4.6}$$

For each $x \in X$, we have

$$|\langle \varphi + \psi, x \rangle| \leq |\langle \varphi, x \rangle| + |\langle \psi, x \rangle| \leq \|\varphi\|_{X'} \|x\| + \|\psi\|_{X'} \|x\|.$$

Dividing both sides by $\|x\|$ ($\|x\| \neq 0$), we easily obtain (4.6). The property for scalar multiplication can be shown similarly.

Let us show that this space is complete. Take a Cauchy sequence

$$\|\psi_n - \psi_m\|_{X'} \to 0, \quad n, m \to \infty.$$

This yields, for every $x \in X$, that

$$|\langle \psi_n - \psi_m, x \rangle| \to 0, \quad n, m \to \infty.$$

Hence the sequence $\{\langle \psi_n, x \rangle\}$ is a Cauchy sequence in $\mathbb{K}$. By the completeness of $\mathbb{K}$, there exists a number (limit) a_x, depending on x, such that

$$\lim_{n \to \infty} \langle \psi_n, x \rangle = a_x.$$

Denote this number a_x by $\psi(x)$. Then we have the correspondence

$$\psi : X \to \mathbb{K} : x \mapsto \psi(x) = a_x.$$

It is easy to see that this gives a linear mapping on X. (Prove this.) Hence we denote $\psi(x)$ by $\langle \psi, x \rangle$. We need to prove that ψ is continuous, and that $\|\psi_n - \psi\|_{X'} \to 0$ as $n \to \infty$.

Take any $\epsilon > 0$. There exists N such that if $n, m \geq N$, then

$$|\langle \psi_n - \psi_m, x \rangle| < \epsilon \|x\|, \quad x \in X.$$

Now let $m \to \infty$ to obtain

$$|\langle \psi_n, x \rangle - \langle \psi, x \rangle| \leq \epsilon \|x\|, \quad x \in X. \tag{4.7}$$

Then setting $n = N$ and using the triangle inequality, we have

$$|\langle \psi, x \rangle| \leq |\langle \psi, x \rangle - \langle \psi_N, x \rangle| + |\langle \psi_N, x \rangle|$$
$$\leq \epsilon \|x\| + \|\psi_N\|_{X'} \|x\| = (\|\psi_N\|_{X'} + \epsilon) \|x\|, \quad x \in X.$$

Hence by Proposition 2.1.11 (p. 43), ψ is continuous.

Also, if $n \geq N$, then by (4.7)

$$\|\psi_n - \psi\|_{X'} \leq \epsilon.$$

That is, $\psi_n \to \psi$ as $n \to \infty$. $\qquad \square$

The following equalities hold for the dual norm.

Proposition 4.1.6. *Let $\phi \in X'$. Then*

$$\|\phi\|_{X'} = \sup_{\|x\|=1, x \in X} |\langle \phi, x \rangle|, \tag{4.8}$$

$$\|\phi\|_{X'} = \inf\{M : |\langle \phi, x \rangle| \le M \|x\|\}. \tag{4.9}$$

Proof. Let us first show (4.8). Let C_0 denote the right-hand side. Then from the definition of the dual norm, $C_0 \le \|\phi\|_{X'}$ clearly follows, and C_0 is finite. Conversely, for $x \ne 0$, put $y := x/\|x\|$. Then we have

$$\frac{|\langle \phi, x \rangle|}{\|x\|} = \left| \frac{\langle \phi, x \rangle}{\|x\|} \right| = \left| \left\langle \phi, \frac{x}{\|x\|} \right\rangle \right| = |\langle \phi, y \rangle| \le C_0;$$

$\|\phi\|_{X'} \le C_0$ follows because $\|y\| = 1$.

For (4.9), first set $M_0 = \|\phi\|_{X'}$, and denote by M_1 the right-hand side of (4.9). Since

$$|\langle \phi, x \rangle| \le M_0 \|x\|$$

for every x by (4.5), we clearly have $M_1 \le M_0$. Conversely, for every $\epsilon > 0$,

$$|\langle \phi, x \rangle| \le (M_1 + \epsilon) \|x\|$$

must hold by definition. This readily implies

$$M_0 \le M_1 + \epsilon.$$

Since M_0, M_1 are constants and $\epsilon > 0$ is arbitrary, we must have $M_0 \le M_1$. Hence $M_0 = M_1$. □

Remark 4.1.7. Equality (4.8) also yields

$$\|\phi\|_{X'} = \sup_{\|x\| \le 1, x \in X} |\langle \phi, x \rangle| \tag{4.10}$$

(note the range of x on the right-hand side). Indeed, the right-hand side is clearly greater than or equal to the right-hand side of (4.8). On the other hand, if $\|x\| < 1$, $x \ne 0$, then $|\langle \phi, x \rangle| \le |\langle \phi, x \rangle|/\|x\| = |\langle \phi, x/\|x\| \rangle|$, and $\|x/\|x\|\| = 1$, so that the right-hand side of (4.8) is no less than the right-hand side of (4.10).

Let us give a typical example of a dual space.

Example 4.1.8. Let $1 \le p < \infty$, and let q to satisfy $1/p + 1/q = 1$, i.e., $q = p/(p-1)$ (when $p = 1$, set $q := \infty$). Then

$$(L^p)' = L^q,$$

and

$$(\ell^p)' = \ell^q.$$

In particular, $(L^1)' = L^\infty$.

We omit the proof; see, for example, [20]. We note only that an element of L^q indeed gives a continuous linear functional on L^p. For example, let $f \in L^p$, $g \in L^q$. Hölder's inequality (A.15) (p. 245) yields

$$\int |f(t)g(t)|dt \leq \|g\|_q \|f\|_p.$$

Hence the functional

$$T_g : L^p \to \mathbb{K} : f \mapsto \int f(t)g(t)dt \tag{4.11}$$

certainly gives a continuous linear functional on L^p. The rest of the assertion of this example amounts to saying that (i) the dual norm in the sense of Definition 4.1.4 (p. 78) coincides with $\|g\|_q$, and (ii) every continuous linear functional on L^p is expressible as an element of L^q via (4.11). We omit the proof, however. For a complete proof, see, for example, [20].

4.2 The Riesz–Fréchet Theorem

We have so far discussed general properties applicable to general normed or Banach spaces, but we here prove a theorem that holds only in Hilbert space.

Let $(X, (,))$ be a Hilbert space. For any fixed $g \in X$, the correspondence

$$T_g : X \to \mathbb{K} : x \mapsto (x, g)$$

is clearly an element of X', i.e., a continuous linear functional on X. The next theorem, called the Riesz–Fréchet theorem, guarantees that the converse is also valid; that is, every element of X' is always expressible as above for some g. It is the most important theorem in the theory of Hilbert space.

Theorem 4.2.1 (Riesz–Fréchet). [46] *Let X be a Hilbert space and ψ a continuous linear functional on it, i.e., $\psi \in X'$. Then there exists a unique element g in X such that*

$$\langle \psi, x \rangle = (x, g), \quad x \in X, \tag{4.12}$$

and

$$\|g\| = \|\psi\|_{X'}. \tag{4.13}$$

Proof. Let us first note that if there exists such a g, then it should be unique. For if there exist two elements g_1, g_2 satisfying (4.12), then we must have $(x, g_1 - g_2) = 0$ for every x. Then by Proposition 3.1.4 (p. 61), we have $g_1 - g_2 = 0$; i.e., $g_1 = g_2$ follows.

Hence we need only find one such g that satisfies the requirement.

Now if $\psi = 0$, i.e., if $\langle \psi, x \rangle = 0$ for all x, then $g = 0$ clearly satisfies the requirement. Hence we may assume that $\psi \neq 0$ in what follows. Define a subspace

$$M := \ker \psi = \{x \in X : \langle \psi, x \rangle = 0\}.$$

---

[46] Also referred to simply as Riesz's theorem or the Riesz representation theorem.

Since $\psi \neq 0$, this is not equal to the whole space. Also, by the continuity of ψ, this is a closed subspace of X. For if $x_n \to x$ and $\langle \psi, x_n \rangle = 0$, then $\langle \psi, x \rangle = 0$ by continuity.

Theorem 3.2.8 (p. 68) implies that there exists the orthogonal complement $M^\perp \neq 0$ of M such that $X = M \oplus M^\perp$. Take any nonzero element $h_0 \in M^\perp$. Then $\langle \psi, h_0 \rangle \neq 0$. Hence by taking $h := h_0 / \langle \psi, h_0 \rangle$, we have $\langle \psi, h \rangle = 1$ by linearity. Now note that every $x \in X$ can be written as

$$x = (x - \langle \psi, x \rangle h) + \langle \psi, x \rangle h.$$

It is easy to see that the first term belongs to M and the second to $M^\perp$. Taking the inner product of h with both sides yields (note that the first term is orthogonal to h)

$$(x, h) = (\langle \psi, x \rangle h, h) = \langle \psi, x \rangle \|h\|^2. \tag{4.14}$$

Now set $g := h / \|h\|^2$. Then (4.14) implies

$$(x, g) = (x, h) / \|h\|^2 = \langle \psi, x \rangle.$$

Hence the first assertion follows.

Now by the Cauchy–Schwarz inequality

$$|\langle \psi, x \rangle| = |(x, g)| \leq \|x\| \|g\|.$$

Hence $\|\psi\|_{X'} \leq \|g\|$ follows. On the other hand, since $\langle \psi, g \rangle = (g, g)$ (put $x := g$ in (4.12)),

$$\|g\| = |(g, g)| / \|g\| = |\langle \psi, g \rangle| / \|g\| \leq \|\psi\|_{X'}.$$

Hence $\|\psi\|_{X'} = \|g\|$ follows.   □

In particular, the correspondence $X' \to X : \psi \mapsto g$ gives a norm-preserving one-to-one correspondence. In this sense, $X' \cong X$. Note, however, that $\alpha \psi$ corresponds to $\overline{\alpha} g$, and not αg.

Remark 4.2.2. Let us summarize the idea of the proof. Since $\psi \neq 0$ is a linear functional, its kernel space M is of codimension 1; i.e., the quotient space X/M is one-dimensional. Since ψ is 0 on M, its value is determined on this one-dimensional space X/M. The projection theorem and the associated direct sum decomposition allow us to represent this quotient as a one-dimensional subspace of X. The rest follows by adjusting a constant factor to make $\langle \psi, x \rangle$ and (x, g) equal.

4.3   Weak and Weak* Topologies

One of the difficulties in dealing with infinite-dimensional spaces is that there is no suitable notion of coordinates; even when there is one, coordinates can have a fairly weak control over the spatial properties of the whole space. For example, let $\{e_k\}$ be an orthonormal system in a Hilbert space X, and suppose that a sequence $\{x_n\}$ be expanded as

$$x_n = \sum_{k=1}^{\infty} (x_n, e_k) e_k.$$

A natural candidate for the coordinate in the e_k-direction is (x_n, e_k). Let us pose the question: *Suppose that each coordinate (x_n, e_k) converges to 0. Does this imply that x_n itself converges to 0?* This is of course valid for finite-dimensional X but is not necessarily so for infinite-dimensional spaces. For this convergence,

$$\|x_n\|^2 = \sum_{k=1}^{\infty} |(x_n, e_k)|^2 \to 0$$

must hold, but this of course depends on the rate of convergence as a function of n. When this decays rather slowly, the convergence of $\{x_n\}$ cannot be concluded.

On the other hand, we occasionally encounter a situation in which the original topology of X is too strong. A typical consequence is that the closed unit ball $\{x : \|x\| \leq 1\}$ is not compact.[47] We may encounter the need to conclude that a continuous function assumes its maximum on the unit ball, but if the space is infinite-dimensional, the standard compactness argument does not work due to the lack of compactness of the unit ball.

One way of circumventing this difficulty is to consider a much weaker topology. This is the *weak topology* we will now introduce.

Definition 4.3.1. A sequence $\{x_n\}$ in X *converges weakly to x* if $\langle f, x_n - x \rangle \to 0$ for every $f \in X'$.

While the definition above is given for normed linear spaces, this definition is also valid for general topological vector spaces (see Appendix A). In particular, it is also valid for spaces of distributions in Chapter 6. On the other hand, the weak topology does not satisfy what is called the first countability axiom [10, 33] (unless X is finite-dimensional); defining such a topology through the notion of convergence of sequences is, rigorously speaking, not right. As we see in Appendix A, we must introduce a system of neighborhoods defining the weak topology. But it is a little specialized, and for many practical purposes convergence for sequences is enough, so for the moment we leave the above definition as it is. For further details, the reader is referred to books on topological vector spaces such as [29, 51, 58, 64].

In contrast to this, we call the standard topology given by the norm the *strong topology*. The following proposition is almost trivial.

Proposition 4.3.2. *Suppose that a sequence $\{x_n\}$ converges strongly to x, i.e., converges to x in the sense of strong topology. Then $\{x_n\}$ also converges weakly to x.*

Proof. The proof is immediate from

$$|\langle x_n - x, f \rangle| \leq \|f\| \|x_n - x\| \to 0, \quad n \to \infty. \qquad \Box$$

However, the converse does not hold. In fact, one can easily give an example of a weakly convergent sequence that does not converge strongly. The following is such an example.

Example 4.3.3. Consider as X the space ℓ^2 given in Example 2.1.17. Let $e_n = (0, 0, \ldots, 0, 1, 0, \ldots)$, where 1 is placed at the nth entry. Since ℓ^2 is a Hilbert space, $X' \cong X$, as already noted in the previous section; the Riesz–Fréchet theorem, Theorem 4.2.1, implies

---

[47] In fact, it is known [51, 58] that if the unit ball in a normed linear space is compact, then that space must be finite-dimensional. See also the discussion following Example 4.3.3.

that every element of X' can be represented by the inner product with an element $y = (y_1, y_2, \ldots)$ of ℓ^2. Then $(e_n, y) = y_n$, and since $\sum |y_n|^2 < \infty$ by $y \in \ell^2$, $y_n \to 0$ as $n \to \infty$ follows. Hence e_n converges weakly to 0. On the other hand, $\|e_n\|$ is always 1, so that it cannot converge to 0 in the strong topology.

We can generalize this construct to an orthonormal system $\{\psi_1, \ldots, \psi_n, \ldots\}$ of a Hilbert space X. For any x in X, Bessel's inequality (Theorem 3.2.15) implies $\sum_{n=1}^{\infty} |(x, \psi_n)|^2 < \infty$. Hence $(x, \psi_n) \to 0$. That is, $\{\psi_n\}$ converges weakly to 0. Needless to say, this sequence is not strongly convergent.

Example 4.3.3 also shows the following: If there exists a strongly convergent subsequence of $\{e_n\}$, then by Proposition 4.3.2 it also converges weakly to the same limit. On the other hand, since $\{e_n\}$ itself converges weakly to 0, this limit must be 0; i.e., this subsequence converges strongly to 0. On the other hand, since $\|e_n\| \equiv 1$, this is impossible. Hence no subsequence of $\{e_n\}$ can be strongly convergent. This shows that *the unit ball of* ℓ^2 *cannot be compact with respect to the strong topology.* In contrast to this, however, the unit ball can be, and actually is, compact with respect to the weak topology (or the weak* topology introduced below); see Theorem 4.3.6 below. This shows a great advantage of the notion of weak topology.

Definition 4.3.4. Let X be a normed linear space and X' its dual. A sequence $f_n \in X'$ is said to converge to $f \in X'$ in the sense of *weak* topology*, or simply *converges weak* to* f, if for every $x \in X$

$$\langle f_n - f, x \rangle \to 0, \quad n \to \infty.$$

Remark 4.3.5. If X is a Hilbert space, then X' can be identified with X as noted in the previous section (Riesz's theorem). Then X can be further identified with the dual of X'. Since X' is also a Hilbert space identified with X, applying Riesz's theorem, Theorem 4.2.1, again to X' yields that X is identified with the dual space of X'. Thus the weak* topology on X agrees with the weak topology on X. As is clear with this argument, this is valid for any normed linear space in which the bidual $X'' := (X')'$ (the dual space of X') coincides with X.

The following theorem guarantees that the unit ball of the dual space of X is always compact with respect to the weak* topology (weak*-compact, in short). This has a wide range of applications. Its proof may be found in most books on functional analysis and is omitted here; see, for example, [29, 31, 48, 51, 58, 64].

Theorem 4.3.6 (Bourbaki–Alaoglu). *Let X be a normed linear space and X' its dual space. The unit ball*

$$B := \{x \in X' : \|x\| \leq 1\}$$

of X' is compact with respect to the weak* topology.

Proposition 4.3.7. *Let X be a Hilbert space. Then every weakly convergent sequence $\{x_n\}$ in X is bounded in the strong topology, i.e., with respect to the norm. That is, there exists a* *constant $M > 0$ such that*

$$\|x_n\| \leq M \quad \forall n.$$

Proof. Consider the following family of continuous linear functionals on X:

$$T_n : X \to \mathbb{K} : x \mapsto (x, x_n). \tag{4.15}$$

For each $x \in X$, $\{(x, x_n)\}$ is convergent because of the weak convergence. Hence it is also bounded in $\mathbb{K}$. Then by the uniform boundedness principle (Theorem 2.5.6), the sequence $\{x_n\}$ is bounded in the sense that

$$\sup_{\|x\|=1} |(x, x_n)| < \infty.$$

By the Riesz–Fréchet theorem, Theorem 4.2.1, the left-hand side gives the norm of x_n, and hence $\{\|x_n\|\}$ is bounded. □

Remark 4.3.8. The proof above also shows that if $\{x_\lambda\}_{\lambda \in \Lambda}$ is bounded with respect to the weak topology, then it is also bounded in the strong topology. We should also note that the notion of convergence in the weak topology is much weaker than that for the strong topology, so it is not a priori clear that this convergence guarantees the boundedness in the strong topology. But this is indeed true when X is a Hilbert space as above.

4.4 Duality Between Subspaces and Quotient Spaces

There is a natural duality between subspaces and quotient spaces. We show some fundamental results in this section.

Let X be a normed linear space and M a subspace of X. Define its *annihilator* by

$$M^\perp := \{f \in X' : \langle f, x \rangle = 0 \text{ for all } x \in M\}. \tag{4.16}$$

We use the same notation $M^\perp$ following the convention but note that $M^\perp$ is a subset of the dual X'. When X is a Hilbert space, then $X' \cong X$, as noted in section 4.2, so this will not induce much confusion.

It is easy to see that $M^\perp$ is a subspace of X'.

Exercise 4.4.1. Prove the above.

Regard M as a normed linear space as a subspace of X, and consider its dual space M'. Then what is M'? The first result says that M' can be identified with $X'/M^\perp$.

Theorem 4.4.2. *Let X be a normed linear space and M a subspace of X. Then there exists a norm-preserving isomorphism from $X'/M^\perp$ onto M'. In this sense, $M' = X'/M^\perp$.*

Proof. Let j_M be the embedding of M into X:

$$j_M : M \to X : x \mapsto x. \tag{4.17}$$

This is clearly a continuous linear mapping. Define $T : X' \to M'$ by

$$T : X' \to M' : f \mapsto f \circ j_M.^{48} \tag{4.18}$$

That is, Tf is nothing but the restriction $f|_M$ of f to M. By the continuity of composed mappings, we clearly have $Tf \in M'$. Also,

$$\|Tf\| = \sup_{x \in M, x \neq 0} \frac{|\langle f, x \rangle|}{\|x\|} \leq \sup_{x \in X, x \neq 0} \frac{|\langle f, x \rangle|}{\|x\|} = \|f\| \quad \text{(by (4.4))}.$$

Hence $\|T\| \leq 1$. Moreover, T is surjective. This is because for every $g \in M'$, there exists $f \in X'$ such that $Tf = g$, by the Hahn–Banach extension theorem, Corollary A.7.4 (p. 245).

Since $f|_M = 0$ if and only if $\langle f, x \rangle = 0$, $x \in M$, the kernel $\ker T$ is precisely $M^\perp$ by the definition (4.16). Observe here that $M^\perp$ is a closed subspace as the kernel of a continuous linear mapping T (Proposition 2.1.28, p. 48).

Let $\pi : X' \to X'/M^\perp$ be the canonical projection (see p. 27), and introduce the quotient norm to $X'/M^\perp$ (Definition 2.3.3, p. 52). By the homomorphism theorem, Theorem 1.3.21 (p. 29), there exists a bijective mapping $S : X'/M^\perp \to M'$ such that $S \circ \pi = T$. Let us show that this S preserves norms.

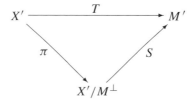

Take any $[f] = f + M^\perp \in X'/M^\perp$. Observing that $Tn = 0$ for every $n \in M^\perp$ and $\|T\| \leq 1$, we have

$$\|S[f]\| = \|Tf\| = \|T(f+n)\| \leq \|T\|\|f+n\| \leq \|f+n\|.$$

It then follows that

$$\|S[f]\| \leq \inf_{n \in M^\perp} \|f+n\| = \|[f]\| \quad \text{(by (2.26), p. 53)}. \tag{4.19}$$

Hence $\|S[f]\| \leq \|[f]\|$.

Conversely, take any $n \in M^\perp$ and set $g := T(f+n) = (f+n)|_M$. Since $n \in M^\perp$, $g = Tf = S[f]$ follows. Then again by the Hahn–Banach extension theorem, Corollary A.7.4, there exists $h \in X'$ such that $Th = g$ and $\|h\| = \|g\|$. This implies $Th = g = T(f+n) = Tf$. Since $T(h-f) = 0$, we have $h - f \in \ker T = M^\perp$. Then $[h] = [f]$, and

$$\|[f]\| = \|[h]\| \leq \|h\| = \|g\| = \|Tf\| = \|S[f]\|$$

follows. Hence $\|S[f]\| \geq \|[f]\|$, and combining this with the reverse inequality (4.19), we have $\|S[f]\| = \|[f]\|$. That is, S is a norm-preserving isomorphism. $\square$

---

[48]According to the definition of dual mappings to be introduced later (section 5.2, p. 90), $T = (j_M)'$.

The next theorem gives a dual of the theorem above. Note, however, that M is assumed to be a closed subspace, in contrast to the above.

Theorem 4.4.3. *Let X be a normed linear space and M a closed subspace of X. Then there exists a norm-preserving isomorphism from $(X/M)'$ to $M^\perp$. In this sense, $M^\perp = (X/M)'$.*

Proof. Let $\pi : X \to X/M$ be the canonical projection, and introduce the quotient norm (Definition 2.3.3) to X/M. Define a mapping $T : (X/M)' \to M^\perp$ by

$$T\xi := \xi \circ \pi, \quad \xi \in (X/M)'.$$

Clearly $\xi \circ \pi$ is a continuous linear functional on X, and since $\pi m = 0$ for $m \in M$, $\langle T\xi, m \rangle = 0$ follows. Hence $T\xi \in M^\perp$. Clearly T is a linear operator, and since $\|T\xi\| \leq \|\xi\| \|\pi\| \leq \|\xi\|$, we have $\|T\| \leq 1$.

Let us first show that T is injective. If $T\xi = 0$, then $0 = \langle T\xi, x \rangle = \langle \xi \circ \pi, x \rangle = \langle \xi, \pi x \rangle$ for every $x \in X$, and hence ξ is 0 on X/M; that is, T is injective.

Showing that T is surjective, $\|\xi\| \leq \|T\xi\|$, and combining these with the reverse inequality obtained above will complete the proof of the norm-preserving property of T. Take any $f \in M^\perp$. Define ξ by $\langle \xi, [x] \rangle := \langle f, x \rangle$. If $[x] = [y]$, then $x - y \in M$, and $f \in M^\perp$ yields $\langle f, x \rangle = \langle f, y \rangle$, so that this ξ is well defined. Since $[x] = \pi(x)$, we see that $f = \xi \circ \pi = T\xi$, and hence T is surjective.

Now for every m in M, we have $|\langle \xi, [x] \rangle| = |\langle f, x + m \rangle| \leq \|f\| \|x + m\|$. It follows that

$$|\langle \xi, [x] \rangle| \leq \|f\| \inf_{m \in M} \|x + m\| = \|f\| \|[x]\|.$$

Hence $\|\xi\| \leq \|f\| = \|T\xi\|$. Combining this with the reverse inequality, we see that T is a norm-preserving isomorphism. □

This duality readily yields some duality results in the following extremal problems.

Proposition 4.4.4. *Let X be a normed linear space, X' its dual space, and x, ψ elements of X, X', respectively. Let also M be a closed subspace of X. Then the following identities hold:*

$$\sup\{|\langle \psi, m \rangle| : m \in M, \|m\| \leq 1\} = \inf\{\|\psi - n\| : n \in M^\perp\}, \tag{4.20}$$

$$\inf\{\|x - m\| : m \in M\} = \sup\{|\langle x, n \rangle| : n \in M^\perp, \|n\| \leq 1\}. \tag{4.21}$$

Proof. The left-hand side of (4.20) represents the norm of ψ considered as a linear functional on the subspace M (a normed linear space in itself) (see (4.8) and Remark 4.1.7, p. 80). On the other hand, the right-hand side represents the norm of $[\psi] = \psi + M^\perp$ considered as an element of the quotient space $X'/M^\perp$ (see Definition 2.3.3, p. 52). By Theorem 4.4.2, these two are equal. The same kind of argument works for identity (4.21). □

Chapter 5

The Space $\mathcal{L}(X,Y)$ of Linear Operators

The theory of linear mappings, as well as that of the spaces themselves, forms an integral part of the theory of finite-dimensional vector spaces. For infinite-dimensional spaces the theory of linear operators is equally, or even more, indispensable. This chapter begins with the introduction of normed linear spaces consisting of linear operators and then discusses their relationships with duality, dual operators, and adjoint operators in Hilbert space. We then give the so-called Hilbert–Schmidt expansion theorem for compact Hermitian operators. All play central roles in classical functional analysis; in particular, adjoint operators appear quite naturally and frequently in many applications, often in connection with optimization problems.

5.1 The Space $\mathcal{L}(X,Y)$

Let X,Y be Banach spaces. We denote by $\mathcal{L}(X,Y)$ the space consisting of all continuous linear operators from X to Y. In particular, when $Y = X$, we write $\mathcal{L}(X)$ instead of $\mathcal{L}(X,Y)$. When $Y = \mathbb{K}$, then $\mathcal{L}(X,\mathbb{K}) = X'$; i.e., it is nothing but the dual space of X. In fact, $\mathcal{L}(X,Y)$ shares many properties with dual spaces by almost the same arguments developed for the latter. First, as in Proposition 4.1.2, $\mathcal{L}(X,Y)$ is closed under addition and scalar multiplication and hence forms a vector space. Also, by defining

$$\|A\| := \sup_{x \neq 0} \frac{\|Ax\|}{\|x\|} \tag{5.1}$$

for $A \in \mathcal{L}(X,Y)$, we see that this satisfies the axiom for norms and also that $\mathcal{L}(X,Y)$ turns out to be a Banach space, as in Theorem 4.1.5.

Let us summarize the above in the form of a theorem.

Theorem 5.1.1. *The space $\mathcal{L}(X,Y)$ of linear operators is a Banach space with respect to the norm defined by (5.1). Also, for $A \in \mathcal{L}(X,Y)$, the following identities hold:*

$$\|A\| = \sup_{\|x\|=1, x \in X} \|Ax\| = \sup_{\|x\| \leq 1, x \in X} \|Ax\| = \inf\{M : \|Ax\| \leq M\|x\| \ \forall x \in X\}.$$

89

The proofs are almost identical to those given in Theorem 4.1.5 and Proposition 4.1.6 and hence are omitted.

Exercise 5.1.2. Complete the proof above.

Remark 5.1.3. The norm (5.1) is called the *operator norm* induced from the norms of X and Y. There exist, however, those norms for linear operators not induced from the norms of the spaces. For example, let $X = Y = \mathbb{R}^n$, and let $A : \mathbb{R}^n \to \mathbb{R}^n$ be an $n \times n$ matrix. Let $\mathbb{R}^n$ be normed by the 2-norm:

$$\|x\|_2 := \sqrt{\sum_{k=1}^{n} |x_k|^2}.$$

The induced norm (5.1) is given by the maximal singular value of A (see Problem 11 at the end of this chapter). On the other hand,

$$\|A\|_2 := \sqrt{\sum_{i,j=1}^{n} |a_{ij}|^2}, \quad A := (a_{ij}) \tag{5.2}$$

and

$$\|A\|_{\max} := \max\{|a_{ij}| : i, j = 1, \ldots, n\} \tag{5.3}$$

both give a norm, but they are not induced norms.

Let X, Y, Z be Banach spaces, and let $B \in \mathcal{L}(X,Y)$, $A \in \mathcal{L}(Y,Z)$. Then the composition AB clearly belongs to $\mathcal{L}(X,Z)$, and

$$\|AB\| \leq \|A\| \|B\|. \tag{5.4}$$

Exercise 5.1.4. Show this inequality. Give an example in which $\|AB\| < \|A\| \|B\|$ holds. (Hint: It suffices to consider matrices. Give two nonzero matrices which become zero when multiplied.)

Examples 1.2.4–1.2.7 given in Chapter 1 are examples of continuous linear operators. Let us give yet another example.

Example 5.1.5. Let $X = C[0,1]$, $\phi(t) \in X$. The multiplication operator defined by

$$M_\phi : C[0,1] \to C[0,1] : \psi(t) \mapsto \phi(t)\psi(t)$$

is a continuous linear operator.

Exercise 5.1.6. Verify the continuity of this operator.

5.2   Dual Mappings

Let X, Y be Banach spaces and $A : X \to Y$ a continuous linear mapping. For any $\phi \in Y'$, define $A'\phi$ by (1.71) (p. 35), or as in (1.73), by

$$A'\phi = \phi \circ A : X \to \mathbb{K} : x \mapsto \langle \phi, Ax \rangle. \tag{5.5}$$

Then this is continuous as a composition of continuous mappings. Hence $A'\phi \in X'$. The correspondence

$$A' : Y' \to X' : \phi \mapsto A'\phi = \phi \circ A \tag{5.6}$$

is, as seen in (1.72), a linear mapping. We denote this mapping by A' in place of A^* in view of its continuity. This A' is called the *dual mapping* or the *dual operator* of A.

Not only is this $A' : Y' \to X'$ linear, but it is also continuous with respect to the respective topologies of X', Y' induced by their dual norms. Let us now show this.

Theorem 5.2.1. *The linear mapping A' defined by (5.6) above gives a continuous mapping from Y' to X'. That is, $A' \in \mathcal{L}(Y', X')$.*

Proof. Let us take any nonzero $\phi \in Y'$ and estimate $\|A'\phi\|$. Theorem 5.1.1, inequality (4.5) (p. 78), and (5.1) together imply

$$\begin{aligned}
\|A'\phi\| &= \sup_{\|x\|=1} |\langle A'\phi, x \rangle| = \sup_{\|x\|=1} |\langle \phi, Ax \rangle| \\
&\leq \sup_{\|x\|=1} \|\phi\|_{X'} \|Ax\| \leq \sup_{\|x\|=1} \|\phi\|_{X'} \|A\| \|x\| \\
&= \|A\| \|\phi\|_{X'}.
\end{aligned}$$

Hence $\|A'\| \leq \|A\|$, so that A' is bounded, i.e., continuous (Proposition 2.1.11, p. 43). $\quad\square$

We have just seen that $\|A'\| \leq \|A\|$ holds in the proof above, but the reverse inequality is also valid. That is, $\|A'\| = \|A\|$.

Proposition 5.2.2. *Under the same notation as above, $\|A'\| = \|A\|$.*

Proof. It suffices to prove $\|A\| \leq \|A'\|$. Take any $x \in X$. By Corollary A.7.3 (p. 244) to the Hahn–Banach extension theorem, for Ax there exists $f \in X'$ with $\|f\|_{X'} = 1$ such that $\langle Ax, f \rangle = \|Ax\|$. It follows that

$$\|Ax\| = \langle Ax, f \rangle = \langle x, A'f \rangle \leq \|A'f\| \|x\| \leq \|A'\| \|x\|.$$

Hence $\|A\| \leq \|A'\|$. $\quad\square$

The following proposition is useful in many situations.

Proposition 5.2.3. *Let X and Y be Banach spaces, $A \in \mathcal{L}(X, Y)$, and $A' \in \mathcal{L}(Y', X')$ be its dual mapping. The subspace $\operatorname{im} A \subset Y$ is dense in Y if and only if A' is injective. In particular, if either X or Y is finite-dimensional, then $\operatorname{im} A$ being dense means that A is surjective.*

Proof. Suppose that $\overline{\operatorname{im} A} = Y$, and suppose $A'f = 0$ for some $f \in Y'$. Then we have

$$\langle A'f, x \rangle = \langle f, Ax \rangle = 0$$

for every $x \in X$. Since $\operatorname{im} A$ is dense in Y, $\langle f, y \rangle = 0$ for every $y \in Y$ by continuity, and this means $f = 0$.

Conversely, suppose $\overline{\operatorname{im} A} \neq Y$. Then there exist $y_0 \in Y \setminus \overline{\operatorname{im} A}$ and $f \in Y'$ such that $\langle f, Ax \rangle = 0$ but $\langle f, y_0 \rangle \neq 0$. Otherwise, by a corollary of the Hahn–Banach theorem, by Corollary A.7.3 (p. 244) in Appendix A, it would imply $y_0 = 0 \in \operatorname{im} A$. It follows that

$$\langle A'f, x \rangle = \langle f, Ax \rangle = 0$$

for every $x \in X$, and hence $A'f = 0$. Since $\langle f, y_0 \rangle \neq 0$, f cannot be 0, and hence A' is not injective, which is a contradiction.

Finally, if either X or Y is finite-dimensional, then $\operatorname{im} A$ becomes a closed subspace of Y by Theorem 2.3.2 (p. 52), and hence $\operatorname{im} A$ being dense means that it is equal to Y, i.e., surjectivity. $\square$

5.3   Inverse Operators, Spectra, and Resolvents

The notion of the spectrum of a linear operator is an important generalization of eigenvalues of a matrix. As we see below, the spectrum of an operator T is defined to be the subset of all λ in $\mathbb{C}$ such that $\lambda I - T$ does not possess a continuous inverse. Since in an infinite-dimensional space it may occur that an operator possesses an inverse but this inverse may not be continuous, a spectrum does not necessarily consist of eigenvalues only. Indeed, there even exist operators that do not possess any eigenvalues at all.

Let X, Y be Banach spaces and $A \in \mathcal{L}(X,Y)$. If there exists $B \in \mathcal{L}(Y,X)$ such that $BA = I_X$ and $AB = I_Y$, then A is said to be *invertible*. B is called the inverse of A and is denoted by A^{-1}. Such a B is clearly uniquely determined. Note that we require continuity for B.

The following theorem is fundamental to the invertibility of operators of type $I - A$ for $A \in \mathcal{L}(X)$.

Theorem 5.3.1. *Let X be a Banach space and $A \in \mathcal{L}(X)$. If $\|A\| < 1$, then $I - A$ is invertible. In fact,*

$$(I - A)^{-1} = \sum_{k=0}^{\infty} A^k, \tag{5.7}$$

where A^0 is understood to be the identity operator I. This series is sometimes called a Neumann series.

Proof. Let $S_n := \sum_{k=0}^{n} A^k$ be the nth partial sum. Using Exercise 5.1.4 and the definition of the operator norm here given by (5.1) or by Theorem 5.1.1, we have, for $m > n$ and $\|x\| \leq 1$,

$$\|S_m x - S_n x\| = \left\| \sum_{k=n+1}^{m} A^k x \right\|$$

$$\leq \sum_{k=n+1}^{m} \left\| A^k x \right\| \leq \sum_{k=n+1}^{m} \|A\|^k \|x\|$$

$$\leq \|x\| \sum_{k=n+1}^{\infty} \|A\|^k \leq \frac{\|A\|^{n+1}}{1 - \|A\|} \to 0 \quad (n \to \infty). \tag{5.8}$$

Thus $\{S_n\}$ forms a Cauchy sequence in $\mathcal{L}(X)$. Hence by the completeness of $\mathcal{L}(x)$ (Theorem 5.1.1), S_n converges to an element in $S \in \mathcal{L}(X)$.

Let us show that this S gives $(I - A)^{-1}$. By the continuity of $I - A$, we have

$$(I - A)Sx = (I - A)\lim_{n\to\infty}(I + A + \cdots + A^n)x$$

$$= \lim_{n\to\infty}(I - A)(I + A + \cdots + A^n)x$$

$$= \lim_{n\to\infty}(x - A^{n+1}x).$$

Since $\|A^{n+1}x\| \leq \|A^{n+1}\|\,\|x\| \leq \|A\|^{n+1}\,\|x\|$ by Exercise 5.1.4, it follows that $A^{n+1}x \to 0$. Hence for every $x \in X$, $(I - A)Sx = x$. Similarly, $S(I - A)x = x$, and $S = (I - A)^{-1}$ thus follows. $\quad\square$

Corollary 5.3.2. *Suppose that $\lambda \in \mathbb{C}$, $\lambda \neq 0$, and also $\|A\| < |\lambda|$. Then $\lambda I - A$ is invertible and is given by*

$$(\lambda I - A)^{-1} = \frac{1}{\lambda}\sum_{k=0}^{\infty}\frac{A^k}{\lambda^k}. \tag{5.9}$$

Proof. The proof is obvious from

$$(\lambda I - A)^{-1} = (\lambda(I - A/\lambda))^{-1},$$

and $\|A/\lambda\| < 1$. $\quad\square$

In general, if we have a sequence $\{A_k\}$ of operators, $A_k \in \mathcal{L}(X,Y)$, such that $\sum_{k=1}^{\infty}\|A_k\| < \infty$, then the sum $\sum_{k=1}^{\infty}A_k$ of these operators exists and indeed belongs to $\mathcal{L}(X,Y)$. The proof is similar to the argument in (5.8) and is left as an exercise.

Exercise 5.3.3. Prove the above fact.

Let us now give the definition of the spectrum of an operator.

Definition 5.3.4. Let X be a Banach space and $A \in \mathcal{L}(X)$. The *spectrum* $\sigma(A)$ of A is the set of all complex numbers λ such that $\lambda I - A$ does not possess a bounded (i.e., continuous) inverse.

When X is a finite-dimensional space, $\lambda I - A$ can be represented by a matrix. Hence its inverse, if it exists at all, can also be represented as the inverse matrix. By a well-known result in matrix theory, $\lambda I - A$ is not invertible if and only if it is not one-to-one—in other words, if and only if there exists a vector $x \neq 0$ such that

$$(\lambda I - A)x = 0.$$

This means that λ is an eigenvalue of A. In this case the spectrum consists only of eigenvalues.

For infinite-dimensional spaces, we still call λ an *eigenvalue* if there exists $x \neq 0$ such that $Ax = \lambda x$, but the situation here is not so simple as it is above. For example, there are many instances where an operator admits no eigenvalues, but its spectrum is still nonempty. The following examples show some of these instances.

Example 5.3.5. Consider $X = \ell^2$. Define the right-shift operator S by

$$S(x_1, x_2, \ldots) := (0, x_1, x_2, \ldots).$$

It is readily seen that for every $\lambda \in \mathbb{C}$, $Sx = \lambda x$ implies $x = 0$. Hence S admits no eigenvalues. On the other hand, it is easy to verify that S is not continuously invertible. Hence $0 \in \sigma(S)$.

Example 5.3.6 (Volterra operator). Let $X = L^2[0,1]$, and let $K : [0,1] \times [0,1] \to \mathbb{R}$ be a square integrable function. Define an operator $A \in \mathcal{L}(X)$ by

$$(Ax)(\xi) := \int_0^\xi K(\xi, \eta) x(\eta) d\eta, \quad 0 \leq \xi \leq 1.$$

Since, for every $x \in X$, Ax is differentiable almost everywhere, this implies that A^{-1} cannot exist, because X contains many functions that are not differentiable. Hence $0 \in \sigma(A)$. On the other hand, it is known that this operator does not in general possess an eigenvalue.

The complement of the spectrum, namely, the set of all $\lambda \in \mathbb{C}$ such that there exists a bounded inverse of $\lambda I - A$, is called the *resolvent set* and is denoted by $\rho(A)$. When $\lambda \in \rho(A)$, the operator $(\lambda I - A)^{-1}$ is called the *resolvent*. The following facts are known (proof omitted; see, e.g., [48, 66]):

1. The spectrum $\sigma(A)$ is a nonempty compact set.

2. $r(A) := \sup\{|\lambda| : \lambda \in \sigma(A)\}$ is called the *spectral radius*. It can be shown that $r(A) \leq \|A\|$.[49]

5.4   Adjoint Operators in Hilbert Space

We have already introduced the notion of dual operators in section 5.2. Let us now introduce a closely related, but slightly different, notion of adjoint operators on Hilbert space.

As we have seen in subsection 1.4.3, dual operators correspond to transposed matrices in finite-dimensional spaces. This is considered over the spaces X, Y along with their duals X', Y'. On the other hand, in Hilbert spaces, as is derived from Riesz's theorem, X itself becomes isomorphic to its dual space, and it is possible to define yet another notion of a related but different concept of *adjoint operators*. For example, in complex Euclidean spaces, there is such a concept as *conjugate transposed matrices*, obtained by taking the complex conjugate of the transposed matrix. This satisfies the following property:

$$\left(x, A^* y\right) = (Ax, y).$$

Note here the difference between $\langle \cdot, \cdot \rangle$ and the inner product $(\cdot, \cdot)$. In what follows, we generalize this to Hilbert space.

Recall that in subsection 1.4.3 we did not make a distinction between dual and adjoint operators. In what follows, however, we make the following distinction to avoid confusion: *When we deal with dual spaces, we use the term* dual mappings (operators), *and when we deal with such concepts in relation with inner products, we use the term* adjoint

---

[49]Some books denote the spectral radius by $\rho(A)$, and the reader should be alerted to the confusion.

mappings (operators) and denote them by A' for the dual operator and A^* for the adjoint operator, respectively. Note, however, that in many books these terms are often used interchangeably. Observe also that when we have a real vector space and the inner product is the standard one, the two notions coincide,[50] as may be seen from the subsequent discussions.

Theorem 5.4.1. *Let X, Y be Hilbert spaces and $A \in \mathcal{L}(X, Y)$. Then there exists a unique mapping $A^* \in \mathcal{L}(Y, X)$ such that*

$$(Ax, y)_Y = (x, A^* y)_X \quad \forall x \in X, \ y \in Y. \tag{5.10}$$

Moreover, $\|A^*\| = \|A\|$.

Proof. Take any $y \in Y$, and consider the following linear functional associated with it:

$$x \mapsto (Ax, y)_Y.$$

This is clearly linear and also continuous because

$$|(Ax, y)_Y| \leq \|Ax\| \, \|y\| \leq \|A\| \, \|x\| \, \|y\|.$$

Hence, by the Riesz–Fréchet theorem, Theorem 4.2.1 there exists a unique $w \in X$ such that $(Ax, y)_Y = (x, w)_X$ holds for every $x \in X$. Regard this w as a function of y and denote it by $A^* y$. Then obviously (5.10) holds.

Let us show that this mapping A^* is linear. For every $x \in X$, $y, z \in Y$, and $\alpha, \beta \in \mathbb{K}$, we have

$$\begin{aligned}
\left(x, A^*(\alpha y + \beta z)\right)_X &= (Ax, \alpha y + \beta z)_Y \\
&= \overline{\alpha}(Ax, y)_Y + \overline{\beta}(Ax, z)_Y \\
&= \overline{\alpha}\left(x, A^* y\right)_X + \overline{\beta}\left(x, A^* z\right)_X \\
&= \left(x, \alpha A^* y + \beta A^* z\right)_X.
\end{aligned}$$

Since x is arbitrary, $A^*(\alpha y + \beta z) = \alpha A^* y + \beta A^* z$ follows from Proposition 3.1.4 on p. 61, and hence A^* is linear.

According to the Riesz–Fréchet theorem, Theorem 4.2.1 (and (4.13)), the norm of any $z \in Y$ equals that of the linear functional induced by it; that is,

$$\|z\| = \sup_{\|y\|=1} |(y, z)|.$$

Hence the continuity of A^* and $\|A^*\| = \|A\|$ follow from the corresponding results in Theorem 5.2.1 and Proposition 5.2.2 for dual operators, by considering these over $\mathcal{L}(Y', X')$. $\qquad\square$

Exercise 5.4.2. Under the notation above, $(A^* y, x)_X = (y, Ax)_Y$ also holds (note that the positions of the operators are interchanged). Prove this identity.

---

[50]They are thus even more easily confused; but the reader should keep in mind that the notion of adjoint operators can be defined only through inner products, while dual operators do not need an inner product.

5.5   Examples of Adjoint Operators

In what follows, we understand that $\mathbb{C}^n$ is equipped with the standard inner product defined by

$$(x,y) := \sum_{i=1}^{n} x_i \overline{y_i} = y^* x, \tag{5.11}$$

where y^* denotes the conjugate transpose of y, i.e., $\overline{y^T} = \overline{y}^T$. Note that the order of y and x is reversed in the last equality sign of (5.11). This is necessary to make the definition compatible with the convention that inner products are conjugate linear in their second variable.[51] Also, $(L^2[a,b])^n$ denotes the space of all $\mathbb{C}^n$-valued functions on $[a,b]$ with square integrable entries in the sense of Lebesgue. Define an inner product on this space by

$$(x,y) := \int_a^b (x(t),y(t))_{\mathbb{C}^n} \, dt = \int_a^b y^*(t) x(t) dt. \tag{5.12}$$

Example 5.5.1. Let $X = \mathbb{C}^n$, and let $A = (a_{ij})$ be an $n \times n$ matrix. It readily follows from the definition (5.11) of the inner product of $\mathbb{C}^n$ that

$$\begin{aligned}
(Ax,y) &= \overline{y}^T A x \\
&= \overline{\left(\overline{A}^T y\right)}^T x \\
&= \left(x, \overline{A}^T y\right).
\end{aligned}$$

Hence $A^* = \overline{A}^T$ in this case, i.e., A^* in this case agrees with the conjugate transpose of A, and is consistent with the standard notation.

Example 5.5.2. Let $X = (L^2[a,b])^n$, $Y = \mathbb{C}^p$, and $K(t)$ be a $p \times n$ matrix whose entries are square integrable in the sense of Lebesgue on $[a,b]$. Define an operator $A : X \to Y$ by

$$Ax := \int_a^b K(t) x(t) dt.$$

It follows that

$$\begin{aligned}
(Ax,y)_{\mathbb{C}^p} &= \left(\int_a^b K(t)x(t)dt, y\right) \\
&= y^* \left(\int_a^b K(t)x(t)dt\right) \\
&= \int_a^b \left(K^*(t)y\right)^* x(t) dt \\
&= \left(x, K^*(\cdot)y\right)_{L^2[a,b]}.
\end{aligned}$$

---

[51] If we were to take $x^* y$ as a definition for (x,y), then it would *not* be linear in the first variable: $(\alpha x, y) = \overline{\alpha}(x,y)$.

Hence by defining $A^* : Y \to X$ by

$$A^* : y \mapsto K^*(\cdot)y,$$

we have $(Ax, y) = (x, A^*y)$.

The next example gives its dual.

Example 5.5.3. Let $X = \mathbb{C}^n$ and $Y = (L^2[a,b])^p$, and let $K(t)$ be the same as in Example 5.5.2 above. Define an operator $A : X \to Y$ by

$$(Ax)(t) := K(t)x.$$

Then we have

$$
\begin{aligned}
(Ax, y)_{L^2[a,b]} &= \int_a^b y^*(t) K(t) x \, dt \\
&= \left\{ \int_a^b K^*(t) y(t) dt \right\}^* x \\
&= \left(x, \int_a^b K^*(t) y \, dt \right)_{\mathbb{C}^n}.
\end{aligned}
$$

Defining $A^* : Y \to X$ by

$$A^* : y(\cdot) \mapsto \int_a^b K^*(t) y(t) dt,$$

we obtain $(Ax, y) = (x, A^*y)$, as expected.

Example 5.5.4 (Fredholm operator). Let $X = (L^2[a,b])^n$ and $Y = (L^2[a,b])^p$, and let $K(\xi, \eta)$ be a $p \times n$ matrix of Lebesgue square integrable functions in two variables on $[a,b] \times [a,b]$. Define an operator $A : X \to Y$ by

$$Ax := \int_a^b K(\cdot, \eta) x(\eta) d\eta.$$

Strictly speaking, we must guarantee that the integrand makes sense almost everywhere in the sense of Lebesgue and whether it is further integrable or not. This can be, and must be, ensured by using Fubini's theorem (e.g., [20, 22, 37]), but we omit the proof here. Of course, when K is continuous, it raises no difficulty, and those who are content with such a case can proceed with this case in mind.

According to the definition, we have

$$
\begin{aligned}
(Ax, y) &= \left(\int_a^b K(\cdot, \eta) x(\eta) d\eta, \, y \right) \\
&= \int_a^b \int_a^b y^*(\xi) K(\xi, \eta) x(\eta) d\eta \, d\xi \\
&= \int_a^b \left\{ \int_a^b \left(K^*(\xi, \eta) y(\xi) \right)^* d\xi \right\} x(\eta) d\eta \\
&= \left(x, \int_a^b K^*(\xi, \cdot) y(\xi) d\xi \right).
\end{aligned}
$$

This yields

$$A^*y = \int_a^b K^*(\xi,\cdot)y(\xi)d\xi.$$

The change of the order of integration in the computation above is guaranteed by Fubini's theorem (e.g., [20, 22, 37]).

Example 5.5.5 (Volterra operator). Let $X = (L^2[a,b])^n$, $Y = (L^2[a,b])^p$, and $K(\xi,\eta)$ be the same as above. Define an operator $A : X \to Y$ by

$$(Ax)(\xi) := \int_a^\xi K(\xi,\eta)x(\eta)d\eta, \quad a \le \xi \le b.$$

The only difference from Example 5.5.4 is in the upper bound of the interval of integration, but this induces much difference.

Following the definition, we have

$$(Ax, y) = \left(\int_a^\bullet K(\cdot,\eta)x(\eta)d\eta, y \right)$$
$$= \int_a^b y^*(\xi) \int_a^\xi K(\xi,\eta)x(\eta)d\eta d\xi.$$

The domain of integration here is the triangular region, as shown in Figure 5.1. It is integrated, as shown on the left of Figure 5.1, first from a to ξ with respect to η and then in ξ from a through b. If we change the order of integration, this must be equal, as shown on the right of Figure 5.1, to that given by integrating in ξ from η to b first, and then integrating in η from a through b afterward. The change of the order of integration is guaranteed by Fubini's theorem, as noted in Example 5.5.4.

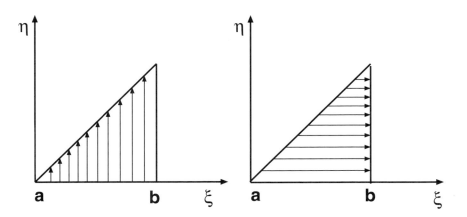

Figure 5.1. *Domain of integration*

Hence we can change the order of integration and obtain

$$\int_a^b y^*(\xi) \int_a^\xi K(\xi,\eta)x(\eta)\,d\eta\,d\xi = \int_a^b \left\{ \int_\eta^b \left(K^*(\xi,\eta)y(\xi) \right)^* d\xi \right\} x(\eta)\,d\eta$$

$$= \left(x, \int_\bullet^b K^*(\xi,\cdot)y(\xi)\,d\xi \right).$$

Therefore,

$$(A^*y)(\eta) = \int_\eta^b K^*(\xi,\eta)y(\xi)\,d\xi.$$

The adjoint operators discussed above appear in various situations in application. While they may take different forms depending on the context, they are often reducible to one of or a combination of the examples given above.

5.6 Hermitian Operators

As we have seen in Examples 5.3.5 and 5.3.6, there can be some bothersome cases of infinite-dimensional operators that do not have any eigenvalues. For compact[52] Hermitian operators to be introduced in this and in the next sections, such anomalies do not occur, and a *spectral resolution theorem* holds as a natural generalization of the corresponding finite-dimensional result. We start with the discussion of Hermitian operators as a generalization of symmetric matrices.

An operator A on a Hilbert space X is said to be *Hermitian* or *self-adjoint* if $A^* = A$, i.e.,

$$(Ax, y) = (x, Ay), \quad x, y \in X.$$

The operator defined in Example 5.5.4 gives a Hermitian operator when $K^*(\xi,\eta) = K(\eta,\xi)$.

Let us begin with the following fact.

Proposition 5.6.1. *Let A be a Hermitian operator on a Hilbert space X. For every $x \in X$, (Ax,x) is always real-valued. The norm of A is given by*

$$\|A\| = \sup_{\|x\|=1} |(Ax,x)|. \tag{5.13}$$

Proof. According to property 2 of inner products (Definition 3.1.1) and the self-adjointness of A, we have

$$\overline{(Ax,x)} = (x, Ax) = (Ax,x).$$

Hence (Ax,x) is a real number.

Let us show (5.13). Note that, by the Cauchy–Schwarz inequality (Theorem 3.1.5), for $\|x\| = 1$,

$$|(Ax,x)| \le \|Ax\|\,\|x\| \le \|A\|\,\|x\|^2 = \|A\|.$$

Hence $\sup_{\|x\|=1} |(Ax,x)| \le \|A\|$.

---

[52]to be defined in the next section.

Let us show the reverse inequality. Let $M := \sup_{\|x\|=1} |(Ax,x)|$, and we show $\|A\| \leq M$. For every $v \neq 0$,

$$(Av,v) = \|v\|^2 (Av/\|v\|, v/\|v\|) \leq M \|v\|^2. \tag{5.14}$$

Now, for every x, y, we have

$$
\begin{aligned}
(A(x \pm y), x \pm y) &= (Ax,x) \pm (Ax,y) \pm (Ay,x) + (Ay,y) \\
&= (Ax,x) \pm (Ax,y) \pm \overline{(x,Ay)} + (Ay,y) \\
&= (Ax,x) \pm (Ax,y) \pm \overline{(Ax,y)} + (Ay,y) \\
&= (Ax,x) \pm 2\,\mathrm{Re}\,(Ax,y) + (Ay,y).
\end{aligned}
$$

Subtracting the respective terms, and using (5.14) and the parallelogram law (Proposition 3.1.7, p. 62), we obtain

$$
\begin{aligned}
4\,\mathrm{Re}\,(Ax,y) &= (A(x+y), x+y) - (A(x-y), x-y) \\
&\leq M \left(\|x+y\|^2 + \|x-y\|^2 \right) \\
&= 2M(\|x\|^2 + \|y\|^2).
\end{aligned}
$$

Put $(Ax,y) = |(Ax,y)| e^{i\theta}$ and replace x by $e^{-i\theta} x$ to obtain

$$\left(A e^{-i\theta} x, y \right) = e^{-i\theta} (Ax,y) = |(Ax,y)|.$$

It follows that

$$|(Ax,y)| = \mathrm{Re} \left(A e^{-i\theta} x, y \right) \leq \frac{M}{2} \left(\left\| e^{-i\theta} x \right\|^2 + \|y\|^2 \right) = \frac{M}{2} (\|x\|^2 + \|y\|^2).$$

If $Ax \neq 0$, then substituting $y = (\|x\| / \|Ax\|) Ax$ for y yields (note $\|y\| = \|x\|$)

$$\|x\| \, \|Ax\| = \frac{\|x\|}{\|Ax\|} |(Ax,Ax)| = \left| \left(Ax, \frac{\|x\|}{\|Ax\|} Ax \right) \right| \leq M \|x\|^2.$$

Hence

$$\|Ax\| \leq M \|x\|.$$

Since this inequality holds trivially for $Ax = 0$, $\|A\| \leq M$ follows, and this completes the proof. $\square$

As is the case with symmetric or Hermitian matrices, the eigenvalues of Hermitian operators are all real, and eigenvectors corresponding to distinct eigenvalues are always mutually orthogonal.

Proposition 5.6.2. *Let $A \in \mathcal{L}(X)$ be a Hermitian operator on a Hilbert space X. All eigenvalues of A are real, and eigenvectors corresponding to distinct eigenvalues are mutually orthogonal.*

Proof. Suppose that λ is an eigenvalue of A and x is a corresponding eigenvector. Since $A = A^*$, we have

$$
\begin{aligned}
0 &= (Ax, x) - (x, Ax) \\
&= (\lambda x, x) - (x, \lambda x) \\
&= (\lambda - \bar{\lambda}) \|x\|^2.
\end{aligned}
$$

Thus $\lambda = \bar{\lambda}$; that is, λ is a real number.

   Now assume that λ and μ are two distinct eigenvalues and that x, y are corresponding eigenvectors, respectively. Then

$$
\begin{aligned}
0 &= (Ax, y) - (x, Ay) = (\lambda x, y) - (x, \mu y) \\
&= (\lambda - \bar{\mu})(x, y) = (\lambda - \mu)(x, y).
\end{aligned}
$$

Since $\lambda \neq \mu$, $(x, y) = 0$ follows. $\qquad\qquad\qquad\qquad\qquad\qquad\qquad\qquad$ □

5.7 Compact Operators and Spectral Resolution

It is well recognized that eigenvalues and eigenvectors play a crucial role in deriving canonical forms of matrices. Does there exist a canonical form, e.g., a diagonal form, for infinite-dimensional operators? In this context, it is called a spectral resolution, but it does not necessarily exist. But for compact Hermitian operators, which we discuss now, a result analogous to that for Hermitian matrices holds. This is called the Hilbert–Schmidt expansion theorem and is also important in applications.

Definition 5.7.1. An operator $T \in \mathcal{L}(X)$ in a Hilbert space X is called a *compact operator* or a *completely continuous operator* if the closure $\overline{T B_1}$ of the image of the unit ball $B_1 :=$ $\{x \in X : \|x\| \leq 1\}$ of X under T is a compact subset[53] of X.

Remark 5.7.2. The definition above is the same if we replace B_1 by its constant multiple cB_1 with $c > 0$. Also, since every bounded set B can be contained in cB_1 for some positive c, the definition remains the same if we replace the requirement that the closure of the image of every bounded set under T be compact.

   This definition may appear somewhat counterintuitive. The following theorem shows that compact operators have a continuity in a very strong sense.

Theorem 5.7.3. *A continuous operator T in a Hilbert space X is compact if and only if for every sequence $\{x_n\}$ weakly convergent to 0, $T x_n$ converges strongly to 0.*

Proof. Let T be compact. Suppose that x_n converges to 0 weakly. Then, for every $y \in X$,

$$
(T x_n, y) = (x_n, T^* y) \to 0.
$$

---

[53]For compactness of a set, see Appendix A, p. 240.

Hence Tx_n also converges to 0 weakly. Since every weakly convergent sequence is bounded according to Proposition 4.3.7 (p. 84), we have

$$\limsup_{n\to\infty} \|Tx_n\| = \alpha < \infty.$$

By taking a suitable subsequence $\{x_{n'}\}$, we can make $\lim_{n'\to\infty} \|Tx_{n'}\| = \alpha$ hold. Furthermore, by the compactness of T, we can take yet a further subsequence $\{x_{n''}\}$ of $\{x_{n'}\}$ such that $Tx_{n''}$ is strongly convergent. Now, since Tx_n is weakly convergent to 0, this limit must be 0. Since $\alpha = \lim_{n''\to\infty} \|Tx_{n''}\|$, this yields $\alpha = 0$. On the other hand, since $\|Tx_n\| \geq 0$ and $\alpha = 0$, $\lim_{n\to\infty} \|Tx_n\| = 0$ follows; that is, Tx_n converges strongly to 0.

Conversely, suppose that T satisfies the condition of this theorem. Pick any sequence $\{Tx_n\}$ from the image of the unit ball under T. It suffices to show that a suitable subsequence is strongly convergent (see Definition A.5.1 on p. 240 in Appendix A). According to the Bourbaki–Alaoglu theorem, Theorem 4.3.6 (p. 84), the unit ball of X is weakly compact.[54] Hence there exists a subsequence $\{x_{n'}\}$ of $\{x_n\}$ and some $x_0 \in X$ such that $x_{n'} \to x_0$ weakly. Then by hypothesis $T(x_{n'} - x_0) \to 0$ strongly; that is, $Tx_{n'} \to Tx_0$.   □

Example 5.7.4. Although we omit the proof, we note that the Fredholm and Volterra operators given in Examples 5.5.4 and 5.5.5 are compact. Note, however, that while the Fredholm operator becomes Hermitian when $K^*(\xi,\eta) = K(\eta,\xi)$, the Volterra operator cannot be Hermitian, as can be seen from the calculations in Example 5.5.5.

Theorem 5.7.5. *Let $T \in \mathcal{L}(X)$ be a compact Hermitian operator in a Hilbert space X. Then either $\|T\|$ or $-\|T\|$ is an eigenvalue of T. Consequently, $\|T\|$ agrees with the maximum of the absolute values of eigenvalues of T.*

Proof. If $T = 0$, then $0 = \|T\|$ is obviously an eigenvalue of T. Hence let us assume that $T \neq 0$. By Proposition 5.6.1 (p. 99),

$$\|T\| = \sup_{\|x\|=1} |(Tx,x)|.$$

Hence there exists a sequence x_n with $\|x_n\| = 1$ such that

$$|(Tx_n,x_n)| \to \|T\|.$$

Since T is Hermitian, each (Tx_n,x_n) is real-valued, and hence by taking a suitable subsequence we can let $(Tx_n,x_n) \to \lambda$ for either $\lambda = \|T\|$ or $\lambda = -\|T\|$. Since λ is real and $\|T\| = |\lambda|$, we have

$$\|Tx_n - \lambda x_n\|^2 = \|Tx_n\|^2 - 2\lambda(Tx_n,x_n) + \lambda^2 \|x_n\|^2$$
$$\leq 2\lambda^2 - 2\lambda(Tx_n,x_n) \to 0 \quad (n \to \infty).$$

Thus $Tx_n - \lambda x_n \to 0$. On the other hand, since T is compact and $\|x_n\| = 1$, there exists a suitable subsequence of $\{Tx_n\}$ (that is, by taking a suitable subsequence $\{x_{n'}\}$ of $\{x_n\}$ and considering $\{Tx_{n'}\}$) that is strongly convergent. Let us denote this sequence also by $\{Tx_n\}$ for brevity. Hence we must have

$$Tx_n \to y, \quad Tx_n - \lambda x_n \to 0.$$

---

[54] Since X is a Hilbert space, weak* and weak topologies coincide.

This implies $\lambda x_n \to y$, and $\|y\| = \lim \|\lambda x_n\| = |\lambda| = \|T\| \neq 0$ yields $y \neq 0$. Now that $\lambda T x_n \to \lambda y$, and on the other hand $\lambda T x_n = T \lambda x_n \to T y$, we must have $T y = \lambda y$; that is, λ is an eigenvalue of T. $\square$

The following facts hold for compact Hermitian operators.

Proposition 5.7.6. *Let T be a compact Hermitian operator on a Hilbert space X. The following properties hold:*

    1. *For every nonzero eigenvalue of T, its corresponding eigenspace is finite-dimensional.*

    2. *Every nonzero eigenvalue of T is an isolated point in the set of eigenvalues.*

Proof. Suppose that the eigenspace corresponding to an eigenvalue $\lambda \neq 0$ is infinite-dimensional. Then there exists an orthonormal system $\{e_1, \ldots, e_n, \ldots\}$ in this subspace (Schmidt orthogonalization (Problems in Chapter 3)). We have seen that this $\{e_n\}$ converges weakly to 0 (Example 4.3.3, p. 83). On the other hand, since T is compact, $T e_n$ converges strongly to 0 (Theorem 5.7.3). In particular, $\|T e_n\| \to 0$. But this contradicts $\|T e_n\| = \|\lambda e_n\| = |\lambda| \neq 0$. Hence such an eigenspace must be finite-dimensional.

    Now suppose that $\lambda_0 \neq 0$ is an eigenvalue such that there exists a sequence $\{\lambda_n\}$ of distinct eigenvalues such that $\lambda_n \to \lambda_0$. Taking the eigenvectors x_n such that $T x_n = \lambda_n x_n$, $\|x_n\| = 1$, $n = 1, 2, \ldots$, we see from Proposition 5.6.2 that they constitute an orthonormal system. Hence, by the same reasoning as above, we have $T x_n \to 0$. But this contradicts $\|T x_n\| = |\lambda_n| \to |\lambda_0| \neq 0$. $\square$

We are now ready to state and prove the Hilbert–Schmidt expansion theorem.

Theorem 5.7.7 (Hilbert–Schmidt expansion theorem). *Let T be a compact Hermitian operator on a Hilbert space X, $\{\lambda_n\}$ its nonzero eigenvalues, and $\{\varphi_n\}$ the corresponding eigenvectors. Let these eigenvalues λ_n be counted according to multiplicities derived by the dimension of each corresponding eigenspace and ordered decreasingly by order of magnitude of their absolute value. Let also $\{\varphi_n\}$ be orthonormalized. Then, for every $x \in X$,*

$$T x = \sum_{n=1}^{\infty} \lambda_n (x, \varphi_n) \varphi_n \tag{5.15}$$

holds. Further, if T is injective (one-to-one), then the orthonormal system $\{\varphi_n\}$ is complete (cf. Definition 3.2.17), and for every $x \in X$,

$$x = \sum_{n=1}^{\infty} (x, \varphi_n) \varphi_n \tag{5.16}$$

holds.

Remark 5.7.8. The orthonormal system above always exists. For distinct eigenvalues their corresponding eigenvectors are mutually orthogonal (Proposition 5.6.2, p. 100), and for a fixed nonzero eigenvalue the corresponding eigenspace is finite-dimensional by Proposition 5.7.6. Then, by restricting the action of T to this finite-dimensional space, it becomes a

Hermitian linear transformation, and hence we can choose an orthonormal basis for this eigenspace by relying on the result on canonical forms for Hermitian matrices.

Proof. Let us first show (5.15). It goes as follows: Let M be the closure of the subspace spanned by $\{\varphi_1,\ldots,\varphi_n,\ldots\}$, decompose X into the direct sum of M and its orthogonal complement $M^\perp$, and then show that (5.15) holds on each of these components. Denote by $\tilde{T}$ the operator defined by the right-hand side of (5.15). Since $T\varphi_i = \tilde{T}\varphi_i$ clearly holds for every i, T and $\tilde{T}$ agree on $\mathrm{span}\{\varphi_1,\ldots,\varphi_n\}$. Since this is valid for every n and since T and $\tilde{T}$ are both continuous, they also agree on the closure $M = \overline{\mathrm{span}\{\varphi_1,\ldots,\varphi_n,\ldots\}}$.

Now the orthogonal complement of M turns out to be

$$M^\perp = \ker T = \{x \in X \, : \, Tx = 0\}.$$

In fact, if $x \in \ker T$, then $0 = (Tx,\varphi_n) = (x,T\varphi_n) = \lambda_n(x,\varphi_n)$ implies $(x,\varphi_n) = 0$, and hence $x \in M^\perp$.

Conversely, if $x \in M^\perp$, then

$$(Tx,\varphi_n) = (x,T\varphi_n) = \lambda_n(x,\varphi_n) = 0$$

also yields $Tx \in M^\perp$. This means that $M^\perp$ is an invariant subspace of T. Hence if $T \neq 0$ on $M^\perp$, then T would become a nonzero compact Hermitian operator there. Then by Theorem 5.7.5 there must exist a nonzero eigenvalue of T and a corresponding eigenvector in $M^\perp$. This contradicts the choice of M. Hence we must have $T|_{M^\perp} = 0$. That is, $M^\perp \subset \ker T$.

On the other hand, since $\tilde{T}$ clearly vanishes on $M^\perp$ by its definition, T and $\tilde{T}$ are both equal to zero on $M^\perp$. Since $X = M \oplus M^\perp$ and T and $\tilde{T}$ are identical on each components, we must have $T = \tilde{T}$. Hence (5.15) is proved.

Now subtracting the right-hand side from the left-hand side in (5.15), we obtain

$$0 = Tx - \sum_{n=1}^\infty \lambda_n(x,\varphi_n)\varphi_n = T\left(x - \sum_{n=1}^\infty (x,\varphi_n)\varphi_n\right).$$

If T is injective, clearly $x - \sum_{n=1}^\infty (x,\varphi_n)\varphi_n = 0$ follows; that is, (5.16) holds. $\quad\square$

Remark 5.7.9. This is clearly a generalization of the diagonalization of Hermitian matrices in finite-dimensional spaces. This is also a special of a more general spectral resolution theorem; see, e.g., [38, 48, 66].

Problems

1. Let X be a Banach space and M a proper closed subspace of X. Show that the norm of the canonical projection $\pi : X \to X/M$ is 1.

2. Let X be a Banach space, and let $A, B \in \mathcal{L}(X)$. Show that if $I - AB$ has a continuous inverse, then so does $I - BA$. (Hint: Consider $I + B(I - AB)^{-1}A$.)

3. Let X be Banach space, and let $A \in \mathcal{L}(X)$. Show that if $\|I - A\| < 1$, then A^{-1} exists and is continuous. (Hint: Write $A = I - (I - A)$, and use Theorem 5.3.1.)

4. Let X be a Hilbert space, and let $A \in \mathcal{L}(X)$. Show that if A is continuously invertible, then so is A^*, and $(A^*)^{-1} = (A^{-1})^*$ holds.

5. Let X and A be as above. Show that

$$\sigma(A^*) = \{\bar{\lambda} : \lambda \in \sigma(A)\}.$$

6. Determine the spectrum of the right-shift operator S given in Example 5.3.5 (p. 94).

7. Define the *left-shift operator* S^* in ℓ^2 by

$$S^*(x_1, x_2, x_3, \ldots) := (x_2, x_3, \ldots).$$

It is easy to see that S^* is the adjoint of S. Find the eigenvalues of S^*, and discuss the relationship with Problem 5 above.

8. Let X be a Banach space, and let $A, B \in \mathcal{L}(X)$. Show that $\sigma(AB)$ and $\sigma(BA)$ share the nonzero elements even if $AB \neq BA$, i.e.,

$$\sigma(AB) \setminus \{0\} = \sigma(BA) \setminus \{0\}.$$

9. Let X be a Hilbert space, let $T, S \in \mathcal{L}(X)$, and suppose that either T or S is compact. Show that TS is also compact.

10. Let X be a Banach space, and let $T \in \mathcal{L}(X)$. Suppose that the range of T is finite-dimensional. Show that T is compact.

11. Let $A : \mathbb{C}^n \to \mathbb{C}^m$ be an $m \times n$ matrix. Endow $\mathbb{C}^n$ and $\mathbb{C}^m$ with the Euclidean norm. Define the *maximal singular value* of A by

$$\sigma_{\max}(A) = \max\{\sqrt{\lambda} : \lambda \in \sigma(A^*A)\},$$

that is, the square root of the maximum of eigenvalues of A^*A. Show that the norm $\|A\|$ of A defined by (5.1) is equal to $\sigma_{\max}(A)$.

12. Let X be a Hilbert space and T a compact operator on X. Prove the following:

   (a) T^*T is a compact Hermitian operator.
   (b) For every $x \in X$, $(T^*Tx, x) \geq 0$. In this sense, T^*T is nonnegative-(positive semi-)definite.
   (c) The norm of T is given by the maximal singular value of T.

13. Let A and B be $n \times n$ and $n \times m$ real matrices, respectively. Consider the differential equation

$$\frac{dx}{dt} = Ax(t) + Bu(t) \qquad (5.17)$$

under the initial condition $x(0) = 0$. Define the operator

$$T : (L^2[0,h])^m \to \mathbb{R}^n : u \mapsto x(h)$$

by mapping u to the value $x(h)$ at $t = h$ of the solution $x(t)$ of (5.17) under the initial condition $x(0) = 0$. Prove the following:

(a) The adjoint operator of T is given by introducing the adjoint differential equation

$$\frac{dp}{dt} = -A^T p(t), \quad p(h) = p_h \tag{5.18}$$

and associating $B^T p(t)$ to p_h as

$$T^* : p_h \mapsto B^T p(t)$$

via the solution of (5.18).

(b) The norm of this operator T is given by the square root of the maximal eigenvalue of the matrix $\int_0^h e^{At} B B^T e^{A^T t} dt$.

Chapter 6

Schwartz Distributions

A generalized function called the δ (delta) function introduced by Dirac appears frequently in physics, dynamical system theory, control, signal processing, and engineering in general. Unfortunately, in many (perhaps too many) engineering textbooks, this is imprecisely defined as a "function" satisfying the following two properties:

$$\delta(t) := \begin{cases} +\infty, & t = 0, \\ 0, & t \neq 0, \end{cases}$$

$$\int_{-\infty}^{\infty} \delta(t) dt = 1.$$

This does not fall into the category of ordinary functions.[55] Moreover, these two conditions are not enough to effectively use the delta function. On the other hand, the introduction of this concept has brought great computational and conceptual advantages to various situations, and there have been quite a few attempts to justify this concept in a mathematically satisfactory way. One of them is the theory of distributions due to Schwartz. This chapter describes some fundamentals, confined to one-variable theory, of this distribution theory, to the extent that can be useful for many engineering applications. This will also greatly simplify our subsequent treatments of Fourier and Laplace transforms.

6.1 What Are Distributions?

6.1.1 Difficulty Concerning Differentiation

Perhaps the first annoying problem we encounter when we try to apply the theory of differential equations seriously to practical problems is that of differentiability. As long as we deal with differential equations, they do certainly contain derivatives of the solution, and hence we must guarantee the differentiability of the solution, often without exhibiting a solution in a concrete form. This, however, is not always easy and is the source of a problem. Moreover, there are some problems in practice, for example in control, where the solutions are not necessarily, or even required to be, differentiable everywhere.

---

[55]For example, the symbol $+\infty$ is not a number.

Similar problems arise also in studying Fourier series. It is often bothersome to guarantee in advance the differentiability of a *candidate* of a solution of a partial differential equation obtained, for example, via separation of variables. Another issue is how a Fourier series behaves around discontinuities. This requires careful treatment and is often quite delicate for the uninitiated.

A chief reason for such problems is that differentiation is not a continuous operation. Of course, as already noted in Chapter 2, the notion of continuity depends on topology, i.e., the concept of "closeness" in the given space, so that we cannot naively say that something is discontinuous without specifying a topology. However, let us consider very intuitively the following function:

$$t \mapsto \epsilon \sin(e^t). \tag{6.1}$$

This function is not only bounded; it in fact does not exceed ϵ in magnitude (so it is very small if ϵ is small). But its derivative

$$t \mapsto \epsilon e^t \cos(e^t) \tag{6.2}$$

diverges as $t \to \infty$ no matter how small ϵ may be. That is, although (6.1) is a very "small" function, differentiation may amplify its oscillation and yield a very "large" function such as (6.2). Hence in a rather naive sense, it may be concluded to be discontinuous.

This is, of course, not very convenient. Our intuition (and most of engineering) is based on continuity. For example, suppose that certain procedures are valid and applicable under some conditions. But in reality there are hardly any cases in which such conditions are satisfied in a mathematically rigorous sense; at best they are satisfied only approximately. We nonetheless usually believe that such procedures can still apply to some reasonable extent by "continuity." In other words, when the difference in conditions is small, the resulting consequence varies by only a small amount. This conviction, however, is not quite valid for differentiation, at least in a naive sense, since small functions may yield a very large function as a result of differentiation.

Thus differentiation introduces a contradictory viewpoint to our ongoing intuition based on continuity, and this raises many technical issues. The concept of distributions to be discussed here gives a solution to this problem. It generalizes the notion of functions and reconsiders the concept of closeness (i.e., topology). With respect to such generalized notions, differentiation becomes a continuous operation, for example. The theory of distributions hence leads to a great simplification of the treatment of various problems indicated above, and, in a sense, it is more compatible with applications than the classical treatments of functions.

Roughly speaking, we can list the following two advantages in considering distributions:

- The δ function gives the identity element in a convolution algebra (to be introduced below); i.e., $f * \delta = \delta * f = f$ holds with respect to convolution. This provides effective machinery of operational calculus, particularly effective in differential and integral equations.

- Every distribution is differentiable to an arbitrary order. Furthermore, differentiation turns out to be a continuous operation in the sense of distributions.

Needless to say, the utility of distributions is not limited only to these. For example, Fourier and Laplace transforms, and their relationships with complex functions become much more transparent. We will witness these in subsequent chapters, but for the moment it is perhaps enough to keep in mind the two properties above and watch for how they are related to applications.

We note that the theory of Schwartz distributions is not the only way of justifying generalized functions (i.e., the generalization of the concept of functions) such as δ functions. Other such approaches are, for example, *Nonstandard Analysis* by Robinson [46], *Hyperfunctions* by Sato [50, 26], and *Analytic Functionals* by Martineau. Each has its own advantages. Among them, hyperfunctions have been quite actively studied and applied to the theory of partial differential equations by many Japanese mathematicians. But we here confine ourselves to Schwartz distributions. For these other concepts, the reader is referred to the literature cited above.

6.1.2 An Approximating Example

The reader must surely be familiar with a device called an electronic flash. This is a device for emitting light so that one can take photographs in the dark. The lapse in time of emission is very short—about the order of 1/several thousands or $1/10,000$ of a second, or even shorter. Why is it possible to take a picture with such a short period of light emission? The reason is that the flash magnifies the luminous intensity to ensure enough total light radiation for the film to be sufficiently exposed to light. Since the total radiation is the integral over the whole range of emission, a sufficient amount of light radiation will be ensured, making the intensity large even if the total lapse time is small.

For simplicity suppose the intensity to be constant over this range. Its graph would look like that in Figure 6.1. The shorter the emitting lapse time becomes, the higher the height of the graph should be in order to ensure enough light. Generally speaking, for the quality of a photograph, as far as precision is concerned, the shorter the emitting period is, the better. If it is long, the subject might move and can cause a blur. The delta function may be regarded as a limit of such a procedure, making the lapse time approach 0. Needless to

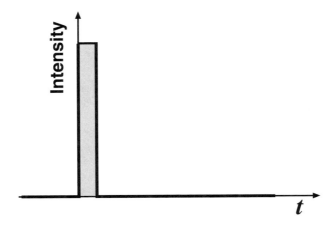

Figure 6.1. *Light emission of an electronic flash*

say, the condition

$$\int f_n(t)dt = 1$$

must be maintained to guarantee a sufficient amount of light (normalized to 1 here). This is why the condition

$$\int \delta(t)dt = 1$$

is required in the "definition" of the δ function. But this is an imprecise way of specifying a new "function." The δ function is not an ordinary function, and its rigorous treatment requires the subsequent developments on distributions.

6.1.3 Functions and Distributions

Let us now introduce the concept of distributions. A distribution in the sense of Schwartz is, in short, a continuous linear functional, i.e., an element of the dual space of a suitable function space. But this may give the impression that it is overly simplistic or, to the contrary, that it is too esoteric. Let us begin by describing how ordinary functions may be regarded as a distribution.

First, what do we (or should we) mean by *ordinary functions?* What class of functions should we take into consideration? We would certainly wish to include continuous functions. We would like to consider not only continuous functions but also some discontinuous functions. It would appear reasonable to consider those in L^1, i.e., integrable functions. But if we were to impose the requirement that they be integrable over the whole real line $(-\infty,\infty)$, it would also turn out to be restrictive. For example, we would not be able to consider a function such as e^t, which does not belong to $L^1(-\infty,\infty)$. The very motivation for generalizing the concept of functions is in relaxing the requirement of local regularity, so that it is more appropriate to confine our attention to the local behavior of functions. We thus take any *locally integrable function f*. A function f is locally integrable if for every bounded interval $[a,b]$ the integral

$$\int_a^b |f(t)|dt$$

converges and is finite; needless to say, the value of the integral varies depending on $[a,b]$. We can then consider functions such as e^t as locally integrable functions. The integral here should be understood in the sense of Lebesgue, but if you are not familiar with Lebesgue integration theory, you can take it in the sense of Riemann integrals without much loss in applications.

Distribution theory takes the standpoint that when we consider such locally integrable functions, we focus our attention on *how such functions act on other functions and how these actions characterize each function*, rather than on the individual value they assume at each point. For example, if ϕ is a continuous function, we consider the functional associated to a locally integrable function f:

$$T_f : \phi \mapsto \int_a^b f(t)\phi(t)dt. \tag{6.3}$$

This is a continuous[56] linear functional defined over the space of continuous functions. The first step of distribution theory is in identifying this functional with function f.

In order for this idea to work, the above-introduced functionals must be fine enough to distinguish different functions. In other words, different functions should induce different functionals in the above sense. While we do not give a proof here because the proof requires Lebesgue integration theory, the following theorem indeed holds.

Theorem 6.1.1. *Let f be an integrable function on an interval $[a,b]$. If the associated functional T_f defined by (6.3) is 0, that is,*

$$\int_a^b f(t)\phi(t)dt = 0$$

for every continuous function ϕ, then f is 0 almost everywhere in the sense of Lebesgue.[57]

The term *almost everywhere* means that a certain property (e.g., a function is zero) holds *except on a "small" set* such as sets consisting of finitely many or countably infinite points. To justify this idea, consider, for example, the function

$$f(t) = \begin{cases} 1, & t = 0, \\ 0, & \text{otherwise.} \end{cases}$$

It would appear reasonable to identify this with 0 by disregarding its value at one point. If we try to physically observe the function above with a certain sensor, such a sensor will always have a limit due to its physical characteristics (often involving integration). The difference above is just too small to be observed in such a process. It is thus more natural also in the engineering sense to disregard a difference at one point. This naturally generalizes to finitely many points. Generalizing this, one may think of ignoring the values at a countably infinite[58] number of points. The idea of Lebesgue goes yet one more step further. He discovered that integration theory becomes much simpler and more transparent if one ignores function values on small sets called "null sets" or sets of measure zero. Let N be a subset of $\mathbb{R}$. N is said to be a *set of measure zero* or a *null set* if for every $\epsilon > 0$ there exists a sequence of intervals $\{(a_i,b_i)\}_{i=1}^{\infty}$ such that

$$N \subset \cup_{i=1}^{\infty}(a_i,b_i), \quad \text{with} \sum_{i=1}^{\infty}(b_i - a_i) < \epsilon.$$

Countably infinite sets are null sets, but there are also some uncountably infinite null sets. For details, the reader is referred to, e.g., [20, 22, 37]. He defines that a certain property holds *almost everywhere* if this property holds except on a set of measure zero. Theorem 6.1.1 states that if two functions f and g induce the same functional T_f ($= T_g$) via the correspondence (6.3), then $f = g$ almost everywhere. Hence it virtually says that the distributional interpretation of functions uniquely determines function values almost everywhere. Let us examine the details more closely but intuitively.

---

[56]The meaning is to be clarified later.

[57]with respect to Lebesgue measure, to be precise.

[58]A set is countably infinite if its elements can be in one-to-one correspondence with natural numbers. The sets $\mathbb{Z}$ of integers and $\mathbb{Q}$ of rational numbers are examples of countably infinite sets.

Suppose that a function f is nonzero, say, positive, in a neighborhood V of $t = t_0$. One can easily find a continuous function ϕ such that it is identically 1 on V and 0 outside a slightly larger neighborhood containing V. Hence for a function which is not identically zero, the integral (6.3) becomes nonzero for some ϕ. However, if f is nonzero only on a small negligible set, then it is indistinguishable from 0 by the nature of integration. This is why we need the term "almost everywhere" in the statement of the theorem.

The functional T_f is easily seen to be linear. Moreover, this correspondence is also "continuous." For example, let us confine ourselves to the interval $[a, b]$. It is easily seen that

$$|T_f(\phi)| \leq \int_a^b |f(t)\phi(t)|dt \leq \left\{ \int_a^b |f(t)|dt \right\} \cdot \sup_{a \leq t \leq b} |\phi(t)|.$$

Hence we see that if $\phi_n(t)$ converges uniformly to 0 on $[a, b]$, then $T_f(\phi_n)$ approaches 0.[59]

A moment of reflection would tell us that such a correspondence mapping continuous functions to $\mathbb{R}$ (or $\mathbb{C}$) does not necessarily arise from a function f. For instance, Dirac's delta function[60] is defined as

$$\delta : \phi \mapsto \phi(0).$$

Schwartz distributions are a generalization of this idea. One takes a function space Z, considers the space of continuous linear functionals on Z, and calls it a space of distributions. That is, a distribution is an element of the dual space of Z. As we have seen above, ordinary locally integrable functions or the δ function satisfy this requirement.

We have some freedom in choosing Z. For example, if we take as Z the space $L^2[a, b]$ of Lebesgue square integrable functions on $[a, b]$, then its dual is (isomorphic to) $L^2[a, b]$ itself by the Riesz–Fréchet theorem, Theorem 4.2.1 (p. 81), and hence this choice would not yield a generalization of the concept of functions. On the other hand, the case of the delta function (and locally integrable functions) amounts to the choice $Z = C[a, b]$, and we may consider $\delta \in (C[a, b])'$ following the arguments above, and we can consider the δ function as a distribution. We can also observe that as Z assumes higher regularity (smoothness), its dual can allow more irregular elements. In view of this, it is advantageous to take as Z a space of infinitely differentiable functions to include objects of maximum possible generality, for example, the space $\mathcal{D}$ to be discussed in the next section. Such a space is called the space of *test functions*.

6.2   The Space of Distributions

We start by first introducing the space of test functions which is crucial to distribution theory.

6.2.1   The Space $\mathcal{D}$ of Test Functions

Let us introduce a space of infinitely differentiable test functions. Take a bounded interval $[a, b]$. Define $\mathcal{D}[a, b]$ as follows:

$$\mathcal{D}[a, b] := \{\phi : \phi \in C^\infty(-\infty, \infty) \text{ and } \phi(t) = 0 \text{ for } t \notin [a, b]\}, \qquad (6.4)$$

---

[59]Of course, the notion of continuity depends on the topology, so we should be careful and should not jump to conclusions. For the moment, let us take this only as an intuitive explanation until a precise definition is given later.

[60]This is in fact not a function, so it should be called the delta distribution to be precise.

namely, the space of infinitely differentiable functions on $(-\infty, \infty)$ and identically 0 outside the interval $[a,b]$. See Figure 6.2.

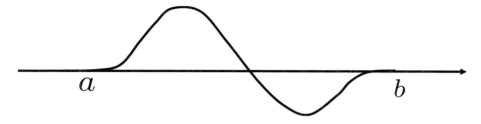

Figure 6.2. *A function in* $\mathcal{D}[a,b]$

This space of course constitutes a vector space with respect to the usual addition and scalar multiplication defined for functions. It is not appropriate, however, to try to introduce a norm topology to this space. It is necessary to consider a topology induced by an *infinite family* of norms here. To see this, consider the following norms $\{\|\cdot\|_m\}_{m=0}^{\infty}$:

$$\|\phi\|_m := \sup_{a \le t \le b, 0 \le k \le m} \left| \frac{d^k \phi}{dt^k}(t) \right|, \quad \phi \in \mathcal{D}[a,b], \quad m = 1, 2, \ldots. \tag{6.5}$$

It is easy to see that each $\|\cdot\|_m$ qualifies as a norm in $\mathcal{D}[a,b]$. The topology of $\mathcal{D}[a,b]$ is defined to be that of a *topological vector space with all these norms* (see Appendix A, p. 241). More concretely, the fundamental system of neighborhoods of the origin 0 is defined to be the sets of the following type:[61]

$$V(m, \epsilon) := \{x \in \mathcal{D}[a,b] \ : \ \|x\|_k < \epsilon, \quad k = 1, \ldots, m\}, \quad m = 1, 2, \ldots, \epsilon > 0.$$

Then a sequence $\{\phi_n\}$ in $\mathcal{D}[a,b]$ is, by definition, said to converge to 0 if for every $V(m, \epsilon)$ there exists N such that $n \ge N$ implies $\phi_n \in V(m, \epsilon)$. In other words, the derivative $\{\phi_n^{(m)}\}$ of $\{\phi_n\}$ must converge uniformly to 0 for every m.

No single norm given by (6.5) can define this topology, for the norm $\|\cdot\|_m$ governs only the behavior of derivatives up to the order m but nothing beyond, and hence there can be a sequence convergent with respect to this norm, but the limit is not infinitely differentiable and hence does not belong to $\mathcal{D}[a,b]$. Thus the topology defined by this norm does not render completeness to $\mathcal{D}[a,b]$.

Let us summarize this in the form of a definition.

Definition 6.2.1. A sequence $\phi_n(t) \in \mathcal{D}[a,b]$ is said to *converge* to 0 if for every m $\|\phi_n\|_m$ converges to 0 as $n \to \infty$.

We are now ready to define the space $\mathcal{D}(\mathbb{R})$ which is fundamental for defining distributions.

---

[61]That is, a subset V is a neighborhood if V contains one of these fundamental neighborhoods for some m and ϵ. See Appendix A, section A.6.

Definition 6.2.2. Define $\mathcal{D}(\mathbb{R})$ as

$$\mathcal{D}(\mathbb{R}) := \bigcup_{a<b} \mathcal{D}[a,b]. \tag{6.6}$$

That is, an element ϕ belongs to $\mathcal{D}(\mathbb{R})$ if and only if

1. it belongs to the C^∞ class, i.e., it is infinitely differentiable, and

2. it has compact support, i.e., it is identically 0 outside some interval $[a,b]$.

Here the *support* suppϕ of a function ϕ means the smallest closed set outside of which $\phi(t) = 0$.

The space of distributions is the dual of this space $\mathcal{D}(\mathbb{R})$. To this end we must suitably introduce a topology on $\mathcal{D}(\mathbb{R})$. We must also make it compatible with each component $\mathcal{D}[a,b]$ which comprises $\mathcal{D}(\mathbb{R})$ as a union. We will explain how this can be done, although rather briefly.

Note first that there exists the following *injection (embedding)*:

$$j_{ab} : \mathcal{D}[a,b] \hookrightarrow \mathcal{D}(\mathbb{R}) : \phi \mapsto \phi \tag{6.7}$$

for each $\mathcal{D}[a,b]$. This embedding may be either continuous or discontinuous depending on the topology on $\mathcal{D}(\mathbb{R})$. If it is too strong (fine), i.e., contains too many open sets (or neighborhoods), then j_{ab} will not be continuous, while if it is too weak (coarse), as in the case with the so-called *indiscrete topology*,[62] then it will not make much sense. It is very natural to require that these embeddings be continuous; were they not, we would certainly encounter all sorts of anomalies.

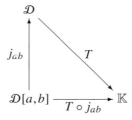

In view of the above, we introduce to $\mathcal{D}(\mathbb{R})$ the strongest (finest) topology that makes all these embeddings continuous. There is always a unique topology satisfying this requirement [10, 33]. This is the *inductive limit* topology [51, 53, 58] introduced by Schwartz.

Although we do not give a proof here, the following facts hold for the inductive limit topology.

Theorem 6.2.3. *Let X be a normed linear space[63] and T a linear map from $\mathcal{D}(\mathbb{R})$ to X. A necessary and sufficient condition for T to be continuous is that for every $\mathcal{D}[a,b]$ the restriction $T \circ j_{ab}$ of T to $\mathcal{D}[a,b]$ be continuous. A sequence $\{\phi_n\}_{n=1}^\infty$ in $\mathcal{D}(\mathbb{R})$ converges to 0 if and only if there exist real numbers a,b such that*

---

[62]where the only open sets consist of the whole space and the empty set $\emptyset$; see Appendix A and [10, 33] for more details.

[63]As will be noted in Appendix A, this can be a locally convex space.

1. *all ϕ_n belong to $\mathcal{D}[a,b]$, and*

2. *ϕ_n converges to 0 in $\mathcal{D}[a,b]$ in the sense of Definition 6.2.1.*

The space $\mathcal{D}(\mathbb{R})$ is nothing but the set of C^∞ functions having compact support. But introducing a suitable topology to this space is, as we have seen, nontrivial. The reason we did not take uniform convergence of every order of derivatives on the whole line $\mathbb{R}$ is that it would not make $\mathcal{D}(\mathbb{R})$ complete. There can be a sequence in $\mathcal{D}(\mathbb{R})$ whose derivatives converge uniformly on $\mathbb{R}$, yet its limit does not belong to $\mathcal{D}(\mathbb{R})$. The topology of $\mathcal{D}(\mathbb{R})$ introduced here is much stronger than this, and it indeed makes $\mathcal{D}(\mathbb{R})$ complete. The theorem above states that the continuity of $T : \mathcal{D}(\mathbb{R}) \to X$ can be determined *locally*. That is, it is continuous if and only if it is continuous on every $\mathcal{D}[a,b]$. This is a great advantage to discuss continuity of distributions to be defined below.

6.2.2 Definition and Examples of Distributions

We are now ready to give the definition of distributions as follows.

Definition 6.2.4. A continuous linear functional T on the space $\mathcal{D}(\mathbb{R})$ is called a *distribution*. The totality of distributions is nothing but the dual space of $\mathcal{D}(\mathbb{R})$. It is denoted by $\mathcal{D}'(\mathbb{R})$ or $\mathcal{D}'$.

Here the continuity of $T : \mathcal{D}(\mathbb{R}) \to \mathbb{K}$ means that $T(\phi_n) \to 0$ as $\phi_n \to 0$. Strictly speaking, the notion of convergence of sequences does *not* determine the topology of $\mathcal{D}(\mathbb{R})$,[64] so this may be a little imprecise; but it is known that for the present situation this notion of convergence of sequences is sufficient.

Since a distribution T is in short a continuous linear form on $\mathcal{D}(\mathbb{R})$, it is more convenient to denote its value at $\phi \in \mathcal{D}(\mathbb{R})$ as

$$\langle T, \phi \rangle, \tag{6.8}$$

exhibiting its linearity explicitly. From here on we will adopt this notation.

Let us list several examples of distributions.

Example 6.2.5. Let us begin with the discussion of locally integrable functions already seen in the previous section. Let f be a locally integrable function on $(-\infty, \infty)$. The correspondence

$$T_f : \mathcal{D}(\mathbb{R}) \to \mathbb{K} : \phi \mapsto \int_{-\infty}^{\infty} f(t)\phi(t)dt$$

defines a distribution. Note that the integral on the right-hand side is always confined to a bounded interval since the support of ϕ is compact, and hence the integral always exists.

From here on, we will always regard a locally integrable function as a distribution when the context so demands.

---

[64]This is, technically, because $\mathcal{D}(\mathbb{R})$ does not satisfy the so-called first countability axiom [10, 33].

Example 6.2.6 (Heaviside unit step function).

$$H(t) := \begin{cases} 1, & t > 0, \\ 0, & t < 0. \end{cases} \qquad (6.9)$$

This is a locally integrable function and is a special case of Example 6.2.5, but it deserves special attention for its importance in applications and for preparing notation for it. Other texts may use $Y(t)$ or $\mathbf{1}(t)$. We did not specify its value at the origin 0 because of its irrelevance since one point is of (Lebesgue) measure zero.

Example 6.2.7 (Dirac's delta). Let $a \in (-\infty, \infty)$. The correspondence

$$\delta_a : \mathcal{D}(\mathbb{R}) \to \mathbb{K} : \phi \mapsto \langle \delta_a, \phi \rangle = \phi(a) \qquad (6.10)$$

defines a distribution. This is the delta distribution with point mass at point a. In particular, we denote δ_0 by δ. In view of the fact that it is a distribution but not a function, we will hereafter avoid the confusing notation $\delta(t)$.

Example 6.2.8. The correspondence

$$T : \mathcal{D}(\mathbb{R}) \to \mathbb{K} : \phi \mapsto \langle T, \phi \rangle = \phi'(0) \qquad (6.11)$$

also defines a distribution. To see this, suppose that $\phi_n \to 0$ in $\mathcal{D}(\mathbb{R})$. Then ϕ_n' converges uniformly to 0, and hence $\phi_n'(0) \to 0$ (see Theorem 6.2.3 and (6.5) defining the topology of $\mathcal{D}[a,b]$).

The relationship of this to differentiation of distributions will be discussed in the next section.

6.3 Differentiation of Distributions

6.3.1 Definition of Derivatives of Distributions

Since distributions should give a generalization of the concept of functions, we naturally want to differentiate and integrate them. Let us now introduce the notion of differentiation. As a natural requirement, if a distribution T agrees with an ordinary function f in the sense of Example 6.2.5, and if f admits a continuous derivative f', then the derivative of T in the sense of distributions must coincide with the distribution defined by this f'. Otherwise we would encounter serious difficulties.

Let us first compute the distribution defined by f' under the conditions above. Take any test function $\phi \in \mathcal{D}$. By definition, ϕ is identically 0 outside some interval $[a,b]$, so integration by parts yields

$$\int_{-\infty}^{\infty} f'(t)\phi(t)\,dt = [f(t)\phi(t)]_a^b - \int_{-\infty}^{\infty} f(t)\phi'(t)\,dt = -\int_{-\infty}^{\infty} f(t)\phi'(t)\,dt. \qquad (6.12)$$

This means that if f is differentiable, then $\langle f', \phi \rangle = -\langle f, \phi' \rangle$. But observe here that the last term of this equation makes sense even when f is not differentiable because ϕ is always differentiable. Generalizing this, we arrive at the following definition.

Definition 6.3.1. Let T be a distribution. Its *derivative* T' in the sense of distributions is defined by

$$\langle T',\phi \rangle := -\langle T,\phi' \rangle. \tag{6.13}$$

To ensure that this makes sense, we must guarantee that T' as defined above is also a distribution. But this is obvious from the definition of distributions: If a sequence ϕ_n converges to 0 in $\mathcal{D}$, then by the definition of the topology of $\mathcal{D}$, ϕ'_n also converges to 0 in $\mathcal{D}$. Hence, by the continuity of T, $\langle T,\phi'_n \rangle$ also converges to 0.

This definition appears quite innocent. But its implication is in fact far-reaching. That is, *every distribution is differentiable to an arbitrary order*. This is in marked contrast to the classical situation, where not all functions are differentiable. This further leads to the following question: *How can a nondifferentiable function in the classical sense become differentiable in the sense of distributions?* We will discuss this in the next section through some examples.

6.3.2   Differentiation of Distributions—Examples

Example 6.3.2. Let $H(t)$ be the Heaviside function introduced in Example 6.2.6. We claim

$$\frac{dH}{dt} = \delta. \tag{6.14}$$

In fact, we have, for every test function $\phi \in \mathcal{D}$,

$$\left\langle \frac{dH}{dt},\phi \right\rangle := -\int_{-\infty}^{\infty} H(t)\phi'(t)dt = -\int_{0}^{\infty}\phi'(t)dt = -[\phi(t)]_0^{\infty} = \phi(0) = \langle \delta,\phi \rangle.$$

If this computation appears too formal, it is helpful to proceed intuitively as follows. Consider the following function $H_\epsilon(t)$:

$$H_\epsilon(t) := \begin{cases} 0, & t < 0, \\ t/\epsilon, & 0 \le t \le \epsilon, \\ 1, & t > \epsilon. \end{cases}$$

It is obvious that $H_\epsilon(\cdot) \to H(\cdot)$ as $\epsilon \to 0$. Furthermore,

$$H'_\epsilon(t) := \begin{cases} \dfrac{1}{\epsilon}, & 0 < t < \epsilon, \\ 0 & \text{elsewhere.} \end{cases}$$

We then obtain

$$\langle H'_\epsilon,\phi \rangle = \int_{-\infty}^{\infty} H'_\epsilon(t)\phi(t)dt = \int_0^{\epsilon}\frac{\phi(t)}{\epsilon}dt.$$

Since $\phi(t)$ is a continuous function, $\phi(t)$ is close to $\phi(0)$ for sufficiently small ϵ. Hence the term above converges to $\phi(0)$ as $\epsilon \to 0$. (This argument is a bit rough. For a rigorous proof, see Example 6.5.2 below. Observe also that this is the same example given in Figure 6.1.)

Generalizing this example, we obtain the following proposition.

Proposition 6.3.3. *Suppose that a function f satisfies the following conditions:*

1. *f is of class C^1 except at point a; i.e., it is continuously differentiable except at a;*

2. *f has first-order discontinuity at a; i.e., the right limit $f(a+) := \lim_{t \downarrow a} f(t)$ and the left limit $f(a-) := \lim_{t \uparrow a} f(t)$ both exist.*

Then the derivative of f in the sense of distributions is given by

$$\frac{df}{dt} = [f'] + (f(a+) - f(a-))\delta_a, \qquad (6.15)$$

where $[f']$ denotes the distribution defined by the derivative of f (for $t \neq a$) as follows:

$$[f'](t) := \begin{cases} \frac{df}{dt}, & t > a \text{ and } t < a, \\ \text{arbitrary}, & t = a, \end{cases} \qquad (6.16)$$

where df/dt is understood in the ordinary sense.

Proof. Take any test function $\phi \in \mathcal{D}$. It follows that

$$
\begin{aligned}
-\langle f, \phi' \rangle &= -\int_{-\infty}^{\infty} f(t)\phi'(t)dt \\
&= -\int_{-\infty}^{a} f(t)\phi'(t)dt - \int_{a}^{\infty} f(t)\phi'(t)dt \\
&= -[f(t)\phi(t)]_{-\infty}^{a} + \int_{-\infty}^{a} [f'](t)\phi(t)dt - [f(t)\phi(t)]_{a}^{\infty} + \int_{a}^{\infty} [f'](t)\phi(t)dt \\
&= -f(a-)\phi(a) + f(a+)\phi(a) + \int_{-\infty}^{\infty} [f'](t)\phi(t)dt \\
&= (f(a+) - f(a-))\langle \delta_a, \phi \rangle + \int_{-\infty}^{\infty} [f'](t)\phi(t)dt \\
&= \langle [f'] + (f(a+) - f(a-))\delta_a, \phi \rangle,
\end{aligned}
$$

and hence the conclusion follows.[65]   □

As a special case of the above, if the function f is of class C^1, then its derivative f' in the sense of distributions agrees with the distribution $[f']$ defined by the ordinary derivative (in the sense of Example 6.2.5).

Example 6.3.4. Consider the function

$$f(t) = |t|.$$

---

[65]As seen from the proof, this proposition holds with suitable modifications when there are finitely many discontinuities. Also, f does not have to be in the C^1 class, and it is enough that f is an absolutely continuous function. For a definition of absolutely continuous functions, see references on real analysis [20, 22, 37].

Take any $\phi \in \mathcal{D}$. Integration by parts yields

$$-\int_{-\infty}^{\infty} |t|\phi'(t)dt = -\int_{-\infty}^{0}(-t)\phi'(t)dt - \int_{0}^{\infty} t\phi'(t)dt$$

$$= [t\phi(t)]_{-\infty}^{0} - \int_{-\infty}^{0}\phi(t)dt - [t\phi(t)]_{0}^{\infty} + \int_{0}^{\infty}\phi(t)dt$$

$$= \int_{-\infty}^{\infty}(\operatorname{sgn} t)\phi(t)dt.$$

Hence $|t|' = \operatorname{sgn} t$, where $\operatorname{sgn} t$ denotes the signature function

$$\operatorname{sgn} t := \begin{cases} 1, & t > 0, \\ -1, & t < 0. \end{cases}$$

Furthermore, $|t|$ is certainly a locally integrable function, and $[|t|'] = \operatorname{sgn} t$ for $t \neq 0$ and also assumes the value 0 at the origin, so that $|t|' = \operatorname{sgn} t$ can also be deduced from Proposition 6.3.3.

6.4 Support of Distributions

Let us introduce the notion of *support* for distributions.

The support of a function is, as defined in subsection 6.2.1, the smallest closed subset outside of which this function is zero. Let us generalize this to distributions.

Given an open set $O \subset \mathbb{R}$, we define the space $\mathcal{D}(O)$ of test functions on it by

$$\mathcal{D}(O) := \bigcup_{[a,b] \subset O} \mathcal{D}[a,b], \tag{6.17}$$

similarly as in Definition 6.2.2 (p. 114) in subsection 6.2.1.

The support of a distribution is then defined as follows.

Definition 6.4.1. Let $T \in \mathcal{D}'$. For an open set O, the *restriction* $T|_O$ of T to O is defined to be 0 if for every $\phi \in \mathcal{D}(O)$

$$\langle T, \phi \rangle = 0$$

holds. The *support* of the distribution T is the set

$$\operatorname{supp} T := \left(\bigcup \{O \, : \, T|_O = 0\} \right)^c, \tag{6.18}$$

that is, the complement of the largest open set on which the distribution T is zero. As a complement of an open set, $\operatorname{supp} T$ is always closed.

Example 6.4.2. The support of the function f shown in Figure 6.3 is $[0,1] \cup [2,2.5]$.

6.5 Convergence of Distributions

When can we say that a sequence $\{T_n\}$ of distributions converges to a distribution T? This is the problem of introducing a topology to the space $\mathcal{D}'$ of distributions, and it is a more complicated problem than that for $\mathcal{D}$. Since $\mathcal{D}$ is not a Banach space, there can be several

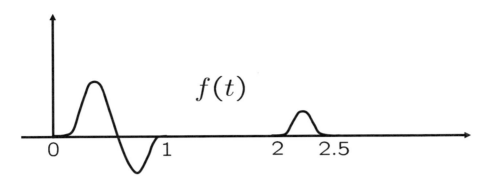

Figure 6.3. *Support of a function f*

plausible and reasonable ways to introduce a topology on its dual space. The discussion of this topic is beyond the scope of this book, and the reader is referred to the existing accounts on topological vector spaces [29, 51, 58, 64].

We here content ourselves with the weak topology, i.e., the topology of simple convergence (convergence at each point, where "point" refers to an element in $\mathcal{D}$).

Definition 6.5.1. A sequence $\{T_n\}$ of distributions is said to converge to T if

$$\langle T_n, \phi \rangle \to \langle T, \phi \rangle$$

for every test function $\phi \in \mathcal{D}$. In other words,

$$\langle T_n - T, \phi \rangle \to 0$$

for every $\phi \in \mathcal{D}$.

How fast $\langle T_n - T, \phi \rangle$ approaches 0 depends of course on each ϕ. But it is known that for the convergence of sequences in $\mathcal{D}'$ the above convergence implies a uniform convergence on a fairly "large" set of ϕ [58, 53, 64]. This is a special situation for the space of distributions.

In particular, convergence to δ has many important applications. We begin with the following example.

Example 6.5.2. The sequence shown in Figure 6.1 is

$$g_n(t) = \begin{cases} n, & 0 \leq t \leq 1/n, \\ 0 & \text{elsewhere.} \end{cases} \tag{6.19}$$

Let us show that this sequence converges to δ in the sense defined above. Take any $\phi \in \mathcal{D}$. Clearly ϕ is continuous at the origin 0. Hence for every $\epsilon > 0$ there exists $\nu > 0$ such that $|t| < \nu$ implies $|\phi(t) - \phi(0)| < \epsilon$. It follows that for $1/n < \nu$

$$\langle g_n - \delta, \phi \rangle = \int_0^{1/n} [n\phi(t) - n\phi(0)]dt < \int_0^{1/n} n\epsilon\, dt = \epsilon,$$

and hence $g_n \to \delta$.

The sequence here satisfies the following properties:

1. The integral of each function is identically 1.

2. The sequence converges uniformly to 0 except in the neighborhoods of the origin; in other words, the mass of each function becomes concentrated at the origin.

This observation suggests that a sequence satisfying these conditions converges to δ. The following proposition indeed guarantees this.

Proposition 6.5.3. *Suppose that a sequence ψ_n of functions satisfies the following conditions:*

1. *There exists $\nu_0 > 0$ such that if $|t| < \nu_0$, then $\psi_n(t) \geq 0$ for every n, and $\lim_{n \to \infty} \int_{-\nu}^{\nu} \psi_n(t)dt = 1$ for every $\nu > 0$.*

2. *ψ_n converges to 0 uniformly outside of any interval $(-\nu, \nu)$.*

Then $\psi_n \to \delta$.

Proof. Take any function $\phi(t) \in \mathcal{D}$. Then by Definition 6.2.2 of $\mathcal{D}(\mathbb{R})$ (p. 114), $\phi \in \mathcal{D}[a,b]$ for some a,b. Hence $\phi(t)$ can be regarded as a continuous function on a closed interval $[a,b]$, so it is bounded; i.e., there exists $M > 0$ such that $|\phi(t)| < M$ for all t.

It suffices to show

$$\int_{-\infty}^{\infty} \psi_n(t)\phi(t)dt \to \langle \delta, \phi \rangle = \phi(0). \qquad (6.20)$$

To this end, take any $\epsilon > 0$. The continuity of ϕ at 0 implies that there exists $\nu > 0$ such that

$$|\phi(t) - \phi(0)| < \epsilon \qquad (6.21)$$

whenever $|t| < \nu$. Condition 2 above also implies that for this ν there exists $N > 0$ such that $n \geq N$ yields

$$|\psi_n(t)| < \epsilon \qquad (6.22)$$

uniformly for $|t| \geq \nu$. We can also assume $\nu \leq \nu_0$ without loss of generality. It then follows that

$$\int_{-\infty}^{\infty} \psi_n(t)\phi(t)dt = \int_{|t|<\nu} \psi_n(t)\phi(0)dt + \int_{|t|<\nu} \psi_n(t)(\phi(t)-\phi(0))dt$$

$$+ \int_{|t|\geq\nu} \psi_n(t)\phi(t)dt.$$

Here the second term is estimated as

$$\left| \int_{|t|<\nu} \psi_n(t)(\phi(t)-\phi(0))dt \right| \leq \int_{|t|<\nu} \psi_n(t)\,|\phi(t)-\phi(0)|\,dt$$

$$\leq \int_{|t|<\nu} \psi_n(t)\epsilon\,dt$$

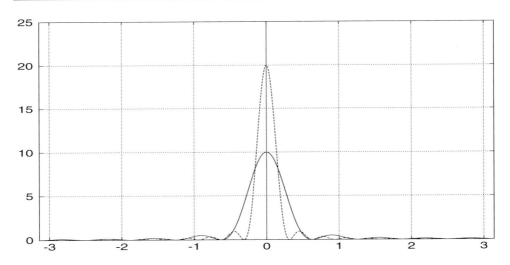

Figure 6.4. *Fejér kernel*

and clearly approaches ϵ by condition 1. For the third term we have from (6.22)

$$\left| \int_{|t| \geq \nu} \psi_n(t)\phi(t)dt \right| \leq \epsilon \int_{|t| \geq \nu} |\phi(t)|\, dt \leq \epsilon M(b-a).$$

On the other hand, the first term is easily seen to converge to $\phi(0)$ by condition 1. Thus (6.20) is proved. $\square$

Let us give some examples.

The following Fejér kernel is known in Fourier analysis. This appears by taking the arithmetic mean[66] instead of the simple n-partial sum of a Fourier series.

Example 6.5.4 (Fejér kernel). This function plays a crucial role in Fourier series:

$$K_n(t) = \begin{cases} \dfrac{1}{n} \cdot \dfrac{\sin^2(n/2)t}{\sin^2(1/2)t}, & -\pi \leq t \leq \pi, \\ 0 & \text{elsewhere.} \end{cases} \tag{6.23}$$

Figure 6.4 shows the graphs of $K_n(t)$ for $n = 10, 20$. We see that the mass becomes concentrated at the origin. Unlike the Dirichlet kernel to be treated below in Example 6.5.6, $K_n(t)$ is everywhere nonnegative.

The functions $K_n(t)/2\pi$ satisfy the conditions of Proposition 6.5.3. This kernel is used to show the completeness of the orthonormal basis $\{e^{int}/\sqrt{2\pi}\}$ appearing in the Fourier expansion (see Chapter 7).

Remark 6.5.5. To see rigorously that $K_n(t)/2\pi$ satisfies the conditions of Proposition 6.5.3, see Lemma 7.1.5 (p. 145).

---

[66]Called the first Cesaro mean.

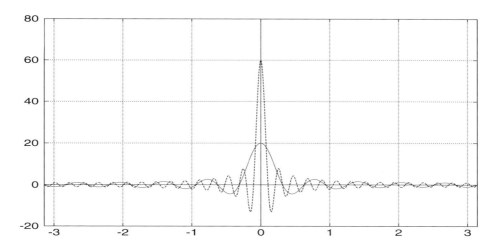

Figure 6.5. *Dirichlet kernel, $n = 10, 30$*

We here give another example of the Dirichlet kernel which is more straightforward from the point of view of Fourier series but more involved in light of Proposition 6.5.3.

Example 6.5.6 (Dirichlet kernel).

$$D_n(t) = \begin{cases} \dfrac{\sin(n+1/2)t}{\sin(1/2)t}, & -\pi \leq t \leq \pi, \\ 0 & \text{elsewhere.} \end{cases} \qquad (6.24)$$

Figure 6.5 shows the graphs for $n = 10, 30$; its mass becomes concentrated to the origin as n increases. Strictly speaking, this sequence does not satisfy the conditions of Proposition 6.5.3. The mass certainly becomes concentrated at the origin, but as this proceeds, there appears a peak in the negative direction around the origin, and this peak also becomes larger as $n \to \infty$. This conflicts with condition 1 of the proposition. But in this case also it can be shown that $D_n(t)/2\pi$ converges to δ. See subsection 6.8.3 below for a rigorous proof.

This function is fundamental in Fourier series. In fact, the nth partial sum $s_n(t)$ of the Fourier series of a function f is nothing but the convolution[67] of f with $D_n(t)/2\pi$. Moreover, δ is the unit for convolution, and, assuming the continuity of convolution, we obtain

$$s_n = \frac{1}{2\pi} D_n * f \to \delta * f = f.$$

This would show the convergence of the Fourier series to f itself. Note, however, that this convergence is only in the sense of distributions, and if one wants a convergence in a stronger sense, e.g., uniform convergence, a stronger restriction needs to be imposed on f. This partly explains why Fourier series theory deals with functions of bounded variation,

---

[67]See the next section and Chapter 7.

first-order derivatives, etc. This, on the other hand, is also due to the fact that the Dirichlet kernel is not well behaved (for example, it has dips to the negative side around the origin). As we saw in the case of the Fejér kernel in Example 6.5.4, the situation is much simpler.

Incidentally,

$$D_n(t) = \sum_{k=-n}^{n} e^{ikt} = 1 + 2\cos t + \cdots + 2\cos nt; \qquad (6.25)$$

each Fourier coefficient satisfies $1 = \langle 2\pi\delta, e^{-ikt}\rangle/2\pi$. This clearly shows, at least formally, that D_n is the nth partial sum of the Fourier expansion of $2\pi\delta$.

The following example is well known to appear as the normal (Gaussian) distribution but also plays a central role in proving the Fourier inversion formula.

Example 6.5.7. Consider the sequence

$$g_n(t) := \sqrt{\frac{n}{2\pi}} e^{-nt^2/2} \to \delta. \qquad (6.26)$$

According to the well-known Gauss integral formula

$$\int_{-\infty}^{\infty} e^{-\alpha t^2} = \sqrt{\frac{\pi}{\alpha}}, \quad \alpha > 0, \qquad (6.27)$$

we have

$$\int_{-\infty}^{\infty} g_n(t)dt = 1$$

for every n. Moreover, $g_n(t) \geq 0$, and condition 2 of Proposition 6.5.3 is easily seen to be satisfied. Hence $g_n \to \delta$. Figure 6.6 shows the graphs of $g_n(t)$ for $n = 0.001, 1, 20, 100$. They clearly exhibit the concentration of mass toward the origin as n increases.

On the other hand, as $\epsilon \to 0$, $e^{-\epsilon t^2}$ converges to the function 1. To see this, take any bounded interval $[a,b]$. Then

$$|1 - e^{-\epsilon t^2}| \leq \max\{1 - e^{-\epsilon a^2}, 1 - e^{-\epsilon b^2}\} \to 0, \qquad (6.28)$$

as $\epsilon \to 0$, so that the left-hand side converges uniformly to 0 on this interval. Hence for $\phi \in \mathcal{D}[a,b]$ we have

$$\left|\langle e^{-\epsilon t^2} - 1, \phi\rangle\right| = \left|\int_a^b (e^{-\epsilon t^2} - 1)\phi(t)dt\right| \leq \int_a^b \left|e^{-\epsilon t^2} - 1\right| \cdot |\phi(t)|dt$$

$$\leq \sup_{a \leq t \leq b}\left|e^{-\epsilon t^2} - 1\right| \int_a^b |\phi(t)|dt,$$

and the last term converges uniformly to 0 by (6.28).

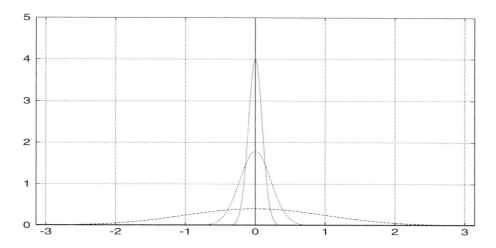

Figure 6.6. *Gaussian distribution functions*

Let us list yet one more important theorem on the convergence of distributions.

Theorem 6.5.8. *Let $\{T_n\}$ be a sequence of distributions. If for every $\phi \in \mathcal{D}$, $\langle T_n, \phi \rangle$ possesses a limit, then there exists $T \in \mathcal{D}'$ such that $T_n \to T$.*

Outline of Proof. If the conditions of the theorem are satisfied, then we can define the correspondence

$$T : \phi \mapsto \lim_{n \to \infty} \langle T_n, \phi \rangle,$$

and this is easily seen to be linear. We have already noted in Remark 2.5.8 that the Banach–Steinhaus theorem holds in a general context such as when X is a so-called barreled space. The space $\mathcal{D}$ is indeed known to be barreled, and hence the conclusion of Corollary 2.5.7 holds, and hence the theorem holds. See [51, 53, 58] for details.   □

Example 6.5.9. The series

$$T = \sum_{n=-\infty}^{\infty} \delta_n \tag{6.29}$$

converges. Indeed, for every $\phi \in \mathcal{D}$, there exist a, b such that $\operatorname{supp}\phi \subset [a, b]$. Then $\langle T, \phi \rangle$ turns out to be a finite sum as

$$\sum_{a \le n \le b} \langle \delta_n, \phi \rangle = \sum_{a \le n \le b} \phi(n).$$

Theorem 6.5.8 yields the following very important result.

Theorem 6.5.10. *Differentiation is a continuous operation on $\mathcal{D}'$. Hence if $T_n \to T$ as $n \to \infty$, then $T_n' \to T'$. In particular, a convergent series (i.e., one whose partial sums converge in the sense of Definition 6.5.1) is always termwise differentiable.*

Proof. Let $T_n \to T$. Then for every $\phi \in \mathcal{D}(\mathbb{R})$, Definition 6.3.1 of differentiation (p. 117) readily yields

$$\langle T_n', \phi \rangle = -\langle T_n, \phi' \rangle \to -\langle T, \phi' \rangle = \langle T', \phi \rangle.$$

Hence $T_n' \to T'$. If we have a convergent series $S = \sum_{k=1}^{\infty} V_k$, apply this to $S_n := \sum_{k=1}^{n} V_k \to S$. $\qquad\qquad\qquad\qquad\qquad\qquad\qquad\qquad\qquad\qquad\qquad\qquad\qquad\qquad\qquad\quad \square$

6.6 Convolution of Distributions

6.6.1 Definition and Basic Properties

Let us now introduce the notion of *convolution* for distributions. To simplify the treatment, we always assume that the supports of distributions are bounded on the left whenever we deal with convolutions, that is, for a distribution f there exists a real number a such that supp $f \subset [a, \infty)$. Of course, this a depends on f. In what follows, we denote by $\mathcal{D}_+'(\mathbb{R})$ or simply $\mathcal{D}_+'$ the space of all distributions with support bounded on the left.[68]

Let us start with the case of locally integrable functions. Let f and g be locally integrable functions with support bounded on the left. Then the convolution $h = f * g$ is usually defined through

$$h(t) := \int_{-\infty}^{\infty} f(t - \tau) g(\tau) d\tau = \int_{-\infty}^{\infty} f(\tau) g(t - \tau) d\tau. \tag{6.30}$$

Observe that by the assumption on the supports of f and g the region of integration is bounded for each fixed t. Then the following facts are known:

- For almost every t, $h(t)$ exists and becomes a locally integrable function.

- Suppose that one (e.g., f) is integrable and the other (respectively, g) is bounded. Then $h(t)$ is a bounded continuous function and satisfies

$$|h(t)| \leq \left\{ \sup_{-\infty < t < \infty} |g(t)| \right\} \int_{-\infty}^{\infty} |f(t)| dt. \tag{6.31}$$

In this case, the condition on the supports of f, g is not necessary to define h.

- If both f and g are integrable, then $h(t)$ is integrable and satisfies

$$\|h\|_{L^1} \leq \|f\|_{L^1} \|g\|_{L^1}. \tag{6.32}$$

In this case, too, the constraint on the supports of f, g becomes unnecessary.

---

[68]This condition can be relaxed a little for defining convolution; for details, see Schwartz [52, 53]. But for most cases this will be enough.

- When the supports of f and g are both contained in $[0, \infty)$, the support of h is also contained in $[0, \infty)$, and the convolution takes the well-known form

$$
h(t) = \begin{cases} 0, & t \leq 0, \\ \displaystyle\int_0^t f(t - \tau)g(\tau)d\tau, & t > 0. \end{cases} \tag{6.33}
$$

The proof depends on Lebesgue integration theory and hence is omitted; see, e.g., [20, 22, 37, 66].

Exercise 6.6.1. Ensure formally the estimates (6.31) and (6.32).

We now extend the definition above to distributions. To this end, let us first see how (6.30) defines a distribution. Take any test function $\phi \in \mathcal{D}$. Then $\langle h, \phi \rangle$ should be given by

$$
\langle h, \phi \rangle = \int_{-\infty}^{\infty} h(t)\phi(t)dt = \iint f(t - \tau)g(\tau)\phi(t)d\tau dt.
$$

The double integral in the last term is guaranteed to exist for almost every t by Fubini's theorem in Lebesgue integration theory [20, 22, 37]. Performing the change of variables $\xi := t - \tau$ and $\eta := \tau$, we obtain that the integral above becomes

$$
\iint f(\xi)g(\eta)\phi(\xi + \eta)d\xi d\eta,
$$

namely, $\langle f_{(\xi)}g_{(\eta)}, \phi(\xi + \eta) \rangle$. Here $f_{(\xi)}, g_{(\eta)}$ denote that f and g are regarded as functions in ξ and η, respectively.

We adopt this formula to define the convolution for distributions, taking advantage of the fact that variables of f and g can be taken independently.

Let T and S be distributions having support bounded on the left. We define their convolution $T * S$ by

$$
\langle T * S, \phi \rangle := \langle T_{(\xi)}S_{(\eta)}, \phi(\xi + \eta) \rangle = \langle T_{(\xi)}, \langle S_{(\eta)}, \phi(\xi + \eta) \rangle \rangle, \quad \phi \in \mathcal{D}. \tag{6.34}
$$

Since the support of ϕ is bounded, $\phi(\xi + \eta)$ is 0 where $|\xi + \eta| > C$ for some $C > 0$. On the other hand, since $T, S \in \mathcal{D}'_+$, $\operatorname{supp} T, \operatorname{supp} S \subset [a, \infty)$ for some $a \in \mathbb{R}$. Now for each fixed ξ, $\phi(\xi + \eta)$ is equal to 0 for $|\eta| > C - \xi$, and hence $\langle S_{(\eta)}, \phi(\xi + \eta) \rangle$ is actually evaluated on a bounded interval. This must then be regarded as a function of ξ. Since T is 0 outside $\xi \geq a$, it suffices to consider the region $a \leq \xi, \eta \leq C - a$ in evaluating $\langle T_{(\xi)}, \langle S_{(\eta)}, \phi(\xi + \eta) \rangle \rangle$. We may thus assume that the pertinent region for ξ, η in (6.34) is bounded. Finally, since $\langle S_{(\eta)}, \phi(\xi + \eta) \rangle$ is infinitely differentiable as a function of ξ, as can be seen in the following exercise, (6.34) is certainly well defined.

Exercise 6.6.2. Show that the function $\psi(\xi) := \langle S_{(\eta)}, \phi(\xi + \eta) \rangle$ is infinitely differentiable as a function of ξ. (Hint: Calculate $\lim_{h \to 0}(\psi(\xi + h) - \psi(\xi))/h$ according to the definition. Then, by the continuity of distributions, this is seen to be equal to $\langle S_{(\eta)}, \frac{\partial}{\partial \xi}\phi(\xi + \eta) \rangle$.)

The notation $T_{(\xi)}S_{(\eta)}$ above is a little imprecise. To be more precise, one should introduce the notion of tensor product $T_{(\xi)} \otimes S_{(\eta)}$ for distributions, and formula (6.34) above should be understood in this sense. But since it is a "product" in regard to two independent variables ξ and η, this abuse does not cause a serious problem. In fact, one can understand its meaning on the far right side of (6.34), and this will not lead to any difficulty. The reader should, however, keep in mind that an arbitrary "product" of two distributions is not necessarily well defined.

Also, the defining equation (6.34) is symmetric in ξ and η, so that it is invariant if we change the roles of T and S. Hence, if the convolution $T * S$ exists, then $S * T$ also exists, and they are equal.

Let us give some examples.

Example 6.6.3. For every distribution T,

$$\delta * T = T. \tag{6.35}$$

In fact,

$$\langle \delta * T, \phi \rangle = \langle T_{(\eta)}, \langle \delta_{(\xi)}, \phi(\xi + \eta) \rangle \rangle = \langle T_{(\eta)}, \phi(\eta) \rangle = \langle T, \phi \rangle$$

for every $\phi \in \mathcal{D}$. That is, δ *acts as the identity with respect to convolution.*

Example 6.6.4.

$$\delta_a * T = \tau_a T, \quad \delta_a * \delta_b = \delta_{a+b}. \tag{6.36}$$

Here $\tau_a T$ is the shift of T, i.e., a distribution defined by

$$\langle \tau_a T, \phi \rangle := \langle T_{(t)}, \phi(t + a) \rangle. \tag{6.37}$$

Obviously by definition

$$\langle \delta_a * T, \phi \rangle = \langle T_{(\eta)}, \langle \delta_a, \phi(\xi + \eta) \rangle \rangle = \langle T_{(\eta)}, \phi(a + \eta) \rangle = \langle \tau_a T, \phi \rangle. \tag{6.38}$$

As a special case, putting $T = \delta_b$ yields the second formula above. In particular, we observe that if T is a locally integrable function, then $(\delta_a * T)(t) = (\tau_a T)(t) = T(t - a)$. This readily follows from expressing (6.37) or (6.38) in integral form.

Example 6.6.5.

$$\delta' * T = T'. \tag{6.39}$$

Here T' is of course understood in the sense of distributions. The definition readily yields

$$\langle \delta' * T, \phi \rangle = \langle T_{(\eta)}, \langle \delta'_{(\xi)}, \phi(\xi + \eta) \rangle \rangle$$
$$= \langle T_{(\eta)}, -\phi'(\eta) \rangle$$
$$= \langle T', \phi \rangle.$$

This shows that convolution with δ' acts as the differentiation. In general, given a polynomial $\sum_{k=1}^{n} a_k z^k$, we denote by $\sum_{k=1}^{n} a_k \delta^{(k)}$ the distribution obtained by substituting δ' for z. Then we have

$$\left(\sum_{k=1}^{n} a_k \delta^{(k)} \right) * T = \sum_{k=1}^{n} a_k T^{(k)}, \tag{6.40}$$

where $\delta^{(k)}$ denotes the kth order derivative of δ, which is also equal to the one obtained by convolving δ' with itself $k - 1$ times.

For three distributions T, S, U belonging to $\mathcal{D}'_+$, the associative law holds among their convolutions.

Proposition 6.6.6. *Let T, S, U belong to $\mathcal{D}'_+$. Then*

$$(T * S) * U = T * (S * U) \tag{6.41}$$

holds. Hence we can denote either term as $T * S * U$ without ambiguity.

The proof is not difficult but requires some preparation of notation and hence is omitted. See [52, 53] for details.

Using Proposition 6.6.6, we obtain the following theorem, which has a fundamental importance in application of convolutions.

Theorem 6.6.7. *Let $S, T \in \mathcal{D}'_+$. Then*

$$(S * T)' = S * T' = S' * T.$$

That is, to differentiate a convolution, it suffices to differentiate one factor and then take the convolution afterward.

Proof. Proposition 6.6.6 and Example 6.6.5 yield

$$(S * T)' = \delta' * (S * T) = (\delta' * S) * T = S' * T$$
$$= (S * \delta') * T = S * (\delta' * T) = S * T'. \qquad \square$$

6.6.2 Continuity of Convolution

The convolution operation as defined over $\mathcal{D}'_+$ satisfies the following continuity property.

Theorem 6.6.8. *Suppose that $S_j \to S$ and $T_j \to T$ in $\mathcal{D}'_+$ as $j \to \infty$. Then*

$$S_j * T \to S * T, \quad S * T_j \to S * T.$$

Furthermore, if there exists c such that $\operatorname{supp} S_j \subset (c, \infty)$ and $\operatorname{supp} T_j \subset (c, \infty)$ for all j, then

$$S_j * T_j \to S * T$$

also holds.

Proof. The proof of this theorem is not elementary and hence is omitted. See, e.g., [53, 58]. $\qquad \square$

6.6.3 Regularization of Distributions

Let us consider the problem of approximating distributions by smooth functions. Take a function ρ satisfying the following conditions:

1. $\rho \in \mathcal{D}$, $\rho(t) \geq 0$ for every t, and $\operatorname{supp} \rho \subset [-1, 1]$.

2. $\int_{-\infty}^{\infty} \rho(t)dt = 1$.

The following function, for example, satisfies the requirements:

$$\rho(t) = \begin{cases} C \exp\left(-\frac{1}{1-t^2}\right), & |t| < 1, \\ 0, & |t| \geq 1, \end{cases} \tag{6.42}$$

where the constant C is to be chosen to satisfy requirement 2 above.

For a positive parameter $\epsilon > 0$, set

$$\rho_\epsilon(t) := \frac{1}{\epsilon} \rho\left(\frac{t}{\epsilon}\right). \tag{6.43}$$

Clearly $\operatorname{supp} \rho_\epsilon \subset [-\epsilon, \epsilon]$, and ρ_ϵ satisfies the conditions of Proposition 6.5.3. Hence $\rho_\epsilon \to \delta$. We then obtain the following proposition.

Proposition 6.6.9. *Let T be an arbitrary distribution, and let ρ_ϵ be as above. Then $T * \rho_\epsilon$ is infinitely differentiable, and $T * \rho_\epsilon \to T$ as $\epsilon \to 0$.*

Approximating a distribution by smooth functions as above is called the *regularization* of a distribution.

Proof. Since $\rho_\epsilon \in \mathcal{D}(\mathbb{R})$, the function $\langle T_\eta, \rho_\epsilon(t - \eta) \rangle$ is well defined at each t and is infinitely differentiable (Exercise 6.6.2,[69] p. 127).

Since definition (6.34) of convolutions is precisely the extension of (6.30) to distributions, which has the same meaning as $\langle T_\eta, \rho_\epsilon(t - \eta) \rangle$, this function certainly agrees with $T * \rho_\epsilon(t)$. Since $\rho_\epsilon \to \delta$ by the discussion above, it follows from Theorem 6.6.8 on the continuity of convolution that

$$T * \rho_\epsilon \to T * \delta = T. \qquad \qquad \square$$

Strictly speaking, the discussions in this section are confined to $\mathcal{D}'_+$, so applying the result above to arbitrary $T \in \mathcal{D}'$ is not yet fully guaranteed. But even if the support of T is not bounded on the left, if its counterpart in the convolution has compact support, then the convolution can be well defined, and the argument above applies without any change.

If, further, T is continuous or an L^p ($p \geq 1$) function, then it is known that $T * \rho_\epsilon$ turns out to be convergent to T uniformly or in the L^p topology, respectively. For details, see [39, 58].

Regularization provides a convenient tool for proving various properties. In particular, it is often employed to extend identities for ordinary functions to distributions. Let us give a proof of the following fact using regularization.

---

[69]Be careful not to leap at Theorem 6.6.7. The theorem guarantees the differentiability only in the sense of distributions.

Proposition 6.6.10. *Let $T \in \mathcal{D}'$ and ϕ be a C^∞ function. Suppose that conditions for $T * \phi$ to be well defined are satisfied. Then $T * \phi \in C^\infty$, and*

$$(T * \phi)(t) = \langle T_{(\tau)}, \phi(t - \tau) \rangle. \tag{6.44}$$

In particular, if we denote the reflection of a variable as $\check{\phi}(t) := \phi(-t)$, then $\langle T, \phi \rangle = (T * \check{\phi})(0)$.

Proof. That $T * \phi$ is a C^∞ function has already been shown in Exercise 6.6.2.

First suppose that T is an ordinary locally integrable function. Then by (6.30), we have

$$(T * \phi)(t) = \int_{-\infty}^{\infty} T(t - \tau) \phi(\tau) d\tau = \langle T_{(\tau)}, \phi(t - \tau) \rangle,$$

and (6.44) clearly holds. Then by Proposition 6.6.9 there exists a sequence $T_n \in C^\infty$ such that $T_n \to T$. From Theorem 6.6.8 on the continuity of convolution, it follows that $T_n * \phi \to T * \phi$. On the other hand, since $\langle (T_n)_{(\tau)}, \phi(t - \tau) \rangle \to \langle T_{(\tau)}, \phi(t - \tau) \rangle$ is clearly valid by the definition of convergence of distributions, (6.44) holds in the limit. $\square$

6.7 System Theoretic Interpretation of Convolution

Let us give an interpretation of convolution from the viewpoint of dynamical system theory. Consider the following simple system defined by the first-order difference equation:[70]

$$\begin{aligned} x_{k+1} &= a x_k + b u_k, \\ y_k &= c x_k, \quad a, b, c \in \mathbb{R}, \quad k = 0, 1, 2, \dots. \end{aligned} \tag{6.45}$$

Here x_k, u_k, and y_k are real-valued. They are called state, input, and output variables, respectively. The above set of equations describes the dynamical behavior of the system in such a way that the internal state changes according to the input u_k, which enters into the system at successive time instants and produces output y_k correspondingly. Let us compute the correspondence from inputs to outputs under the assumption that the initial state x_0 is 0. Apply formula (6.45) to $k = 0, 1, 2, \dots$ successively to obtain

$$\begin{aligned} x_1 &= b u_0, \\ x_2 &= a x_1 + b u_1 = a b u_0 + b u_1, \\ x_3 &= a x_2 + b u_2 = a^2 b u_0 + a b u_1 + b u_2, \\ &\vdots \\ x_n &= \sum_{k=0}^{n-1} a^{n-k-1} b u_k. \end{aligned}$$

Hence

$$y_n = \sum_{k=0}^{n-1} c a^{n-k-1} b u_k. \tag{6.46}$$

---

[70] We will encounter a more general continuous-time system in Chapter 10.

Setting $w_k := ca^{k-1}b$, we can express (6.46) as

$$y_n = \sum_{k=0}^{n-1} w_{n-k}u_k.$$

Its continuous-time limit (counterpart) is precisely the convolution equation:

$$y(t) = \int_0^t w(t-\tau)u(\tau)d\tau = (w * u)(t). \qquad (6.47)$$

That is, $w(t)$ represents the instantaneous response (output) against a stimulus (input) at time t, and the convolution represents the overall cumulative effect in the output during the period while an input function is applied. The weight in the integral is given by the function w.

Recall that δ acts as the identity with respect to convolution (Example 6.6.3, p. 128). Substituting δ for u in (6.47) produces the output $(w * \delta)(t) = w(t)$. In other words, the *weighting function* $w(t)$ is obtained as the output against the impulse (i.e., delta distribution δ) input. In this sense, $w(t)$ in the input-output relation (6.47) described by convolution is called the *impulse response*. Needless to say, $w(t)$ should be by no means restricted to be the special one derived from (6.45).

The relation

$$w = w * \delta$$

is very often (imprecisely) written in the integral form as

$$w(t) = \int w(t-\tau)\delta(\tau)d\tau. \qquad (6.48)$$

But since δ is not a function but only a distribution, this is an inaccurate expression that can easily lead to confusion and misunderstanding. The very introduction of δ by Dirac as a "function" satisfying

$$\int_{-\infty}^{\infty} \delta(t)dt = 1, \quad \delta(t) = 0, \, t \neq 0, \qquad (6.49)$$

was motivated by the need to justify (6.48). In fact, if there were such a function, its "mass" would be concentrated at the origin $\{0\}$, and if the total integral were 1, then by these two properties the weight $w(t)$ would come out of the integral (because "δ" is concentrated at 0 we set $\tau = 0$ in $w(t-\tau)$), and hence it would appear as if (6.48) were obtained. But we should repeat that such a "function" does not exist (if it is 0 except at the origin, then its integral must also be 0 by Lebesgue integration theory), and we must understand that (6.49) is no more than a symbolic interpretation for the delta "function."

But there can be *sequences of functions* that asymptotically realize (6.49). Those given in Examples 6.5.2, 6.5.6, and 6.5.4 serve as examples.

6.8 Application of Convolution

6.8.1 Convolution Equations

The primary importance of the delta distribution lies in the fact that it gives the identity element with respect to convolution. Moreover, the convolution with its derivative δ' acts as the differentiation operator as seen in Example 6.6.5. This suggests that the use of δ and its derivative δ' is effective in solving and discussing the properties of differential equations, particularly of constant coefficients.

Consider the following general convolution equation in $\mathcal{D}'_+$:

$$A * f = g. \tag{6.50}$$

Here A is an element of $\mathcal{D}'_+$, and our problem here is to find a solution $f \in \mathcal{D}'_+$ for a given $g \in \mathcal{D}'_+$. If we require that (6.50) be uniquely solvable for an arbitrary g, it will become the problem of finding a convolutional inverse of A.

The space $\mathcal{D}'_+$ is clearly a vector space and is also endowed with a notion of product, i.e., convolution; it satisfies the condition of what is called an *algebra*[71] in mathematics. However, it is not guaranteed that every nonzero element admits an inverse. In fact, it is known (and can be easily shown) that infinitely differentiable functions never admit an inverse [53].

On the other hand, it is also known that there exist no zero-divisors in $\mathcal{D}'_+$. That is, if $\alpha * \beta = 0$, $\alpha, \beta \in \mathcal{D}'_+$, then either $\alpha = 0$ or $\beta = 0$. In other words, $\mathcal{D}'_+$ is a so-called integral domain. This is the content of Titchmarsh's convolution theorem, only the result of which we now state [66].

Theorem 6.8.1 (Titchmarsh's theorem). *The convolution algebra $\mathcal{D}'_+$ has no zero-divisors. That is, if $\alpha * \beta = 0$, $\alpha, \beta \in \mathcal{D}'_+$, then either $\alpha = 0$ or $\beta = 0$.*

The original form of Titchmarsh's theorem is given for L^1 functions, but the above generalization is also possible. This provides a basis for operational calculus, particularly that by Mikusiński [36]. But we will not go into detail here.

It is to be noted, however, that this implies that an inverse of an element $\alpha \in \mathcal{D}'_+$ is, if it exists, unique. For if $\alpha\beta_1 = \alpha\beta_2 = \delta$, then $\alpha * (\beta_1 - \beta_2) = 0$, so that $\beta_1 = \beta_2$ by Titchmarsh's theorem. Hence from here on we will speak of *the* inverse of an element instead of *an* inverse.

We also note that Titchmarsh's theorem does not necessarily hold for choices of spaces other than $\mathcal{D}'_+$. For example, take $\alpha = \delta'$ and $\beta = 1$, i.e., the function that takes the value 1 identically. The latter does not belong to $\mathcal{D}'_+$ (the condition on support is not satisfied). Note that $\alpha * \beta = (d/dt)1 = 0$.

We now discuss a special class of convolution operators. Take the ordinary differential operator of constant coefficients:

$$D := \frac{d^n}{dt^n} + a_{n-1}\frac{d^{n-1}}{dt^{n-1}} + \cdots + a_1\frac{d}{dt} + a_0. \tag{6.51}$$

Let us find the inverse of $D\delta$. (The following treatment is based on that given in [52].)

Consider the ordinary differential equation

$$D\delta * x = Dx = \left(\frac{d^n}{dt^n} + a_{n-1}\frac{d^{n-1}}{dt^{n-1}} + \cdots + a_1\frac{d}{dt} + a_0\right)x = 0. \tag{6.52}$$

---

[71]A vector space that simultaneously constitutes a ring with compatible binary operations is called an algebra.

By the existence and uniqueness of solutions for ordinary differential equations, this equation admits a unique solution for any initial condition [45]:

$$x(0) = x_0,\ x'(0) = x_1, \ldots,\ x^{(n-1)}(0) = x_{n-1}. \tag{6.53}$$

Denote by $E(t)$ a particular solution corresponding to the initial condition

$$x_0 = 0,\ x_1 = 0, \ldots,\ x_{n-2} = 0,\ x_{n-1} = 1. \tag{6.54}$$

The function $E(t)$ is an infinitely differentiable function in the ordinary sense. We claim that the product $HE(t)$ with the Heaviside function $H(t)$ gives the convolutional inverse of $D\delta$.

The function $E(t)$ is infinitely differentiable, whereas $H(t)$ possesses the only discontinuity at the origin, and the jump there is 1. Thus the product $HE(t)$ admits the origin as a discontinuity of the first kind,[72] and its jump is given by $E(0)$. Hence by Proposition 6.3.3 (p. 118) we have

$$
\begin{aligned}
(HE)' &= HE' + \delta E(0), \\
(HE)'' &= HE'' + \delta E'(0) + \delta' E(0), \\
&\ \ \vdots \\
(HE)^{(n)} &= HE^{(n)} + \sum_{k=1}^{n} \delta^{(k-1)} E^{(n-k)}(0).
\end{aligned}
\tag{6.55}
$$

In view of the initial condition (6.54) on $E^{(k)}(0)$, we obtain

$$
\begin{aligned}
(HE)^{(k)} &= HE^{(k)}, \quad 0 \le k \le n-1, \\
(HE)^{(n)} &= HE^{(n)} + \delta.
\end{aligned}
$$

It now follows that

$$D\delta * (HE) = D(HE) = HDE + \delta = \delta.$$

That is, HE is the convolutional inverse of $D\delta$.

Such an inverse HE is called the *fundamental solution* to the differential operator D (or $D\delta$), because one can then construct a solution for the inhomogeneous equation (6.50) with an arbitrary initial condition using this solution. Let us now see this.

Consider the inhomogeneous equation

$$D\delta * x = Dx = \left(\frac{d^n}{dt^n} + a_{n-1}\frac{d^{n-1}}{dt^{n-1}} + \cdots + a_1\frac{d}{dt} + a_0 \right) x = f. \tag{6.56}$$

Let us find the solution $x(t)$ satisfying the initial condition (6.53) with the aid of the fundamental solution. We confine ourselves to $t \ge 0$ and attempt to find $H(t)x(t)$. Successively differentiating similarly to (6.55), we must have

$$(D\delta) * Hx = D(Hx) = HDx + \sum_{k=0}^{n-1} e_k \delta^{(k)} \tag{6.57}$$

---

[72] A function f has a discontinuity of the first kind at a if both $\lim_{\epsilon \to 0, \epsilon > 0} f(a + \epsilon)$ and $\lim_{\epsilon \to 0, \epsilon > 0} f(a - \epsilon)$ exist and are not equal.

for some constants e_k, $k = 0, \ldots, n-1$. Since $x(t)$ must be a solution, we must have $Dx = f$. Substituting x for E in (6.55), stacking and adding the coefficients of $\delta^{(i)}$, $i = 0, \ldots, n-1$, multiplied by a_k's together, and then substituting these into the initial condition (6.53), we obtain

$$e_k = x_{n-1-k} + a_{n-1}x_{n-2-k} + \cdots + a_{k+1}x_0.$$

Applying $(D\delta)^{-1} = HE$ on both sides of (6.57), we have

$$Hx = (D\delta)^{-1} * (D\delta) * Hx = HE * \left(Hf + \sum_{k=0}^{n-1} e_k \delta^{(k)} \right). \tag{6.58}$$

In other words, we obtain

$$x(t) = \int_0^t E(t-\tau)f(\tau)d\tau + \sum_{k=0}^{n-1} e_k E^{(k)}(t) \quad \text{for} \quad t > 0. \tag{6.59}$$

Here we have used

$$(HE * Hf)(t) = \int_0^t E(t-\tau)f(\tau)d\tau.$$

(Observe that either of $H(t-\tau), H(\tau)$ vanishes outside $[0,t]$.) Thus we can solve the ordinary differential equation (6.56) for any initial condition using the fundamental solution.

Let us list the following lemma concerning the convolutional inverse. This is also fundamental to the operational calculus in the next section.

Lemma 6.8.2. *Assume that $A_1, A_2 \in \mathcal{D}'_+$ both admit the inverses A_1^{-1} and A_2^{-1} in $\mathcal{D}'_+$, respectively. Then $A_1 * A_2$ is also invertible in $\mathcal{D}'_+$, and*

$$(A_1 * A_2)^{-1} = A_2^{-1} * A_1^{-1} = A_1^{-1} * A_2^{-1}.$$

The proof is left to the reader as an exercise.

6.8.2 Operational Calculus

As we have seen above, solving the ordinary differential equation $Dx = f$ reduces to finding the fundamental solution to D. Since $Dx = (D\delta) * x$, finding the fundamental solution is nothing but finding the convolutional inverse of $D\delta$. To this end, we can use partial fraction expansion for this, making use of the fact that $\mathcal{D}'_+$ constitutes an algebra.

For example, it is easy to see $(\delta' - \lambda\delta)^{-1} = H(t)e^{\lambda t}$ (see Problem 4 on p. 139). In general, if

$$D\delta = (\delta' - \lambda_1\delta) * \cdots * (\delta' - \lambda_n\delta), \tag{6.60}$$

then by Lemma 6.8.2 we have

$$(D\delta)^{-1} = H(t)e^{\lambda_1 t} * \cdots * H(t)e^{\lambda_n t}.$$

Instead of executing this computation directly, we give yet another method via partial fraction expansion. For this purpose, we will now denote, only in this subsection, δ' by p and δ

by 1 and treat convolution as an ordinary product. Since δ is the identity for convolution, this will not yield confusion. For simplicity, assume that

$$Q(p) = (p - \lambda_1) \cdots (p - \lambda_n)$$

and that the λ_j's are distinct (this is not an essential assumption, though). The partial fraction expansion of $1/Q(p)$ takes the form

$$\frac{1}{Q(p)} = \sum_{k=1}^{n} \frac{c_k}{p - \lambda_k}.$$

Then $(D\delta)^{-1} = Q(\delta')^{-1}$ and $(p - \lambda_k)^{-1} = (\delta' - \lambda_k \delta)^{-1} = H(t) e^{\lambda_k t}$ yield

$$(D\delta)^{-1} = \sum_{k=1}^{n} c_k H(t) e^{\lambda_k t}. \tag{6.61}$$

That is, by calculating the partial fraction expansion coefficients c_k, we can obtain the fundamental solution $(D\delta)^{-1}$. For example, the inverse of

$$\frac{1}{p^2 + \omega^2} = \frac{1}{2\omega i} \left(\frac{1}{p - i\omega} - \frac{1}{p + i\omega} \right)$$

turns out to be

$$\frac{1}{2\omega i} \left(H(t) e^{i\omega t} - H(t) e^{-i\omega t} \right) = \frac{1}{\omega} H(t) \sin \omega t.$$

Remark 6.8.3. The reader may notice that the computation above is precisely parallel to one by Laplace transforms. Of course, there is an important difference that here the symbol p represents the differentiation operator, while it corresponds to a complex variable in Laplace transforms, but as far as the algebraic structure is concerned, the two treatments are identical. However, let us also note that functions of a complex variable appears in Laplace transforms, and hence their function-theoretic properties play various important roles, whereas in the operational calculus in the scope above such a relationship is not particularly emphasized.

6.8.3   Convolution and Fourier Transforms

Let us present an application of convolution to Fourier analysis. Although a more complete treatment is to be developed in Chapter 7, we here show that its essence is the convolution with a kernel function (e.g., Dirichlet kernel) approximating δ.

Let us confirm the fact stated in Example 6.5.6 (p. 123). Namely, we prove that the Dirichlet kernel $D_n(t)/2\pi$ converges to δ as $n \to \infty$. Take a periodic C^1 function $f(t)$ on $(-\infty, \infty)$ having period 2π. By (6.24) and (6.25), we have

$$D_n(t) = \frac{\sin(n + 1/2)t}{\sin(1/2)t} = \sum_{k=-n}^{n} e^{ikt}.$$

Taking the convolution of f with $D_n(t)/2\pi$ on the interval $[-\pi,\pi]$, we obtain

$$\frac{1}{2\pi}\int_{-\pi}^{\pi} D_n(t-\tau)f(\tau)d\tau = \frac{1}{2\pi}\sum_{k=-n}^{n}\int_{-\pi}^{\pi} f(\tau)e^{-ik\tau}e^{ikt}d\tau = \sum_{k=-n}^{n} c_k e^{ikt},$$

$$c_k := \frac{1}{2\pi}\int_{-\pi}^{\pi} f(\tau)e^{-ik\tau}d\tau.$$

This is nothing but the n-partial sum of the Fourier expansion of f. Let us show that this indeed converges uniformly to f itself. (Then by Proposition 6.6.10 $\langle T, f\rangle = (T * \check{f})(0)$, and this readily yields $D_n/2\pi \to \delta$.)

Theorem 6.8.4. *Let f be a C^2 function of period 2π, and $S_n(t)$ its n-partial sum of its Fourier series:*

$$S_n(t) := \sum_{k=-n}^{n} c_k e^{ikt}, \quad c_k := \frac{1}{2\pi}\int_{-\pi}^{\pi} f(\tau)e^{-ik\tau}d\tau.$$

Then S_n converges to f uniformly on $(-\infty,\infty)$.

Proof. In view of the periodicity of f, it suffices to prove the uniform convergence on $[-\pi,\pi]$. Since $\int_{-\pi}^{\pi} D_n(t)dt/2\pi = 1$,

$$\begin{aligned}
S_n(t) - f(t) &= \frac{1}{2\pi}\int_{-\pi}^{\pi} [f(t-\tau) - f(t)]D_n(\tau)d\tau\\
&= \frac{1}{\pi}\int_{-\pi}^{\pi}\left\{\frac{f(t-\tau)-f(t)}{2\sin\tau/2}\right\}\sin\left(n+\frac{1}{2}\right)\tau\, d\tau\\
&= -\frac{1}{\pi}\frac{\cos\left(n+\frac{1}{2}\right)\tau}{n+\frac{1}{2}}g(t,\tau)\Bigg|_{\tau=-\pi}^{\tau=\pi} + \frac{1}{\pi}\int_{-\pi}^{\pi}\frac{\partial g(t,\tau)}{\partial\tau}\frac{\cos\left(n+\frac{1}{2}\right)\tau}{n+\frac{1}{2}}d\tau,
\end{aligned}$$

$$(6.62)$$

where g is defined by

$$g(t,\tau) := \frac{f(t-\tau)-f(t)}{2\sin\tau/2}.$$

Since $f \in C^2$, $|f(t-\tau) - f(t)|$ is of the same order as $|\tau|$ (as $\tau \to 0$), and hence the same order of $\sin\tau/2$. Hence $g(t,\tau)$ is bounded. Now differentiate $g(t,\tau)$ with respect to τ to obtain

$$\frac{\partial g(t,\tau)}{\partial\tau} = \frac{-2f'(t-\tau)\sin\frac{\tau}{2} - [f(t-\tau) - f(t)]\cos\frac{\tau}{2}}{(2\sin\frac{\tau}{2})^2}. \tag{6.63}$$

The first term $-2f'(t-\tau)\sin\frac{\tau}{2}$ in the numerator is to be of the same order as $-\tau(f'(t) - \tau f''(\theta))$ by applying the mean value theorem to $f'(t-\tau) - f'(t)$, $t - \tau < \theta < t$. The second term $-[f(t-\tau) - f(t)]\cos\frac{\tau}{2}$ is of order $\tau f'(t)$ (and higher order terms). Adding these two terms, the first order term in τ vanishes, and the remaining terms are of the same order of τ^2 or higher. Since the denominator $(2\sin\frac{\tau}{2})^2$ is of order τ^2, this means that $\partial g/\partial\tau$

is bounded. Hence by the expression (6.62), there exists a constant $C > 0$ independent of t such that

$$|S_n(t) - f(t)| \le \frac{C}{n + \frac{1}{2}}.$$

Thus S_n converges to f uniformly.   □

We thus see that the Fourier series expansion of f converges to f itself. That is,

$$\frac{1}{2\pi} \int_{-\pi}^{\pi} D_n(t - \tau) f(\tau) d\tau = \frac{1}{2\pi} D_n * f \to \delta * f = f.$$

This looks very simple. But we should realize that this has been achieved by the combination of the following fundamental ingredients in convolution of distributions:

- convolution algebra $\mathcal{D}'_+$ and δ as the identity,

- convergence of $D_n(t)/2\pi$ to δ in the sense of distributions,

- continuity of convolution.

Putting all of these together, we may say that they are not technically simple or easy. However, the simplicity of recognizing that the foundation of Fourier analysis lies in the fact that the Dirichlet kernel approaches δ, the identity for convolution, cannot be dispensed with.

Of course, the convergence of $D_n(t)/2\pi$ to δ is in the sense of distributions, and it is not yet clear how this guarantees the convergence of Fourier series in the ordinary sense with respect to different topologies, depending on the regularity of the target function f. For example, when f belongs to L^p, does its Fourier series converge to f in L^p? We will discuss this question for $p = 2$ in the next chapter. On the other hand, if we are content with convergence in the sense of distributions, then f does not need to be even a function, and the convergence still holds. In other words, the Fourier expansion of a distribution T on a bounded interval can always be defined and it always converges to T in the sense of distributions. While this convergence may be too weak, this is a strikingly simple result compared to the classical Fourier theory.

We conclude this chapter by noting that the discussion above carries over to the Fejér kernel $K_n(t)$ as well. The only difference is that the Fejér kernel represents the Cesaro mean of the Fourier expansion in contrast with the Dirichlet kernel which represents the Fourier expansion itself. The details will be discussed in the next chapter.

Problems

1. Show that the distributions

$$\frac{1}{h} \{\delta_{-h} - \delta\}$$

   converge to δ' as $h \to 0$.

2. Show the identity (6.25) on p. 124.

3. Verify that function (6.59) gives a solution of (6.56).

4. Find the convolutional inverse of each of the following ordinary differential operators:

   (a) $(\delta' - \lambda\delta)^{-1}$;

   (b) $(\delta'' + \omega^2\delta)^{-1}$ $(\omega \in \mathbb{R})$.

5. Prove Lemma 6.8.2.

Chapter 7

Fourier Series and Fourier Transform

The importance of Fourier analysis needs no justification. The crux underlying its idea is in expanding a function into a sum of stationary waves and then capturing how it behaves in terms of each basis (i.e., sinusoid) component.

A characteristic feature in Fourier series expansion is that the basis functions are mutually orthogonal, and they give a typical example of an orthonormal basis in Hilbert space. Hence it is quite compatible with the framework of Hilbert space theory. We will show that Fourier series expansion and Fourier transformation indeed give the desired expansions, first in the framework of L^2 theory and then for distributions.

We should also note that such an expansion is very much based on stationary waves and hence more suited to analyzing those properties closely related to stationary waves. For example, the sinusoid $\sin \omega t$ is periodic and not quite adequate for encoding where or when a specific incident happens. For example, when we expand a function via sinusoids, it is very difficult to tell where a certain phenomenon (e.g., discontinuity) occurs. In other words, expansion via stationary functions such as sinusoids (i.e., Fourier analysis) is not appropriate for nonstationary phenomena. For such purposes, other expansion methods, such as wavelet theory, based on nonstationary basis functions become more adequate. But we here confine ourselves to Fourier analysis. For wavelets, see, e.g., [5, 32, 61].

7.1 Fourier Series Expansion in $L^2[-\pi,\pi]$ or of Periodic Functions

Let us start by motivating the Fourier series expansion. Suppose that we are given a function f which is expressed in terms of a trigonometric series $f(t) = \sum_{n=-\infty}^{\infty} c_n e^{int}$. What property should c_n satisfy?

Let us view this series without worrying about its convergence issues. The right-hand side is periodic of period 2π, so we see that f should also be periodic with the same period. There is a further constraint on c_n in order that this expansion be valid. Multiplying by e^{-int} on both sides and performing termwise integration, we obtain

$$\int_{-\pi}^{\pi} f(t)e^{-int}dt = 2\pi c_n, \tag{7.1}$$

that is,

$$c_n = \frac{1}{2\pi} \int_{-\pi}^{\pi} f(t)e^{-int}\,dt. \tag{7.2}$$

Given any integrable f, we can define its *Fourier series* as the series

$$\sum_{n=-\infty}^{\infty} c_n e^{int}, \tag{7.3}$$

where the c_n's are defined by (7.2). Naturally we ask the following questions:

- Does this series converge?

- Provided it does converge, does the series represent the original function f?

It is not straightforward to answer these questions. For example, if we artificially alter the value of $f(t)$ at a point, the integral (7.1) remains the same, and so does the Fourier series. If we further change the values on a set of measure 0, it still remains invariant. This suggests that the Fourier series is not quite an adequate tool for discussing pointwise convergence.

Of course, when there is a stronger condition imposed on f, e.g., C^1, or it is of bounded variation, etc., pointwise convergence of its Fourier series can be proven. But generally speaking, under the circumstances above, it is easier and more transparent to discuss convergence under a more relaxed condition.

We thus discuss the convergence of the Fourier series in L^2 space within the framework of Hilbert space theory. This leads to the discussion of completeness of the orthonormal system of functions in $L^2[-\pi, \pi]$:

$$\{e_n\}_{n=-\infty}^{\infty} = \left\{ \frac{1}{\sqrt{2\pi}} e^{int} \right\}_{n=-\infty}^{\infty}. \tag{7.4}$$

Remark 7.1.1. If we consider $L^2[-L, L]$ instead of $L^2[-\pi, \pi]$, then we should take as e_n

$$e_n = \frac{1}{\sqrt{2L}} \exp\left(\frac{n\pi i}{L} t \right). \tag{7.5}$$

The constant $1/\sqrt{2L}$ is a normalization factor to make $\|e_n\| = 1$.

First note that

$$(e_n, e_m) = \frac{1}{2\pi} \int_{-\pi}^{\pi} \overline{e^{imt}} e^{int}\,dt = \frac{1}{2\pi} \int_{-\pi}^{\pi} e^{i(n-m)t}\,dt = \begin{cases} 1 & (n = m), \\ 0 & (n \neq m). \end{cases}$$

Hence $\{e_n\}$ forms an orthonormal system. We have to show its completeness, that is, whether every function in $L^2[-\pi, \pi]$ can be expanded into an (infinite) linear combination or can be approximated by a linear combination with arbitrary accuracy.

Remark 7.1.2. According to the notation above, the Fourier series of f can be written as $\sum_{n=-\infty}^{\infty} (f, e_n)e_n$. This is precisely the abstract Fourier series introduced in Chapter 3, subsection 3.2.2 (p. 69).

Remark 7.1.3. When we want an expansion of a real-valued function via real-valued functions, we impose a restriction $\alpha_{-n} = \overline{\alpha_n}$ on the expansion $\sum \alpha_n e^{int}$. Let $\alpha_n = a_n + ib_n$, $a_n, b_n \in \mathbb{R}$. Then by

$$\cos nt = \frac{e^{int} + e^{-int}}{2}, \quad \sin nt = \frac{e^{int} - e^{-int}}{2i},$$

we have $\alpha_n e^{int} + \alpha_{-n} e^{-int} = 2a_n \cos nt - 2b_n \sin nt$, and we obtain an expansion via real-valued functions. The completeness of the obtained family is clearly reduced to the completeness of $\{e_n\}$.

Let us now adopt the following fact from integration theory.

Theorem 7.1.4. *The set of continuous functions f such that $f(-\pi) = f(\pi)$ constitutes a dense subset of $L^2[-\pi,\pi]$.*

In other words, for every $f \in L^2[-\pi,\pi]$, there exists a sequence $\{f_n\}$ of continuous functions with $f_n(-\pi) = f_n(\pi)$ such that

$$\lim_{n \to \infty} \|f - f_n\|_2 = 0.$$

To this end, it suffices to show that the set of continuous functions with compact support contained in $[-\pi,\pi]$ is dense in $L^2[-\pi,\pi]$; details may be found in [20, 22, 37].

Once this fact is accepted, the proof of completeness of $\{e_n\}$ in $L^2[-\pi,\pi]$ reduces to showing that every continuous function can be approximated by its partial Fourier sum with arbitrary precision. The approximation measure must be taken of course with respect to the L^2-norm, but we here give a stronger result by showing that every continuous function on $[-\pi,\pi]$ can be approximated uniformly by a trigonometric series derived from its Fourier series. Suppose that f_n (not necessarily a Fourier series) converges to f uniformly. As we easily see that

$$\|f - f_n\|_2 = \left\{ \int_{-\pi}^{\pi} |f(t) - f_n(t)|^2 dt \right\}^{1/2}$$

$$\leq \sup_{-\pi \leq t \leq \pi} |f(t) - f_n(t)| \left\{ \int_{-\pi}^{\pi} dt \right\}^{1/2}$$

$$= \sqrt{2\pi} \sup_{-\pi \leq t \leq \pi} |f(t) - f_n(t)|,$$

the claimed uniform convergence establishes the convergence in the L^2-norm.[73]

---

[73]There is a subtlety here. We will show that a trigonometric series obtained by taking the mean of the partial sum of the Fourier series converges to f, but we are *not* claiming that for every continuous function its Fourier series converges uniformly to itself. What converges here is a modified trigonometric series called the Cesaro mean. It is known that taking the average as (7.6) improves the convergence property of a sequence. Indeed, if a sequence f_n converges to f, then its Cesaro mean (7.6) also converges to f. But the converse is not necessarily true. There exists a continuous function whose Fourier series actually diverges. But, as we see below, the convergence of such a modified series is enough to guarantee the convergence of the Fourier series in the L^2-norm.

Take any continuous $f \in L^2[-\pi, \pi]$, and let f_n be its n-partial sum of its Fourier series expansion:

$$f_n := \sum_{k=-n}^{n} (f, e_k) e_k.$$

Now, instead of showing $f_n \to f$ directly, we now prove

$$F_n := \frac{1}{n+1}(f_0 + f_1 + \cdots + f_n) \to f. \tag{7.6}$$

This is called the Cesaro mean of f_n. With this change F_n still belongs to the subspace spanned by $\{e_k\}_{k=-n}^{n}$, so this shows that f can be approximated by a linear combination of $\{e_k\}$.

First observe that

$$f_n(t) = \sum_{k=-n}^{n} \frac{1}{\sqrt{2\pi}} \left(f, \frac{1}{\sqrt{2\pi}} e^{ikt} \right) e^{ikt}$$

$$= \sum_{k=-n}^{n} \frac{1}{2\pi} \left(\int_{-\pi}^{\pi} f(\eta) e^{-ik\eta} d\eta \right) e^{ikt}$$

$$= \frac{1}{2\pi} \int_{-\pi}^{\pi} f(\eta) \left(\sum_{k=-n}^{n} e^{ik(t-\eta)} \right) d\eta.$$

Hence

$$F_n(t) = \frac{1}{n+1} \sum_{k=0}^{n} f_k(t)$$

$$= \frac{1}{n+1} \sum_{k=0}^{n} \frac{1}{2\pi} \int_{-\pi}^{\pi} f(\eta) \left(\sum_{\ell=-k}^{k} e^{i\ell(t-\eta)} \right) d\eta$$

$$= \frac{1}{2\pi} \int_{-\pi}^{\pi} f(\eta) \left(\frac{1}{n+1} \sum_{k=0}^{n} \sum_{\ell=-k}^{k} e^{i\ell(t-\eta)} \right) d\eta. \tag{7.7}$$

Set

$$K_n(t) := \frac{1}{n+1} \sum_{k=0}^{n} \sum_{\ell=-k}^{k} e^{i\ell t}. \tag{7.8}$$

Then

$$F_n(t) = \frac{1}{2\pi} \int_{-\pi}^{\pi} f(\eta) K_n(t - \eta) d\eta. \tag{7.9}$$

In fact, K_n is precisely the Fejér kernel introduced in Example 6.5.4 on page 122. Let us now show this.

Note that

$$(n+1)K_n(t) - nK_{n-1}(t) = \sum_{k=-n}^{n} e^{ikt}$$

$$= \sum_{k=0}^{n} e^{ikt} + \sum_{k=1}^{n} e^{-ikt}$$

$$= \frac{1 - e^{i(n+1)t}}{1 - e^{it}} + \frac{1 - e^{-i(n+1)t}}{1 - e^{-it}} - 1$$

$$= \frac{\cos nt - \cos(n+1)t}{1 - \cos t}.$$

Since clearly $K_0(t) = 1$, adding these terms yields

$$K_n(t) = \frac{1}{n+1}\left[\frac{1 - \cos(n+1)t}{1 - \cos t}\right]$$

$$= \frac{1}{n+1}\left[\frac{\sin\dfrac{n+1}{2}t}{\sin\dfrac{1}{2}t}\right]^2. \tag{7.10}$$

This is precisely the Fejér kernel introduced in Example 6.5.4 (except for the difference in the index $n \to n+1$).

As a basic property of the *Fejér kernel*, we have the following.

Lemma 7.1.5. *The following properties hold for K_n:*

1. $K_n \geq 0$.

2. $\frac{1}{2\pi}\int_{-\pi}^{\pi} K_n(t)dt = 1$.

3. *For every $\delta > 0$, $\lim_{n\to\infty} \sup_{|t|\geq\delta} |K_n(t)| = 0$, $(|t| \leq \pi)$.*

Proof. Property 1 is obvious from representation (7.10) for K_n. Property 2 means that the nth Cesaro mean F_n of the constant function 1 is 1. This is obvious from the definition (7.6) of F_n. Now since $(\sin \delta/2)^2 \leq (\sin t/2)^2$ for $0 < \delta \leq |t| \leq \pi$, property 3 follows from the inequality

$$|K_n(t)| \leq \frac{1}{(n+1)\sin^2 \delta/2} \quad \text{for} \quad |t| \geq \delta. \qquad \square$$

We are now ready to prove the following theorem.

Theorem 7.1.6. *Let f be a continuous function on $[-\pi,\pi]$ such that $f(-\pi) = f(\pi)$. Then F_n defined by (7.7) converges uniformly to f.*

Proof. Since $f(-\pi) = f(\pi)$, $f(t)$ can easily be extended to a continuous function of period 2π on $(-\infty, \infty)$. Then (7.9) yields by change of variables

$$F_n(t) = \frac{1}{2\pi} \int_{-\pi}^{\pi} f(t-\eta) K_n(\eta) d\eta.$$

Since $(1/2\pi) \int K_n(t) dt = 1$ for every n, we have

$$F_n(t) - f(t) = \frac{1}{2\pi} \int_{-\pi}^{\pi} [f(t-\eta) - f(t)] K_n(\eta) d\eta. \tag{7.11}$$

Since f is a continuous function on a closed interval $[-\pi, \pi]$, it is uniformly continuous and remains so after it is extended to the whole real line $(-\infty, \infty)$. That is, for any given $\epsilon > 0$, there exists $\delta > 0$ such that

$$|f(t) - f(s)| \leq \epsilon \tag{7.12}$$

whenever $|t - s| \leq \delta$, irrespective of where t, s are. For such $\delta > 0$, Lemma 7.1.5 implies that there exists N such that $n \geq N$ yields

$$\sup_{|t| \geq \delta} |K_n(t)| < \epsilon.$$

This, along with (7.11), implies

$$|F_n(t) - f(t)| \leq \frac{1}{2\pi} \int_{-\delta}^{\delta} |f(t-\eta) - f(t)| K_n(\eta) d\eta$$

$$+ \frac{1}{2\pi} \int_{|\eta| \geq \delta} |f(t-\eta) - f(t)| K_n(\eta) d\eta$$

$$\leq \sup_{|\eta| \leq \delta} |f(t-\eta) - f(t)| + 2 \cdot \sup_{|t| \leq \pi} |f(t)| \cdot \sup_{|t| \geq \delta} K_n(t)$$

$$\leq \epsilon \left(1 + 2 \sup_{|t| \leq \pi} |f(t)|\right)$$

provided $n \geq N$.[74] Hence $F_n(t)$ converges to $f(t)$ uniformly on $[-\pi, \pi]$. $\qquad\square$

As we already noted, this is enough to show that the space spanned by $\{e_n\}$ constitutes a dense subspace of $L^2[-\pi, \pi]$. Then by Proposition 3.2.19 on page 73, it is complete. Hence the Fourier series of f converges to f in the L^2-norm. We have thus obtained the following theorem.

Theorem 7.1.7. *The family*

$$\left\{\frac{1}{\sqrt{2\pi}} e^{int}\right\}_{n=-\infty}^{\infty}$$

constitutes a complete orthonormal system in $L^2[-\pi, \pi]$. That is, as noted in Chapter 3, subsection 3.2.2 (p. 69), the Fourier series of $f \in L^2[-\pi, \pi]$ converges to f in L^2. More

---

[74]Here we used $\sup_{|\eta| \geq \delta} |f(t-\eta) - f(t)| \leq 2 \sup_{|t| \leq \pi} |f(t)|$.

concretely,

$$f(t) = \sum_{n=-\infty}^{\infty} c_n e^{int}, \tag{7.13}$$

$$c_n = \frac{1}{2\pi} \int_{-\pi}^{\pi} f(t) e^{-int} dt. \tag{7.14}$$

The series converges in $L^2[-\pi,\pi]$.

The expansion (7.13) is immediate from Theorem 3.2.18 on page 72.
As a corollary, we obtain Parseval's identity.

Corollary 7.1.8. *For $f \in L^2[-\pi,\pi]$,*

$$\frac{1}{2\pi} \int_{-\pi}^{\pi} |f(t)|^2 dt = \sum_{n=-\infty}^{\infty} |c_n|^2, \tag{7.15}$$

where c_n is as given by (7.14).

Proof. The proof is immediate from (3.21) on page 72.   □

7.2   Fourier Series Expansion of Distributions

We have seen that the Fourier series of a function in L^2 converges to itself with respect to the L^2-norm. The Fourier series is applicable to a much wider class of functions, however, such as distributions, and can be used effectively to derive various useful results. We here consider the Fourier series expansion of distributions. To this end, we first introduce the notion of periodic distributions.

Definition 7.2.1. A distribution T is called a *periodic distribution* (with period L) if there exists $L > 0$ such that $\delta_L * T = T$.

In view of Example 6.6.4 (p. 128), $\delta_L * T$ represents the shifted distribution $\tau_L T$ obtained by shifting T by L to the right, so the definition above is reasonable. In what follows, for consistency of notation, we fix L to be equal to 2π. For other cases, Remark 7.1.1 equally applies. As with the Fourier series of functions, it is convenient to make use of periodicity and confine our consideration to $[-\pi,\pi]$. But for distributions we must fix a space of test functions, but elements of $\mathcal{D}(\mathbb{R})$ are not necessarily periodic, and so this requires some preliminaries.

Let $\mathbb{T}$ denote the unit circle with the origin as its center. Regarding $\mathbb{T}$ as a subset of the complex plane, $\mathbb{T}$ can be identified with $[-\pi,\pi)$ via the polar coordinate $t \leftrightarrow e^{it}$. Denote by $\mathcal{D}(\mathbb{T})$ the space of infinitely differentiable functions on $\mathbb{T}$. Since $\mathbb{T}$ is a circle, this means that for each element $\phi \in \mathcal{D}(\mathbb{T})$ its derivatives should agree at the two points $-\pi$ and π. Hence every element of $\mathcal{D}(\mathbb{T})$ can be extended to an infinitely differentiable periodic function on $\mathbb{R}$. Let us introduce a topology into $\mathcal{D}(\mathbb{T})$ in exactly the same way as with $\mathcal{D}[a,b]$ (see subsection 6.2.1). In other words, ϕ_n converges to 0 if and only if each derivative $\phi_n^{(m)}$ converges to 0 uniformly for every m.

We denote by $\mathcal{D}'(\mathbb{T})$ the space of distributions defined for $\mathcal{D}(\mathbb{T})$, i.e., the space of continuous linear functionals on it. This indeed agrees with the space of periodic distributions restricted to $[-\pi,\pi)$, but its proof is a little delicate and is hence omitted here. See [53] for details. Of course, for locally integrable functions, this identification leads to no difficulty.

Now define the convolution in $\mathcal{D}'(\mathbb{T})$ by

$$\langle T * S, \phi \rangle := \langle T_{(\xi)} S_{(\eta)}, \phi(\xi + \eta) \rangle = \langle T_{(\xi)}, \langle S_{(\eta)}, \phi(\xi + \eta) \rangle \rangle, \quad \phi \in \mathcal{D}(\mathbb{T}), \tag{7.16}$$

as in (6.34). The difference here is that the action of T is confined to $\mathbb{T}$.

By analogy with the Fourier series for functions, define the Fourier coefficients of $T \in \mathcal{D}'(\mathbb{T})$ by

$$c_n := \left\langle T, \frac{1}{\sqrt{2\pi}} e^{-int} \right\rangle = \frac{1}{\sqrt{2\pi}} \langle T, e^{-int} \rangle. \tag{7.17}$$

Its Fourier series is then defined by

$$\sum_{n=-\infty}^{\infty} c_n \cdot \frac{1}{\sqrt{2\pi}} e^{int} = \sum_{n=-\infty}^{\infty} \frac{1}{2\pi} \langle T, e^{-int} \rangle e^{int}. \tag{7.18}$$

Note here that e^{-int} belongs to $\mathcal{D}(\mathbb{T})$, and hence T can be applied to e^{-int}. Of course, when T belongs to $L^2[-\pi,\pi]$, (7.18) agrees with the standard Fourier series for functions in $L^2[-\pi,\pi]$.

Exercise 7.2.2. Verify the facts above for $L^2[-\pi,\pi]$. Note here that $(T, e^{int}) = \langle T, \overline{e^{int}} \rangle = \langle T, e^{-int} \rangle$, where the left-hand side denotes the inner product in $L^2[-\pi,\pi]$, whereas the right-hand side denotes the action of T on an element e^{-int} with respect to the duality between $\mathcal{D}'(\mathbb{T})$ and $\mathcal{D}(\mathbb{T})$.

The following fact holds in $\mathcal{D}'(\mathbb{T})$.

Theorem 7.2.3. *The Fourier series of δ converges to δ in $\mathcal{D}'(\mathbb{T})$, i.e.,*

$$\delta = \frac{1}{2\pi} \sum_{n=-\infty}^{\infty} e^{int}. \tag{7.19}$$

That the right-hand side of (7.19) gives the Fourier series of δ is obvious from $\langle \delta, e^{-int} \rangle = 1$. As we noted in (6.25) on page 124, its n-partial sum converges to the Dirichlet kernel, and it does converge to δ on $[-\pi,\pi]$, as shown in Example 6.5.6 (p. 123). Hence (7.19) follows.

We can extend this result from $\mathbb{T}$ (or $[-\pi,\pi)$) to the whole real line $(-\infty,\infty)$ by making use of the periodicity of e^{int}. Then the left-hand side of (7.19) must be likewise extended to the whole line, and this yields a periodic distribution $\sum_{n=-\infty}^{\infty} \delta_{2n\pi}$ of period 2π (cf. Figure 7.1). This yields the following corollary.

Corollary 7.2.4 (Poisson's summation formula).

$$\sum_{n=-\infty}^{\infty} \delta_{2n\pi} = \frac{1}{2\pi} \sum_{n=-\infty}^{\infty} e^{int}. \tag{7.20}$$

Observing that e^{-int} is the Fourier transform of δ_n (cf. (7.34)), this can also be rewritten as

$$\sum_{n=-\infty}^{\infty} \delta_{2n\pi} = \frac{1}{2\pi} \mathcal{F}\left(\sum_{n=-\infty}^{\infty} \delta_n\right). \tag{7.21}$$

The latter is usually called *Poisson's summation formula.*

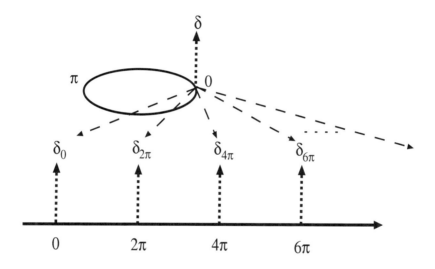

Figure 7.1. *Poisson's summation formula: δ on $[-\pi, \pi]$ appears periodically on the real line*

Theorem 7.2.3 now yields the following theorem on the convergence of the Fourier series of distributions.

Theorem 7.2.5. *The Fourier series of T in $\mathcal{D}'(\mathbb{T})$ converges to T. If T is extended to a periodic distribution, its Fourier series also converges to T in $\mathcal{D}'(\mathbb{R})$:*

$$T = \frac{1}{2\pi} \sum_{n=-\infty}^{\infty} \langle T, e^{-int}\rangle e^{int}. \tag{7.22}$$

Proof. Identity (7.19) implies $s_n = \sum_{k=-n}^{n} e^{ikt}/2\pi \to \delta$. Hence by the continuity of convolution (Theorem 6.6.8, p. 129), $T * s_n \to T * \delta$ follows. Since δ is the identity for convolution ((6.35) on page 128),

$$T = T * \delta = \lim_{n \to \infty} T * s_n = \frac{1}{2\pi} \lim_{n \to \infty} \sum_{k=-n}^{n} T * e^{ikt} = \frac{1}{2\pi} \sum_{n=-\infty}^{\infty} (T * e^{int}).$$

On the other hand, Proposition 6.6.10 (p. 131) implies

$$T * e^{int} = \langle T_{(\tau)}, e^{in(t-\tau)} \rangle = e^{int} \langle T_{(\tau)}, e^{-in\tau} \rangle.$$

Combining these together, we obtain the conclusion (7.22).   □

The derivation above also shows that the Fourier series can be represented as a convolution with a kernel distribution. Let us note that this is in full agreement with (7.9) (except for the difference between the Fejér kernel and the Dirichlet kernel here).

7.3   Fourier Transform

When a function f is not periodic, the Fourier series is no longer effective. When f possesses a period L, the discrete frequencies $\omega_n = 2n\pi/L$, $n = 0, \pm 1, \pm 2, \ldots$, give basis sinusoids $\exp(i\omega_n t)$, $n = 0, \pm 1, \pm 2, \ldots$, for expansion, but this is not applicable if f is not periodic. On the other hand, since this is intuitively a limiting case of $L \to \infty$, the resolution in the frequencies would become infinitely finer and eventually give an expansion in the limit. This forces us to consider an integral transform rather than a series expansion. This is the *Fourier transform*:

$$\mathcal{F}[f](\omega) = \int_{-\infty}^{\infty} f(t)e^{-i\omega t} dt. \tag{7.23}$$

Let us first consider the simplest case. Consider a function $f \in L^1$ integrable on the whole line. Since $|e^{i\omega t}| = 1$, we readily obtain

$$|\mathcal{F}[f](\omega)| \leq \int_{-\infty}^{\infty} |f(t)| dt = \|f\|_{L^1} < \infty.$$

That is, for functions in L^1 its Fourier transform is bounded. Also, by Lebesgue's convergence theorem [20, 22, 37] we can interchange the order of integration and the limit for $\omega \to \omega_0$ to obtain

$$\lim_{\omega \to \omega_0} \mathcal{F}[f](\omega) = \lim_{\omega \to \omega_0} \int_{-\infty}^{\infty} f(t)e^{-i\omega t} dt = \int_{-\infty}^{\infty} \lim_{\omega \to \omega_0} f(t)e^{-i\omega t} dt = \mathcal{F}[f](\omega_0).$$

Hence $\mathcal{F}[f](\omega)$ is a continuous function of ω.

$\mathcal{F}f$ is also denoted as $\hat{f}$. In a dual way, we define the *conjugate Fourier transform* by

$$\overline{\mathcal{F}} f(\omega) = \frac{1}{2\pi} \int_{-\infty}^{\infty} f(t)e^{i\omega t} dt. \tag{7.24}$$

Fundamental properties of $\mathcal{F}$ are also valid for $\overline{\mathcal{F}}$ without any change.

We list in the following theorem some basic properties for the classical Fourier transform as defined above. We omit the proof since it is available in many standard textbooks.

Theorem 7.3.1. *The Fourier transform*

$$\mathcal{F}[f](\omega) = \int_{-\infty}^{\infty} f(t)e^{-i\omega t}\,dt$$

of an integrable function f always exists and is a bounded continuous function of ω. If further $f \in C^m$ and its derivatives up to the mth order are integrable, then

$$\mathcal{F}[f^{(m)}](\omega) = (i\omega)^m \mathcal{F}[f](\omega). \tag{7.25}$$

On the other hand, if $t^m f(t)$ is integrable, then $\mathcal{F}f(\omega)$ belongs to C^m and

$$\mathcal{F}[(-it)^m f] = \hat{f}^{(m)}. \tag{7.26}$$

How should we generalize the Fourier transform to distributions? First suppose $f \in L^1, \phi \in \mathcal{D}$. Then we have

$$\langle \mathcal{F}[f], \phi \rangle = \int_{-\infty}^{\infty} \phi(\omega) \int_{-\infty}^{\infty} f(t)e^{-i\omega t}\,dt\,d\omega = \int_{-\infty}^{\infty}\int_{-\infty}^{\infty} \phi(\omega)f(t)e^{-i\omega t}\,dt\,d\omega.$$

The double integral on the right converges and is finite because

$$\iint |\phi(\omega)f(t)e^{-i\omega t}|\,dt\,d\omega \le \int_{-\infty}^{\infty} |\phi(\omega)|\,d\omega \cdot \int_{-\infty}^{\infty} |f(t)|\,dt < \infty.$$

Hence by Fubini's theorem [20, 22, 37] in Lebesgue integration theory, we can interchange the order of integration, and we have

$$\langle \mathcal{F}[f], \phi \rangle = \int_{-\infty}^{\infty} f(t) \int_{-\infty}^{\infty} \phi(\omega)e^{-i\omega t}\,d\omega\,dt = \langle f, \mathcal{F}[\phi]\rangle.$$

This suggests the definition of the Fourier transform as

$$\langle \mathcal{F}[T], \phi \rangle := \langle T, \mathcal{F}[\phi]\rangle$$

for a distribution T. Here we wish to understand $\mathcal{F}[T]$ as a distribution. In fact, the Fourier transform of ϕ is well defined as a function. Unfortunately, this is not the whole story: $\mathcal{F}\phi$ does not necessarily qualify as a test function. More precisely, $\mathcal{F}\phi$ does not necessarily belong to $\mathcal{D}$ even if $\phi \in \mathcal{D}$. In fact, according to the Paley–Wiener theorem [53, 58], the Fourier transform of $\phi \in \mathcal{D}$ is always an analytic function of ω and cannot belong to $\mathcal{D}$.[75] This would imply that $\mathcal{F}\phi$ may not be appropriate as a test function.

Actually, there is a complementary relationship between the behavior of a function as $t \to \pm\infty$ and the regularity of its Fourier transform as shown in Table 7.1. If we take ϕ in $\mathcal{D}$, then its decay as $t \to \pm\infty$ becomes too fast (indeed it becomes identically zero outside a compact interval) for its Fourier transform $\mathcal{F}\phi$ to belong to $\mathcal{D}$. To circumvent this

---

[75] $\mathcal{D}$ forces its elements to be identically 0 outside some interval, and this contradicts analyticity.

Table 7.1. *Complementary properties of the Fourier transform*

f	$\mathcal{F} f$
Rapid decay	Higher regularity
Compact support	Analytic

problem, we must relax the requirement on the test function ϕ so that $\mathcal{F}\phi$ is not so regular as to be analytic.

But even then we cannot expect $\mathcal{F}[\phi] \in \mathcal{D}$. Since the same relations hold for the conjugate Fourier transform

$$\overline{\mathcal{F}}[f](\omega) = \frac{1}{2\pi} \int_{-\infty}^{\infty} f(t) e^{i\omega t} dt, \tag{7.27}$$

in order that $\mathcal{F}[\phi]$ belong to $\mathcal{D}$, ϕ itself must be an analytic function. This in turn disqualifies ϕ from being a test function.

To remedy this inconvenience, we should abandon the hope of taking the space of test functions to be $\mathcal{D}$, and attempt to take a new space to which both ϕ and $\mathcal{F}[\phi]$ belong. This is the space $\mathcal{S}$ of rapidly decreasing functions to be introduced in the next section.

7.4   Space $\mathcal{S}$ of Rapidly Decreasing Functions

Let us start with the following definition.

Definition 7.4.1. A function ϕ defined on the real line $(-\infty, \infty)$ is said to be a *rapidly decreasing function* if it satisfies the following conditions:

1. $\phi \in C^\infty$; i.e., it is infinitely differentiable.

2. For every pair of nonnegative integer N, k,

$$\lim_{|t| \to \infty} |t^N \phi^{(k)}(t)| = 0.$$

The set of all rapidly decreasing functions is denoted by $\mathcal{S}$. $\mathcal{S}$ is clearly a vector space. To state it differently, $\mathcal{S}$ is the space of all C^∞ functions that converge to 0 as $|t| \to \infty$ faster than any polynomials.

For example, function e^{-t^2} belongs to $\mathcal{S}$.

Endow $\mathcal{S}$ with the topology defined by the following family of norms (see Appendix A):

$$\|\phi\|_{N,k} := \sup_{-\infty < t < \infty} \left| t^N \phi^{(k)}(t) \right|, \tag{7.28}$$

where N, k are arbitrary nonnegative integers. In other words, ϕ_n converges to 0 in $\mathcal{S}$ if and only if $t^N \phi^{(k)}(t)$ converges uniformly to 0 on $(-\infty, \infty)$ for arbitrary nonnegative integers N, k. Clearly $\mathcal{D} \subset \mathcal{S}$.

The following theorem holds.

Theorem 7.4.2. $\mathscr{F}\mathscr{S} \subset \mathscr{S}$; *i.e., if* $\phi \in \mathscr{S}$, *then* $\mathscr{F}[\phi] \in \mathscr{S}$. *Further, if* $\phi_n \to 0$ *in* $\mathscr{S}$, *then* $\mathscr{F}[\phi_n] \to 0$ *in* $\mathscr{S}$. *That is, the Fourier transform gives a continuous linear mapping from* $\mathscr{S}$ *into itself.*

We omit the proof. See [53, 58].

Denote by $\mathscr{S}'$ the dual space of $\mathscr{S}$. Elements of $\mathscr{S}'$ are called *tempered distributions*. Since $\mathscr{D} \subset \mathscr{S}$, $\mathscr{S}' \subset \mathscr{D}'$. For example, integrable functions and functions of at most polynomial increasing order are elements of $\mathscr{S}'$. Also, if $T \in \mathscr{S}$, its derivative T' also belongs to $\mathscr{S}'$. On the other hand, functions that increase very fast, such as e^t, do not belong to $\mathscr{S}'$.

Exercise 7.4.3. Verify the fact $e^t \notin \mathscr{S}'$.

We can now define the Fourier transform for elements in $\mathscr{S}'$ as follows.

Definition 7.4.4. Let $T \in \mathscr{S}'$. We define the *Fourier transform* $\mathscr{F}T$ and the *conjugate Fourier transform* $\overline{\mathscr{F}}T$ by

$$
\begin{aligned}
\langle \mathscr{F}T, \phi \rangle &:= \langle T, \mathscr{F}\phi \rangle, \\
\langle \overline{\mathscr{F}}T, \phi \rangle &:= \langle T, \overline{\mathscr{F}}\phi \rangle,
\end{aligned}
\tag{7.29}
$$

where $\phi \in \mathscr{S}$. $\mathscr{F}$ and $\overline{\mathscr{F}}$ both give continuous linear mappings from $\mathscr{S}'$ into itself. $\mathscr{F}T$ is also denoted by $\hat{T}$.

If the support of T is compact, then its Fourier transform takes a particularly simple form.

Theorem 7.4.5. *Let* T *be a distribution with compact support. Then* $(\mathscr{F}T)(\omega)$ *becomes a function analytic on the whole plane, i.e., an* entire *function, with respect to* ω, *and it admits the representation*

$$
[\mathscr{F}T](\omega) = \langle T, e^{-i\omega t} \rangle.
\tag{7.30}
$$

Proof. For simplicity, consider the case of locally integrable T. From the definition, it follows that

$$
\begin{aligned}
\langle \mathscr{F}T, \phi \rangle &= \langle T, \mathscr{F}\phi \rangle \\
&= \int_{-\infty}^{\infty} T(t) \int_{-\infty}^{\infty} \phi(\omega) e^{-i\omega t} d\omega \, dt \\
&= \iint \phi(\omega) T(t) e^{-i\omega t} d\omega \, dt \\
&= \int_{-\infty}^{\infty} \phi(\omega) \langle T_{(t)}, e^{-i\omega t} \rangle d\omega \\
&= \langle \langle T_{(t)}, e^{-i\omega t} \rangle, \phi \rangle.
\end{aligned}
$$

Thus we obtain (7.30). In this procedure we have used Fubini's theorem to interchange the order of integration on the third line.

The analyticity of $\mathcal{F}T$ is a consequence of the differentiability of $\langle T, e^{-i\omega t}\rangle$ (because $e^{i\omega t}$ is analytic) inside the integral up to an arbitrary order. A precise argument requires Lebesgue integration theory, and the reader is referred to standard references such as [20, 22, 37].

The general case for distributions follows similarly by extending the above argument, or by regularizing T following Proposition 6.6.9, and then extending the identity (7.30) by continuity. $\qquad\qquad\qquad\qquad\qquad\qquad\qquad\qquad\qquad\qquad\qquad\qquad\qquad\square$

For distributions with compact support, their Fourier transforms admit a simple expression as above, and they also possess the remarkable property of being entire, i.e., analytic on the whole complex plane. This relies only on the boundedness of support and does not depend on the local behavior of distributions no matter how irregular they may be. For example, as we will see below, the Fourier transform of δ is the constant function 1, and $\mathcal{F}\delta' = i\omega$, etc. They are both entire functions. On the other hand, the Fourier transform of an ordinary function does not necessarily remain an ordinary (locally integrable) function. For example, the Fourier transform of the constant function 1 is $2\pi\delta$ (Lemma 7.4.9, p. 156). This is of course not a locally integrable function. As we saw already in Table 7.1 (p. 152), the Fourier transform acts to exchange the behavior of a function at infinity and the regularity in the transformed domain. If a function exhibits a very regular behavior at infinity such as being identically zero there (i.e., having compact support) or rapidly decreasing, then its Fourier transform shows a high regularity such as being analytic or infinitely differentiable, etc. On the other hand, if a distribution has a high regularity, its Fourier transform exhibits a regular behavior at infinity. For example, the constant function 1 is entire, and its Fourier transform $\mathcal{F}1 = 2\pi\delta$ has compact support.

Let us give some examples of Fourier transforms.

Example 7.4.6. Applying Theorem 7.4.5, we easily obtain the following:

$$\mathcal{F}\delta = \langle \delta_{(t)}, e^{-i\omega t}\rangle = 1, \tag{7.31}$$

$$\mathcal{F}\delta' = \langle \delta'_{(t)}, e^{-i\omega t}\rangle = i\omega, \tag{7.32}$$

$$\mathcal{F}\delta^{(m)} = \langle \delta^{(m)}_{(t)}, e^{-i\omega t}\rangle = (i\omega)^m, \tag{7.33}$$

$$\mathcal{F}\delta_a = \langle \delta_a, e^{-i\omega t}\rangle = e^{-i\omega a}. \tag{7.34}$$

Example 7.4.7. The Fourier transform of

$$f(t) := \begin{cases} 1 & (|t| \le T_0), \\ 0 & (|t| > T_0) \end{cases} \tag{7.35}$$

is given by

$$\hat{f}(\omega) = \int_{-T}^{T} e^{-i\omega t}\, dt = \frac{e^{i\omega T} - e^{-i\omega T}}{i\omega} = \frac{2\sin\omega T}{\omega}. \tag{7.36}$$

The pole $\omega = 0$ is only superficial and is removable by cancellation with the same zero in the numerator. In fact, expanding the numerator into the Taylor series, we see that this function is indeed analytic. This is in accordance with the claim of Theorem 7.4.5 above.

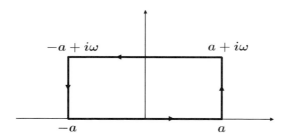

Figure 7.2. *Integral path*

The following example plays a crucial role in proving the Fourier inversion formula to be given later.

Example 7.4.8. The Fourier transform of the Gaussian distribution function $e^{-t^2/2}$ is $\sqrt{2\pi}\,e^{-\omega^2/2}$. Observe first that

$$\int_{-a}^{a} e^{-t^2/2} e^{-i\omega t}\,dt = e^{-\omega^2/2} \int_{-a}^{a} e^{-(t+i\omega)^2/2}\,dt. \tag{7.37}$$

Put $z := t + i\omega$, and consider the contour integral of the analytic function $e^{-z^2/2}$ along the closed path shown in Figure 7.2. By Cauchy's integral theorem this is 0. Hence by noting

$$\int_{-a}^{a} e^{-(t+i\omega)^2/2}\,dt = \int_{-a+i\omega}^{a+i\omega} e^{-z^2/2}\,dz,$$

we obtain

$$\int_{-a}^{a} e^{-(t+i\omega)^2/2}\,dt = \int_{-a}^{a} e^{-t^2/2}\,dt + \int_{-a+i\omega}^{-a} e^{-z^2/2}\,dz + \int_{a}^{a+i\omega} e^{-z^2/2}\,dz.$$

Here the second and third terms on the right-hand side approach 0 as $a \to \infty$. In fact, for the third term, we have

$$\left| \int_{a}^{a+i\omega} e^{-z^2/2}\,dz \right| = \left| \int_{0}^{\omega} e^{-(a^2-y^2)/2-iay}\,i\,dy \right| \le |\omega| \left| e^{-(a^2-\omega^2)/2} \right|,$$

and it approaches 0 as $a \to \infty$ for each fixed ω. The same is true for the second term. Hence, by (7.37), we obtain

$$\int_{-\infty}^{\infty} e^{-t^2/2} e^{i\omega t}\,dt = e^{-\omega^2/2} \int_{-\infty}^{\infty} e^{-t^2/2}\,dt.$$

On the other hand, the last integral is equal to the Gaussian integral (6.27) on page 124:

$$\int_{-\infty}^{\infty} e^{-t^2/2}\,dt = \sqrt{2\pi}.$$

Hence

$$\mathcal{F}[e^{-t^2/2}] = \sqrt{2\pi}\,e^{-\omega^2/2}. \tag{7.38}$$

Note also that we easily obtain

$$\mathcal{F}[e^{-\alpha t^2/2}] = \sqrt{\frac{2\pi}{\alpha}} e^{-\omega^2/(2\alpha)}, \quad \alpha > 0, \tag{7.39}$$

by a change of variables.

Formula (7.39) now yields the following lemma, which will be used in the proof of the Fourier inversion formula.

Lemma 7.4.9.

$$\mathcal{F}[1] = 2\pi\delta. \tag{7.40}$$

Proof. From (7.39) we obtain

$$\mathcal{F}[e^{-\alpha t^2/2}] = \sqrt{\frac{2\pi}{\alpha}} \exp\left(-\frac{\omega^2}{2\alpha}\right) = 2\pi\sqrt{\frac{\beta}{2\pi}} \exp\left(-\frac{\beta\omega^2}{2}\right), \quad \beta := 1/\alpha.$$

According to Example 6.5.7 (p. 124), $e^{-\alpha t^2/2}$ approaches 1 as $\alpha \to 0$. On the other hand, the right-hand side approaches $2\pi\delta$ as $\alpha \to 0$ (hence $\beta \to \infty$) again by Example 6.5.7. Admitting the continuity of the Fourier transform, we obtain $\mathcal{F}[1] = 2\pi\delta$. $\qquad\square$

7.5   Fourier Transform and Convolution

A remarkable property of the Fourier transform (also shared with the Laplace transform to be discussed in the next chapter) is that *it converts convolution to the ordinary product*. This enables us to solve linear differential equations by algebraically inverting the pertinent differential operators in the transformed domain, which is otherwise to be inverted in the sense of convolutional product.

Let us roughly outline how it goes. In view of the definition (6.34) (p. 127) of convolution, we have

$$\begin{aligned}
\mathcal{F}[T * S](\omega) &= \langle T * S, e^{-i\omega t}\rangle \\
&= \langle T_{(\xi)} S_{(\eta)}, e^{-i\omega(\xi+\eta)}\rangle \\
&= \langle T_{(\xi)} S_{(\eta)}, e^{-i\omega\xi} e^{-i\omega\eta}\rangle \\
&= \langle T_{(\xi)}, e^{-i\omega\xi}\rangle \langle S_{(\eta)}, e^{-i\omega\eta}\rangle \\
&= \mathcal{F}[T](\omega)\mathcal{F}[S](\omega).
\end{aligned}$$

It is easy to justify this for $T, S \in L^1$. When both T and S have compact support, it is still easy to guarantee. Actually, if one of T, S has compact support and the other belongs to $\mathcal{S}'$, then the above equality is still known to hold. But this equality is not necessarily always valid. In general, we must impose some restrictions on the regularity of distributions for their product to be well defined; there can be such cases where $\mathcal{F}[T] \cdot \mathcal{F}[S]$ is not well defined. For example, the product of δ with itself is not quite well defined, and for such cases the formula above will be meaningless.

Summarizing the above, we obtain the following theorem.

Theorem 7.5.1. *The Fourier transform converts convolution to multiplication. Dually, multiplication is converted to convolution. That is,*

$$\mathcal{F}[T * S] = \mathcal{F}[T] \cdot \mathcal{F}[S],$$
$$\overline{\mathcal{F}}[T * S] = 2\pi \overline{\mathcal{F}}[T] \cdot \overline{\mathcal{F}}[S],$$
$$\mathcal{F}[T S] = \frac{1}{2\pi} \mathcal{F}[T] * \mathcal{F}[S],$$
$$\overline{\mathcal{F}}[T S] = \overline{\mathcal{F}}[T] * \overline{\mathcal{F}}[S].$$

Proof. The first formula is obtained as described above. For the second assertion on multiplication, we use the *Fourier inversion formula* (to be proven in Theorem 7.5.4 below) as follows: Apply $\overline{\mathcal{F}}$ to both sides of $\mathcal{F}[T * S] = \mathcal{F}[T] \cdot \mathcal{F}[S] = \hat{T}\hat{S}$. Using $\overline{\mathcal{F}}[\mathcal{F}[T]] = T$, we obtain $\overline{\mathcal{F}}[\hat{T} \cdot \hat{S}] = T * S = \overline{\mathcal{F}}[\hat{T}] * \overline{\mathcal{F}}[\hat{S}]$. Now $\mathcal{F}[T S] = \frac{1}{2\pi}\mathcal{F}[T] * \mathcal{F}[S]$ follows similarly. □

Remark 7.5.2. A remark on notational convention may be necessary here. We have defined the Fourier transform by $\mathcal{F}[T] := \langle T, e^{-i\omega t} \rangle$ and the inverse (conjugate) Fourier transform by $\overline{\mathcal{F}}[T] := \frac{1}{2\pi}\langle T, e^{i\omega t} \rangle$, leaving the normalization factor $1/2\pi$ only to the inverse Fourier transform. This unfortunately induces an asymmetry in the formulas above (and various others, too). To avoid such asymmetry, we can either distribute the factor $1/\sqrt{2\pi}$ evenly to the Fourier and inverse Fourier transforms or, as per Schwartz, include this factor into the exponent of the exponential as $e^{-2\pi i \omega t}$ in place of $e^{-i\omega t}$. This is attractive but inconsistent with the convention in Laplace transform theory. For signals in the time domain the Laplace transform is more frequently used, and it is rather common practice to put this factor $1/2\pi$ on the inverse Fourier transform. We here adopt this convention for consistency with such practice. The reader should note, however, that this induces asymmetry in some formulas.

Theorem 7.5.1 now yields the following corollary.

Corollary 7.5.3. *Let* $T \in \mathscr{S}'$. *Then*

$$\mathcal{F}[T^{(m)}] = (i\omega)^m \mathcal{F} T, \tag{7.41}$$
$$\mathcal{F}[(-it)^m T] = [\mathcal{F} T]^{(m)}. \tag{7.42}$$

Proof. Theorem 7.5.1 together with Example 7.4.6 yields

$$\mathcal{F}[T^{(m)}] = \mathcal{F}[\delta^{(m)} * T] = \mathcal{F}[\delta^{(m)}]\mathcal{F}[T] = (i\omega)^m \mathcal{F}[T].$$

A similar formula holds for $\overline{\mathcal{F}}$. Applying $\mathcal{F}$ to both sides of the above and using the Fourier inversion formula $\mathcal{F}\overline{\mathcal{F}}[T] = T$ (Theorem 7.5.4), we obtain (7.42).[76] □

---

[76]We note here that while the first formula (7.41) is used in the proof of Theorem 7.5.4, the second formula (7.42) is not used, so this is free from tautology.

7.5.1   Fourier Inversion Formula

Let us now prove that $\mathcal{F}$ and $\overline{\mathcal{F}}$ give the inverse of each other in $\mathscr{S}'$. This will also be used for the derivation of the inversion formula for the Laplace transform.

Theorem 7.5.4. *In space $\mathscr{S}'$, $\mathcal{F}$ and $\overline{\mathcal{F}}$ give the inverse of each other. That is, $\overline{\mathcal{F}}\mathcal{F} = \mathcal{F}\overline{\mathcal{F}} = I$. This also shows that the Fourier transform is a bijection.*

In this sense, the conjugate Fourier transform $\overline{\mathcal{F}}$ is also referred to as the *inverse Fourier transform*.

Proof. Let us start by showing this for elements in $\mathscr{S}$. We prove $\overline{\mathcal{F}}\mathcal{F}\phi = \phi$. Put $\psi(\omega) := \mathcal{F}\phi$. From the definition, we have

$$\overline{\mathcal{F}}[\psi](\eta) = \frac{1}{2\pi} \int_{-\infty}^{\infty} e^{i\omega\eta} \psi(\omega) d\omega. \tag{7.43}$$

Observe that

$$e^{i\omega\eta} \psi(\omega) = \int_{-\infty}^{\infty} e^{-i\omega(t-\eta)} \phi(t) dt$$

$$= \int_{-\infty}^{\infty} e^{-i\omega t} \phi(t+\eta) dt = \mathcal{F}[\phi(\cdot+\eta)](\omega).$$

Substitute this into (7.43), and use $\mathcal{F}[1] = 2\pi\delta$ (Lemma 7.4.9, p. 156) to obtain

$$\overline{\mathcal{F}}[\psi](\eta) = \frac{1}{2\pi} \int_{-\infty}^{\infty} \mathcal{F}[\phi(\cdot+\eta)] d\omega$$

$$= \frac{1}{2\pi} \langle 1, \mathcal{F}[\phi(\cdot+\eta)] \rangle$$

$$= \frac{1}{2\pi} \langle \mathcal{F}[1], \phi(\cdot+\eta) \rangle$$

$$= \frac{1}{2\pi} \langle 2\pi\delta, \phi(\cdot+\eta) \rangle$$

$$= \phi(\eta).$$

This shows $\overline{\mathcal{F}}\mathcal{F}\phi = \phi$ for $\phi \in \mathscr{S}$. Similarly, $\mathcal{F}\overline{\mathcal{F}}\phi = \phi$.

We need only dualize this to obtain the result for $\mathscr{S}'$. That is, from the definition of the Fourier transform for $\mathscr{S}'$, we obtain for $T \in \mathscr{S}'$, $\phi \in \mathscr{S}$

$$\langle \overline{\mathcal{F}}\mathcal{F}[T], \phi \rangle = \langle \mathcal{F}[T], \overline{\mathcal{F}}[\phi] \rangle = \langle T, \mathcal{F}\overline{\mathcal{F}}[\phi] \rangle = \langle T, \phi \rangle.$$

Hence $\overline{\mathcal{F}}\mathcal{F}T = T$. The proof for $\mathcal{F}\overline{\mathcal{F}}T = T$ is entirely analogous.   □

We are ready to give the following Plancherel–Parseval theorem, which generalizes Parseval's identity (7.15) to the Fourier transform.

Theorem 7.5.5 (Plancherel–Parseval). *Let* $f, g \in L^2(-\infty, \infty)$, *and let* $\hat{f}$ *and* $\hat{g}$ *denote their Fourier transforms. Then*

$$\int_{-\infty}^{\infty} f(t)\overline{g}(t)dt = \frac{1}{2\pi}\int_{-\infty}^{\infty} \hat{f}(\omega)\overline{\hat{g}}(\omega)d\omega. \tag{7.44}$$

In particular,

$$\|f\|_{L^2} = \frac{1}{2\pi}\|\hat{f}\|_{L^2}. \tag{7.45}$$

Proof. Let us first assume that f, g both belong to $\mathcal{S}$. Then we have

$$\begin{aligned}
\int_{-\infty}^{\infty} f(t)\overline{g}(t)dt &= \frac{1}{2\pi}\int_{-\infty}^{\infty}\overline{g}(t)dt\int_{-\infty}^{\infty}\hat{f}(\omega)e^{it\omega}d\omega \\
&= \frac{1}{2\pi}\int_{-\infty}^{\infty}\hat{f}(\omega)d\omega\int_{-\infty}^{\infty}\overline{g}(t)e^{it\omega}dt \\
&= \frac{1}{2\pi}\int_{-\infty}^{\infty}\hat{f}(\omega)d\omega\int_{-\infty}^{\infty}\overline{g(t)e^{-it\omega}}dt \\
&= \frac{1}{2\pi}\int_{-\infty}^{\infty}\hat{f}(\omega)\overline{\hat{g}}(\omega)d\omega.
\end{aligned}$$

Here we have used Fubini's theorem to change the order of integration, and we have also used $\overline{\mathcal{F}}\mathcal{F} = I$ by Theorem 7.5.4. Hence (7.45) holds for $\mathcal{S}$. Since $\mathcal{S}$ is dense[77] in L^2, (7.44) holds by continuity for the whole L^2.

Finally, putting $g := f$, we obtain (7.45). $\qquad\qquad\qquad\qquad\qquad\qquad\Box$

Remark 7.5.6. The identity (7.45) is often also referred to as Parseval's identity.

7.6 Application to the Sampling Theorem

As an application, let us prove the so-called sampling theorem in signal processing.

A typical problem in digital signal processing is one of the following type: Suppose that for a continuous-time signal f, its sampled values $\{f(nh)\}_{n=-\infty}^{\infty}$ with a fixed sampling period h are observed. Reconstruct the original signal f precisely from this sampled signal. Of course, it is hardly expected that f can be uniquely recovered from these sampled values $\{f(nh)\}_{n=-\infty}^{\infty}$ under weak hypotheses such as f being continuous. On the other hand, if we impose a stronger condition on f, it may be expected to be uniquely recoverable from the sampled data. A primitive example is Lagrange interpolation, where one gives n sampled points and reconstructs a unique polynomial of degree $n - 1$. Here we deal with infinitely many, equally spaced, sampled points. The theorem we will now give is the *sampling theorem* usually attributed to Shannon.[78]

---

[77]This is easy to show. In fact, $\mathcal{D}$ is dense in L^2. To see this, just invoke the regularization as shown in Proposition 6.6.9, but show the convergence in L^2 rather than in $\mathcal{D}'$.

[78]As is quite normal for theorems of such a well-known name, there are a number of people involved in the discovery of this theorem—for example, Whittaker and Nyquist. For details, see [71]. (Shannon [54] himself quotes Whittaker and Nyquist.)

Let us state and prove this theorem as an application of Poisson's summation formula. Recall now Poisson's summation formula (Corollary 7.2.4, p. 149):

$$\sum_{n=-\infty}^{\infty} \delta_{2n\pi} = \frac{1}{2\pi} \mathcal{F}\left(\sum_{n=-\infty}^{\infty} \delta_n\right).$$

Let us first convert this into a formula for functions on $[-\pi/h, \pi/h]$ rather than on $[-\pi, \pi]$ as given before. The Fourier series expansion on this interval takes the form

$$\psi(\omega) = \sum_{n=-\infty}^{\infty} c_n e^{inh\omega},$$

$$c_n = \frac{h}{2\pi} \int_{-\pi/h}^{\pi/h} \psi(\omega) e^{-inh\omega} d\omega.$$

Here we have changed the variable to ω for convenience. As in Example 6.5.6 on page 123,

$$\frac{h}{2\pi} \sum_{n=-\infty}^{\infty} e^{inh\omega} = \delta$$

holds on $[-\pi/h, \pi/h]$. Extending this to the whole real line $(-\infty, \infty)$ with period $2\pi/h$, we obtain

$$\sum_{n=-\infty}^{\infty} \delta_{2n\pi/h} = \frac{h}{2\pi} \sum_{n=-\infty}^{\infty} e^{inh\omega}.$$

Using $e^{inh\omega} = \mathcal{F}[\delta_{-nh}]$ ((7.34), p. 154) and changing the index as $n \leftrightarrow -n$, we obtain Poisson's summation formula in the general form:

$$\sum_{n=-\infty}^{\infty} \delta_{2n\pi/h} = \frac{h}{2\pi} \mathcal{F}\left(\sum_{n=-\infty}^{\infty} \delta_{nh}\right). \tag{7.46}$$

Let now f be a continuous function that is Fourier transformable. Take the product $f(t)\sum_{n=-\infty}^{\infty} \delta_{nh}$ of f with $\sum_{n=-\infty}^{\infty} \delta_{nh}$ and consider its Fourier transform. This corresponds to the series of delta functions $\sum_{n=-\infty}^{\infty} f(nh)\delta_{nh}$ generated by the sampled values $\{f(nh)\}$ of f at discrete points nh, $n = 0, \pm 1, \pm 2, \dots$. This is often referred to as an *impulse modulated sequence* of f. Since the Fourier transform of a product becomes the convolution of the Fourier transform of each component (Theorem 7.5.1, p. 157), we obtain

$$\mathcal{F}\left(f(t)\sum_{n=-\infty}^{\infty} \delta_{nh}\right) = \frac{1}{h}\hat{f}(\omega) * \left(\sum_{n=-\infty}^{\infty} \delta_{2n\pi/h}\right)$$

$$= \frac{1}{h}\sum_{n=-\infty}^{\infty} \hat{f} * \delta_{2n\pi/h}$$

$$= \frac{1}{h}\sum_{n=-\infty}^{\infty} \hat{f}(\omega - 2n\pi/h) \tag{7.47}$$

(by Example 6.6.4, p. 128).

In other words, the original spectrum[79] of f is repeated periodically with period $2n\pi/h$. This is not surprising because the sinusoid $\sin\omega t$ takes the values $\sin n\omega h$ at sampled points nh, $n = 0,\pm1,\pm2,\ldots$, but this situation is no different for $\sin(\omega + 2m\pi/h)t$ whose frequency domain component is shifted by $2m\pi/h$ from that of f; that is, the latter signals take the same values $\sin n\omega h$ at sampled points. Hence it is impossible to distinguish these sinusoids based on the sampled values at these points. In other words, the sampling with period h does not have a resolution fine enough to distinguish these two sinusoids that are apart by $2m\pi/h$ in the frequency domain. Thus without a further condition, we cannot recover f based on the information at sampled points.

A condition to avoid this problem is to require that $\hat{f}(\omega)$ not overlap its shifts. For example, if f is fully band-limited and its spectrum $\hat{f}(\omega)$ does not overlap its shifts by $2m\pi/h$, then this condition is satisfied. That is, we require $\mathrm{supp}\,\hat{f} \subset (-\pi/h, \pi/h)$ (if we are to limit $\hat{f}$ to the lowest frequency range). So let us now assume

$$\mathrm{supp}\,\hat{f} \subset [-\omega_0, \omega_0] \subset (-\pi/h, \pi/h). \tag{7.48}$$

However, in contrast to this assumption, formula (7.47) still contains high frequency components $\hat{f}(\omega + 2n\pi/h)$. Hence it is not possible to uniquely recover $f(t)$ out of $\sum_{n=-\infty}^{\infty} \hat{f}(\omega - 2n\pi/h)$ containing such high frequency components. So it suffices to extract only the fundamental frequency range $\hat{f}(\omega)$, $\omega \in [-\omega_0, \omega_0]$, by applying some "filter" to the data. For this, we need only multiply $\sum_{n=-\infty}^{\infty} \hat{f}(\omega - 2n\pi/h)$ by the function

$$\alpha(\omega) := \begin{cases} 1 & (|\omega| \le \pi/h), \\ 0 & (|\omega| > \pi/h), \end{cases} \tag{7.49}$$

and then only the fundamental frequency components remain. Apply the inverse Fourier transform to this, and use Theorem 7.5.1, (7.47), and (7.34) to obtain

$$f(t) = \overline{\mathcal{F}}[\hat{f}(\omega)] = \overline{\mathcal{F}}\left[\alpha(\omega) \sum_{n=-\infty}^{\infty} \hat{f}(\omega - 2n\pi/h)\right] \tag{7.50}$$

$$= h\overline{\mathcal{F}}[\alpha] * \left(\sum_{n=-\infty}^{\infty} f(nh)\delta_{nh}\right) \tag{7.51}$$

$$= \sum_{n=-\infty}^{\infty} f(nh)\,\mathrm{sinc}(t - nh). \tag{7.52}$$

Here $\mathrm{sinc}\,t := h\overline{\mathcal{F}}[\alpha]$. Example 7.4.7 (p. 154) readily yields

$$\mathrm{sinc}\,t = \frac{\sin(\pi t/h)}{\pi t/h}.$$

---

[79] not rigorously defined, but let us understand it as $\hat{f}(\omega)$.

Substituting this into (7.52) yields the *sampling formula*

$$f(t) = \sum_{n=-\infty}^{\infty} f(nh) \frac{\sin \pi (t/h - n)}{\pi (t/h - n)}. \tag{7.53}$$

Summarizing, we obtain the following *sampling theorem*.

Theorem 7.6.1. *Let $f(t)$ be a continuous function belonging to $\mathcal{S}'$. Suppose that its Fourier transform $\hat{f} = \mathcal{F}f$ is band-limited as $\operatorname{supp} \hat{f} \subset (-\pi/h, \pi/h)$. Then*

$$f(t) = \sum_{n=-\infty}^{\infty} f(nh) \frac{\sin \pi (t/h - n)}{\pi (t/h - n)}.$$

Remark 7.6.2. Formula (7.47) appears often in signal processing and sampled-data control theory. It means that when we sample a continuous-time signal $f(t)$ and form the so-called impulse modulated signal[80] $\sum f(nh)\delta_{nh}$, its Fourier transform becomes the sum of all the shifted copies moved by $2n\pi/h$ of the original spectrum $\hat{f}(\omega)$. This phenomenon, namely, when the original spectrum in a high frequency region folds back down to a lower frequency range as above, is called *aliasing*. When one tries to recover (i.e., *decode*) the original signal from the sampled one having aliasing, the result will be different due to such extra frequency components. The distortion due to such an effect is called the aliasing distortion. To avoid the aliasing distortion it is necessary to filter out the aliased extra frequency components. The ideal filter for this is the Shannon filter, which ideally extracts the relevant frequency components, employed in the sampling theorem above. On the other hand, this filter is not causal, i.e., it requires future sampled values for recovering the present signal value, and hence it is physically not realizable. Thus it requires various practical modifications. For details, see the literature on signal processing, e.g., [12, 54, 59, 63].

The aliasing phenomenon can be observed in various situations. For cinemas, the frame-advance speed is 24 (or 30) frames/sec, so we may regard them as having the sampling period $1/24$ (or $1/30$, respectively) seconds. The reader may have seen a carriage in a movie where the wheels deceptively look to be spinning more slowly than they really are, or even stopping. This is due to aliasing: the actual turning period of the wheel spokes is much faster than $1/24$ seconds, but with aliasing it looks to match or is closer to this speed, and hence it appears to synchronize with this speed. Likewise, if it is actually once synchronized and then reduces its speed, it looks to be turning backward, since our eyes may deceptively attempt to track the motion from the aliased frequency.

Another interesting occurrence is a device called stroboscope. This device has cocentric black and white stripe patterns, and it is utilized to observe high-frequency rotations in a stationary motion, taking advantage of aliasing. It is usually attached to analog record-player turntables as an accessory to check the right turning speed. In this case, it makes use of the facts that the alternating power supply has the frequency (either 50 or 60 Hz) and electric light bulbs switch on and off in accordance with the same frequency, thereby inducing an sampling effect. If the rotating speed of the turntable is correct (say, 33 and

---

[80]In signal processing, multiplying the target signals by a particular fixed signal, especially sinusoids, is called *modulation*. Since this fixed signal is taken to be the train of impulses $\sum \delta_{nh}$ here, it is called the *impulse modulation*.

1/3 rotations/min.), one of the cocentric striped belts should look stationary because of aliasing.[81]

Problems

1. Extend the function

$$f(t) := \begin{cases} -t - \pi & (-\pi < t < 0), \\ -t + \pi & (0 < t < \pi) \end{cases}$$

   to a function of period 2π, and expand it to a Fourier series.

2. Differentiate (in the sense of distributions) termwise the Fourier series found in problem 1 above.

---

[81]But this device disappeared as quartz-locked controlled turntables were introduced since their precision of rotation had exceeded the precision of the frequency of the home electric power supply.

Chapter 8

Laplace Transform

Let us now introduce the Laplace transform for distributions. As we have seen in Chapter 6, as far as the operational calculus is concerned for linear ordinary differential equations with constant coefficients, the Laplace transform is not necessarily an indispensable tool. Nonetheless, what makes the Laplace transformation so important is that it provides not only a convenient symbolic tool for exhibiting the algebraic structure of analytical operations but also the complex-analytic properties the transformed functions induces in theory and applications.

Extending the Laplace transform to distributions, we witness the full power of the notion of distributions. For example, consider the differential operator δ' and its Laplace transform s. While the complex function $f(s) = s$ is a very ordinary polynomial function, it cannot be the Laplace transform of any ordinary function. As long as we stay within the Laplace transforms of ordinary functions, we will encounter the inconvenience that $1/s$ can be well handled as a Laplace transform, whereas its inverse function s cannot. Introducing the Laplace transform of distributions, we can fully overcome this difficulty and freely use these objects.

8.1 Laplace Transform for Distributions

Let us introduce the Laplace transform for distributions. Let T be a distribution with support contained in $[0, \infty)$. The function

$$\mathcal{L}[T](s) := \langle T, e^{-st} \rangle, \tag{8.1}$$

regarded as a function in the complex variable s, where it exists,[82] is called the *Laplace transform* of T. To express this formula more precisely, we introduce the notation $\exp_\lambda(t) := e^{\lambda t}$. Then formula (8.1) can be rephrased more precisely as

$$\mathcal{L}[T] : \mathbb{C} \to \mathbb{C} : s \mapsto \langle T, \exp_{-s}(\cdot) \rangle. \tag{8.2}$$

---

[82]As we see below, this is not necessarily well defined for all s, nor is it guaranteed to be well defined for any T.

In what follows, $\mathcal{L}[T](s)$ may also be denoted by $\hat{T}(s)$.[83]

Some remarks are in order: If f is a locally integrable function having support contained in $[0, \infty)$, then the right-hand side of (8.1) becomes

$$\int_0^\infty f(t) e^{-st} dt, \tag{8.3}$$

and hence this definition is certainly a generalization of the classical definition of the Laplace transformation. Note, however, that (8.1) is taken over the whole real line $(-\infty, \infty)$, in contrast to (8.3), where the domain of integration is the half line $[0, \infty)$. But the latter is merely a consequence of the fact that the support of f is contained in $[0, \infty)$. It is indeed more convenient and advantageous to consider Laplace transforms over the whole real line $(-\infty, \infty)$ rather than over $[0, \infty)$, for example, when we consider transform formulas concerning differentiations, etc.; when there exists a discontinuity at the origin, there arises indeed a difference in the obtained formulas, and the approach here gives a simpler and more transparent formula. This will be discussed in subsection 8.1.2.

Definition (8.1) does not necessarily make sense for every distribution T. For example, for the function $f(t) = e^{t^2}$, (8.1) never converges for any s. For this definition to make sense, the growth order of T should not be too large—in fact, at most of exponential order. We will not delve into a detailed discussion of this subject, as it is beyond the scope of this book. In many practical cases, T may be a function of at most exponential growth plus distributional singularities in a bounded region, and we will not encounter difficulty ensuring the existence of Laplace transforms.

In particular, when distribution T has bounded support, its Laplace transform always exists. Also, if $T(t)$ is a locally integrable function satisfying

$$|T(t)| \le C e^{\sigma t}$$

for some $C, \sigma > 0$, then we readily see that its Laplace transform exists for $\operatorname{Re} s > \sigma$.

In general, there is not much sense in restricting the support of T to $[0, \infty)$. In what follows, we assume only that the support of T is bounded on the left; that is, $\operatorname{supp} T \subset [a, \infty)$ for some real a (a can be negative). The Laplace transform defined over $(-\infty, \infty)$ as given here is called the *two-sided Laplace transform* or the *bilateral Laplace transform*, but we simply use the term Laplace transform.

8.1.1   Examples of Laplace Transforms

The following examples can be easily seen.

1. $\mathcal{L}[\delta] = 1$.

2. $\mathcal{L}[\delta^{(m)}] = s^m$.

3. [Advance–delay operator] $\mathcal{L}[\delta_a] = e^{-as}$.

4. [Heaviside function] $\mathcal{L}[H(t)](s) = \frac{1}{s}$.

---

[83]The Fourier transform $\mathcal{F}[T]$ of T is also denoted by $\hat{T}$. However, usually the context demands which is meant, and they both obey the same kind of algebraic rules. When a precise distinction is necessary, we write $\mathcal{F}[T]$ and $\mathcal{L}[T]$.

5. $\mathcal{L}\left[\frac{t^{n-1}}{(n-1)!}H(t)\right](s) = \frac{1}{s^n}$.

6. $\mathcal{L}\left[H(t)e^{\lambda t}\right](s) = \frac{1}{s-\lambda}$.

For example,

$$\mathcal{L}[\delta'](s) = \langle \delta', e^{-st} \rangle = -\langle \delta, -se^{-st} \rangle = s.$$

Also,

$$\mathcal{L}[H(t)](s) = \int_0^\infty e^{-st}\,dt = \left[-\frac{1}{s}e^{-st}\right]_0^\infty = \frac{1}{s}.$$

From these we see that the Laplace transform $1/s$ of the Heaviside function $H(t)$ is the inverse (with respect to the ordinary product of complex functions) of the Laplace transform s of δ'. On the other hand, since we know from Example 6.3.2 (p. 117) that

$$H(t) * \delta' = H' = \delta,$$

$H(t)$ gives the inverse of δ' with respect to convolution. In the usual treatment of Laplace transforms, while $1/s$ (or convolution with $H(t)$) represents the integration operator, and s represents the differentiation, the corresponding convolution kernel (which is to be understood as δ') is missing since it is not a function, and hence the correspondence is not complete. The advantage of considering distributions here is that we can recognize δ' as a concrete entity and then s can be briefly regarded as the Laplace transform of δ'.

Example 8.1.1. Let us compute the Laplace transform of the sinusoid $\sin \omega t$. We must note, however, that this Laplace transform is usually considered over $[0, \infty)$, and hence in view of our convention of adopting the bilateral Laplace transform, it should be understood as the Laplace transform of $H(t)\sin \omega t$.

Since by Euler's formula,

$$\sin \omega t = \frac{e^{i\omega t} - e^{-i\omega t}}{2i},$$

it follows from fact 6 above that

$$\mathcal{L}[H(t)\sin \omega t] = \frac{1}{2i}\left(\frac{1}{s-i\omega} - \frac{1}{s+i\omega}\right) = \frac{\omega}{s^2+\omega^2}.$$

8.1.2 Laplace Transforms and Convolution

Let S and T be two distributions with support bounded on the left, i.e., elements of $\mathcal{D}'_+$, and suppose that $\hat{S}(s)$ and $\hat{T}(s)$ exist for $\mathrm{Re}\,s > a_S$ and $\mathrm{Re}\,s > a_T$, respectively. Then $S * T$ is well defined by the hypothesis on the supports, and in the common region $\mathrm{Re}\,s > \max\{a_S, a_T\}$, where both Laplace transforms exist, the following identities hold:

$$\begin{aligned}
\widehat{S * T}(s) &= \langle S_\xi, \langle T_\eta, e^{-s(\xi+\eta)} \rangle \rangle \\
&= \langle S_\xi, e^{-s\xi}\langle T_\eta, e^{-s\eta} \rangle \rangle \\
&= \langle S_\xi, e^{-s\xi} \rangle \langle T_\eta, e^{-s\eta} \rangle \\
&= \hat{S}(s)\hat{T}(s).
\end{aligned}$$

This means that *the Laplace transform of a convolution is the product of Laplace transforms*. That is, we have the following theorem.

Theorem 8.1.2. *Suppose that $\hat{S}(s)$ and $\hat{T}(s)$ exist for $\operatorname{Re} s > a_S$ and $\operatorname{Re} s > a_T$, respectively. Then $\widehat{S * T}(s)$ exist for $\operatorname{Re} s > \max\{a_S, a_T\}$ and $\widehat{S * T}(s) = \hat{S}(s)\hat{T}(s)$.*

The following corollary is a remarkable consequence of introducing Laplace transforms for distributions.

Corollary 8.1.3. *Suppose that $\mathcal{L}[T](s)$ exists on a half plane $\operatorname{Re} s > a$, and $T^{(m)}$ denotes the mth order derivative in the sense of distributions. Then*

$$\mathcal{L}[T^{(m)}](s) = s^m \mathcal{L}[T](s), \tag{8.4}$$

where the right-hand side exists also for $\operatorname{Re} s > a$.

Proof. Recall that $T^{(m)} = T * \delta^{(m)}$ (Example 6.6.5, p. 128) and also that $\mathcal{L}[\delta^{(m)}](s) = s^m$ (subsection 8.1.1, p. 166). Then Theorem 8.1.2 readily yields

$$\mathcal{L}[T^{(m)}](s) = \mathcal{L}[\delta^m](s)\mathcal{L}[T](s) = s^m \mathcal{L}[T](s). \qquad \square$$

Example 8.1.4. Let us again consider the Heaviside function $H(t)$. Recall that

$$H'(t) = \delta$$

in the sense of distributions (Example 6.3.2, p. 117). It follows that $\mathcal{L}[H'] = s \cdot (1/s) = 1 = \mathcal{L}[\delta]$, as expected, consistent with Corollary 8.1.3. The reader should be careful not to be trapped by the following confusion: since $H(t)$ is nonzero only on $[0, \infty)$, one can consider its Laplace transform on $[0, \infty)$, and since $H'(t) \equiv 0$ there, one should have $0 = \mathcal{L}[H'] = s \cdot (1/s) = 1$, which is a contradiction.[84]

In order to avoid this difficulty, the conventional treatment of Laplace transforms of functions introduces the correction term $f(0)$ as follows:

$$\mathcal{L}[f'] = s\mathcal{L}[f](s) - f(0). \tag{8.5}$$

Here f' is understood in the ordinary sense on the positive real axis. But this correction term $f(0)$ arises from no objects other than the discontinuity of $f(t)$ at the origin when we extend it as 0 for $t < 0$, and this discontinuity $f(0)\delta$ should be absorbed into the derivative f' taken in the sense of distributions; recall Proposition 6.3.3 on page 118. Taking the Laplace transform $\mathcal{L}[f(0)\delta]$ and moving it to the right-hand side, we obtain nothing but (8.5). The reader will no doubt appreciate the beauty of simplification of such computations by introducing the differentiation of distributions and considering the Laplace transforms over the whole real line $(-\infty, \infty)$.

---

[84]The Laplace transform can be taken on $[0, \infty)$, but it is not justified to differentiate it on $[0, \infty)$; what we are dealing with is the bilateral Laplace transform, and one should differentiate functions in the sense of distributions over the whole real line.

Example 8.1.5. Let us apply Corollary 8.1.3 to $\mathcal{L}[H(t)\sin\omega t] = \omega/(s^2+\omega^2)$, considered in Example 8.1.1. Proposition 6.3.3 implies

$$\frac{d}{dt}(H(t)\sin\omega t) = H(t)\omega\cos\omega t + (H(0)\sin(\omega\cdot 0))\delta = \omega H(t)\cos\omega t.$$

Hence

$$\mathcal{L}[H(t)\cos\omega t] = \frac{s}{\omega}\mathcal{L}[H(t)\sin\omega t] = \frac{s}{s^2+\omega^2}.$$

Example 8.1.6. Let us solve the ordinary differential equation

$$\dot{x}(t) = Ax(t) + Bu(t), \quad t > 0, \tag{8.6}$$

$$x(0) = x_0. \tag{8.7}$$

Here $x(t)$ and $u(t)$ are vector-valued functions with their values in $\mathbb{R}^n$ and $\mathbb{R}^m$, respectively, A and B are constant matrices of appropriate size, and $u(t)$ has growth order of at most exponential order and hence is Laplace transformable.

Suppose now that there exists a solution $x(t)$ that is Laplace transformable, and we attempt to find it using Laplace transformation. We first extend $x(t)$ as being identically 0 for $t < 0$, so that we can consider the bilateral Laplace transform; we do the same for $u(t)$. Let us temporarily denote by $[\dot{x}]$ the derivative of $x(t)$ for $t > 0$ in the ordinary sense. Then by Proposition 6.3.3 (p. 118) the derivative $\dot{x}$ of $x(t)$ in the sense of distributions is $[\dot{x}] + x_0\delta$.

On the other hand, since (8.6) is considered only for $t > 0$, the derivative on the left-hand side should of course be understood as $[\dot{x}]$. Thus, if we extend (8.6) to the whole real line, then we see that the Laplace transform of the left-hand side should become $\mathcal{L}([\dot{x}]) = \mathcal{L}(\dot{x} - x_0\delta) = s\hat{x}(s) - x_0$. Here $\mathcal{L}[x](s) = \hat{x}(s)$ and $\mathcal{L}[u](s) = \hat{u}(s)$, and we have used Corollary 8.1.3. We thus have

$$s\hat{x}(s) = A\hat{x}(s) + B\hat{u}(s) + x_0.$$

It follows that

$$\hat{x}(s) = (sI - A)^{-1}(x_0 + B\hat{u}(s)). \tag{8.8}$$

Note here that $(sI - A)^{-1}$ is a rational function and is well defined except at the eigenvalues of A. According to the formula $\mathcal{L}[H(t)e^{At}] = (sI - A)^{-1}$ to be shown below in Example 8.1.7 and by Theorem 8.1.2, we have

$$x(t) = H(t)e^{At}x_0 + (H(t)e^{At} * Bu)(t).$$

Since $u(t) \equiv 0$ for $t < 0$, we obtain the well-known formula

$$x(t) = e^{At}x_0 + \int_0^t e^{A(t-\tau)}Bu(\tau)d\tau, \quad t \geq 0.$$

In view of the uniqueness of linear ordinary differential equations, the solution above is unique, and hence as long as the growth order of u is at most of exponential order, there always exists a Laplace transformable solution that is unique. This guarantees the validity of solving (8.6) by taking the Laplace transforms of both sides. Needless to say, this process does not yield a valid solution where there can be a solution that is not Laplace transformable.

Example 8.1.7.

$$\mathcal{L}[H(t)e^{At}] = (sI - A)^{-1}. \tag{8.9}$$

This gives an extension of the scalar case $\mathcal{L}[e^{\lambda t}] = 1/(s - \lambda)$.

Let us first consider the case where $\operatorname{Re} s$ is greater than the maximum of the real part of the eigenvalues of A. Then $(sI - A)^{-1}$ certainly exists. Integrating both sides of

$$\frac{d}{dt}e^{-(sI-A)t} = -(sI - A)e^{-(sI-A)t}$$

on $[0, \infty)$, noting that $\lim_{t \to \infty} e^{-(sI-A)t} = 0$, we have

$$-I = \int_0^\infty \frac{d}{dt}e^{-(sI-A)t}dt = -(sI - A)\int_0^\infty e^{-(sI-A)t}dt = -(sI - A)\mathcal{L}[H(t)e^{At}].$$

Hence (8.9) follows.

On the other hand, if $\operatorname{Re} s$ does not satisfy the condition above, the Laplace integral does not converge. But the right-hand side of (8.9) is a rational function in s and is well defined everywhere except at the eigenvalues of A. Hence it is convenient and customary to extend the Laplace integral via analytic continuation to there and understand the extended function as the Laplace transform. It should be noted, however, that the obtained transform makes sense as the Laplace integral only on the half plane with real parts greater than the maximum of the real part of the eigenvalues of A.

Using this fact, we can compute e^{At} or find solutions of ordinary differential equations, i.e., by computing $(sI - A)^{-1}$ and finding its inverse Laplace transform.

8.2   Inverse Laplace Transforms

As we have seen in the previous section, the Laplace transform (as well as the Fourier transform) converts convolution to products. Hence convolution equations as discussed in section 6.8 (p. 132) reduce to algebraic equations in functions of s and become easier to deal with.

This method, however, requires the inverse Laplace transform to obtain a solution in the time domain from the one obtained in the Laplace domain. Usually one can easily obtain such inverse Laplace transforms via the examples in section 8.1.1 or from tables of Laplace transforms. However, we may sometimes need a more theoretical consideration and a formal inversion formula, which we now present.

Let us begin by noting the injectivity of the Laplace transform.

Theorem 8.2.1. *Let T be a Laplace transformable distribution, and suppose that there exists $\sigma \in \mathbb{R}$ such that $\hat{T}(s) \equiv 0$ for $\operatorname{Re} s > \sigma$. Then $T = 0$.*

We omit the proof; see, e.g., Schwartz [52, 53].

Hence if a complex function $F(s)$ is the Laplace transform of a distribution, then there can be only one[85] T such that $\mathcal{L}[T] = F(s)$.

Let us consider the case where the preimage T is a locally integrable function $f(t)$ with support contained in $[0, \infty)$. Set $s = \sigma + i\omega$. Then we obtain

$$\hat{f}(\sigma + i\omega) = \int_0^\infty \left(f(t)e^{-\sigma t} \right) e^{-i\omega t} dt. \tag{8.10}$$

This is the Fourier transform of $f(t)e^{-\sigma t}$ if we regard the left-hand side as a function of ω for each fixed σ.

From the Fourier inversion formula Theorem 7.5.4, we obtain

$$f(t)e^{-\sigma t} = \frac{1}{2\pi} \int_{-\infty}^\infty \hat{f}(\sigma + i\omega)e^{i\omega t} d\omega. \tag{8.11}$$

Then it follows that

$$f(t) = \frac{1}{2\pi} \int_{-\infty}^\infty \hat{f}(\sigma + i\omega)e^{(\sigma + i\omega)t} d\omega. \tag{8.12}$$

Or substitute $s = \sigma + i\omega$ back into this formula to obtain

$$f(t) = \frac{1}{2\pi i} \int_{\sigma - i\infty}^{\sigma + i\infty} \hat{f}(s)e^{st} ds. \tag{8.13}$$

This is what is known as the *Bromwich integral* formula.

8.3   Final-Value Theorem

In many problems in control theory, one encounters the need to compute the steady-state value of a signal $f(t)$, i.e., $\lim_{t \to \infty} f(t)$ from its Laplace transform. The *final-value theorem* answers this question, for example, as follows.

Example 8.3.1. Let $f(t) = H(t)(1 - e^{-t})$. Then $\hat{f}(s) = 1/s(s+1)$, and

$$\lim_{t \to \infty} f(t) = 1 = \lim_{s \to 0} \frac{s}{s(s+1)}.$$

That is, the behavior of $f(t)$ as $t \to \infty$ can be deduced from that of $s\hat{f}(s)$ as $s \to 0$.

Let us first consider the following simple case.

Theorem 8.3.2 (final-value theorem). *Suppose that $f(\cdot)$ has support contained in $[0, \infty)$ and satisfies the following conditions:*

1. *The right-limit $f(0+) := \lim_{t \to 0} f(t)$ exists and is finite.*

2. *The ordinary derivative $f'(t)$ of $f(t)$ exists for $t > 0$ and is locally integrable.*

---

[85] in the sense of distributions, of course.

3. *There exists $c > 0$ such that the Laplace integrals of $f(t)$ and $f'(t)$ both converge for* $\operatorname{Re} s > -c$.

4. *The limit $\lim_{s \to 0} s \hat{f}(s)$ exists.*

Then $\lim_{t \to \infty} f(t)$ exists, and

$$\lim_{t \to \infty} f(t) = \lim_{s \to 0} s \hat{f}(s) \qquad (8.14)$$

holds.

Proof. For every $T > 0$,

$$f(T) - f(0+) = \int_0^T f'(t) dt.$$

Since $\mathcal{L}[f'](0)$ exists, we can take the limit as $T \to \infty$ to conclude that $\lim_{T \to \infty} f(T)$ exists, and we have

$$\int_0^\infty f'(t) dt = \lim_{T \to \infty} f(T) - f(0+). \qquad (8.15)$$

On the other hand, it follows from (8.5)[86] that

$$\mathcal{L}[f'](s) = s \hat{f}(s) - f(0+).$$

By condition 3, $s = 0$ belongs to the interior of the domain of convergence of the Laplace integral. Then by continuity[87]

$$\lim_{s \to 0} \mathcal{L}[f'](s) = \int_0^\infty f'(t) dt.$$

Combining this with (8.15), we obtain (8.14). □

The above conditions can be restrictive and may not be satisfied in some applications. In fact, Example 8.3.1 satisfies the condition on $f'(t)$, but it does not satisfy condition 3 on f since $\hat{f}(s) = 1/s(s+1)$ has a pole at the origin.

Actually these conditions can be relaxed quite substantially. For example, it is known [72] that we can replace the conditions of Theorem 8.3.2 by

1. there exists a real number a such that $f(t) = 0$ for $t < a$;

2. f is a locally integrable function; and

3. $|f(t) e^{-\alpha t}|$ is integrable for some real number α.

Example 8.3.1 satisfies these conditions.

Furthermore, even if f' is not a function over the whole domain, there is still a possibility that the result holds. For instance, the following example provides such a case.

---

[86]We are not assuming the right-continuity of $f(t)$ at the origin, so we have written $f(0+)$ in place of $f(0)$.

[87]Or invoke the Lebesgue dominated convergence theorem [20, 22, 37].

Example 8.3.3. Let $f(t) := H(t-1)$. Its Laplace transform is e^{-s}/s. On the other hand, according to Proposition 6.3.3 on page 118, $f' = \delta_1$, and this does not satisfy condition 2 above. But the conclusion of the theorem is still valid as

$$\lim_{t \to \infty} f(t) = 1 = \lim_{s \to 0} s \cdot \frac{e^{-s}}{s}.$$

For generalizations including such cases, the reader is referred to [72].

Not only final values but also initial values can be known from Laplace transforms. We just state a result without a proof.

Theorem 8.3.4 (initial-value theorem). *Let $f(t)$ have support in $[0, \infty)$, and suppose that $f(0+) = \lim_{t \to 0} f(t)$ exists and is finite. Suppose also that conditions 2 and 3 above are satisfied. Then*

$$\lim_{t \to 0} f(t) = \lim_{s \to \infty} s\hat{f}(s) \tag{8.16}$$

holds.

For a proof, the reader is again referred to [72].

Problems

1. Differentiate $H(t)\cos\omega t$ in the sense of distributions, and derive

$$\mathcal{L}[H(t)\sin\omega t] = \omega/(s^2 + \omega^2)$$

from $\mathcal{L}[H(t)\cos\omega t] = s/(s^2 + \omega^2)$. (Hint: See Proposition 6.3.3, page 118; note also that $\cos 0 = 1$.)

2. Let

$$A = \begin{bmatrix} 0 & 1 \\ -2 & -3 \end{bmatrix}.$$

Solve the ordinary differential equation

$$\frac{dx}{dt} = Ax, \quad x_0 = x_0.$$

Chapter 9

Hardy Spaces

The theory of Hardy spaces lies in an interdisciplinary area where function space theory, complex function theory, and Fourier analysis find beautiful interactions. The interplay of these theories leads to very interesting and deep results; moreover, it also gives rise to important applications,[88] particularly H^∞ optimization theory, which has advanced remarkably in the past three decades and has brought about revolutionary advances in the theory of automatic control. We will briefly touch upon this interesting subject along with some introductory materials in Chapter 10.

This chapter assumes some knowledge of complex function theory and aims at giving the theorem of generalized interpolation and its proof, which had a strong impact on the solution of H^∞ optimization theory. Some well-known results such as the Nevanlinna–Pick interpolation theorem and the Carathéodory–Fejér theorem will be derived as corollaries to this general result. It should be noted, however, that, due to the nature of the problem and the space limitations, the treatment here cannot be as elementary as that in previous chapters.

9.1 Hardy Spaces

Let $\sum c_n z^n$ be a given power series analytic on the unit disc $\mathbb{D} := \{z \in \mathbb{C} : |z| < 1\}$. The radius of convergence of this power series is at least 1, but let us here temporarily assume that it also makes sense on the unit circle $\mathbb{T} = \{z : |z| = 1\}$. Then we can execute the substitution $z = e^{i\theta}$, $-\pi \leq \theta \leq \pi$, in this series to obtain $\sum c_n e^{in\theta}$. This is a Fourier series without the terms corresponding to $n < 0$. Hence it gives rise to a function on the circle $\mathbb{T}$ (or one on $[-\pi, \pi]$ via the change of variables $\theta \leftrightarrow e^{i\theta}$). As seen in Chapter 7 on Fourier analysis, a natural space arising in such a situation is $L^2[-\pi, \pi]$. Another possible candidate is the space L^∞ of (almost everywhere) bounded functions on $[-\pi, \pi]$. Under such circumstances, the following questions appear to be, at least mathematically, fairly natural:

---

[88]Perhaps even beyond the original desire of Hardy himself: *I have never done anything 'useful'. No discovery of mine has made, or is likely to make, directly or indirectly, for good or ill, the least difference to the amenity of the world.*—G. H. Hardy [18]. He did not want it to be useful.

- When the function of "boundary values" $\sum c_n e^{in\theta}$ of $f(z) = \sum c_n z^n$ belongs to L^2 or L^∞, to what extent are the properties and behavior of $f(z)$ governed by such boundary values?

- For a given function $f(\theta)$ in L^2 or in L^∞, does there exist a function $F(z)$ analytic on the unit disc such that $f(\theta) = F(e^{i\theta})$?

We begin by defining the Hardy space H^p.
Let f be a function analytic on the disc $\mathbb{D}$, and define

$$M_p(f,r) := \left\{ \frac{1}{2\pi} \int_0^{2\pi} |f(re^{i\theta})|^p d\theta \right\}^{1/p}, \quad 0 < p < \infty, \tag{9.1}$$

$$M_\infty(f,r) := \max_{0 \le \theta < 2\pi} |f(re^{i\theta})|. \tag{9.2}$$

Since analytic functions are continuous, these values are clearly finite for each fixed $r < 1$. We then have the following definition.

Definition 9.1.1. Define H^p, H^∞ spaces by

$$H^p := \{ f \ : \ M_p(f,r) \text{ is bounded as } r \to 1 \}, \tag{9.3}$$

$$H^\infty := \{ f \ : \ M_\infty(f,r) \text{ is bounded as } r \to 1 \}. \tag{9.4}$$

In view of Lemma A.8.5 (p. 246) derived from Hölder's inequality, $M_p(f,r) \le M_q(f,r)$ for $p < q$, and hence $H^q \subset H^p$ for $p < q$. Another important property of $M_p(f,r)$ is its monotonicity.

Lemma 9.1.2. *Let $f \in H^p$, $1 \le p \le \infty$. Then for $r_1 < r_2$*

$$M_p(f,r_1) \le M_p(f,r_2). \tag{9.5}$$

We omit the proof. See, for example, [47], [11]. Incidentally, when $p = \infty$ this is an easy consequence of the maximum modulus principle [47].

Using the lemma above, we now show that Hardy spaces are Banach spaces.

Theorem 9.1.3. *H^p ($1 \le p \le \infty$) is a Banach space with respect to the norm*

$$\|f\|_p := \sup_{r<1} M_p(f,r) = \lim_{r \to 1} M_p(f,r). \tag{9.6}$$

Proof. The second identity of (9.6) follows easily from (9.5); i.e., $M_p(f,r)$ is monotone nondecreasing in r.

Let us first verify that (9.6) gives a norm. One need only check the triangle inequality. Minkowski's inequality (Appendix A, Theorem A.8.4, p. 246) readily yields

$$M_p(f+g,r) \le M_p(f,r) + M_p(g,r).$$

Taking the supremum in r on the right-hand side, we have

$$M_p(f+g,r) \leq \|f\|_p + \|g\|_p.$$

Then taking the supremum in r on the left-hand side, we obtain the triangle inequality

$$\|f+g\|_p \leq \|f\|_p + \|g\|_p. \tag{9.7}$$

Thus $\|\cdot\|_p$ satisfies the axiom of norms, and hence H^p is a normed linear space.

Let us show the completeness. Let $\{f_n\}$ be a Cauchy sequence in H^p. Suppose $|z| \leq r < R < 1$, and apply the Cauchy integral formula to $f_n - f_m$ on the circle of radius R to obtain

$$|f_n(z) - f_m(z)| = \left| \frac{1}{2\pi i} \oint_{|\lambda|=R} \frac{f_n(\lambda) - f_m(\lambda)}{\lambda - z} d\lambda \right|$$

$$\leq \frac{1}{2\pi(R-r)} \oint_{|\lambda|=R} |f_n(\lambda) - f_m(\lambda)| d\lambda.$$

It follows that[89]

$$(R-r)|f_n(z) - f_m(z)| \leq M_1(f_n - f_m, R) \leq M_p(f_n - f_m, R) \leq \|f_n - f_m\|_p.$$

Hence f_n constitutes a Cauchy sequence in the sense of uniform convergence on every disc $\{z : |z| \leq r\}$. Hence there exists an analytic function f on $\mathbb{D}$ such that f_n converges uniformly to f on every compact subset in $\mathbb{D}$. Now, for every $\epsilon > 0$, there exists m such that $\|f_n - f_m\|_p < \epsilon$ for all $n > m$. Hence, for every $r < 1$,

$$M_p(f - f_m, r) = \lim_{n \to \infty} M_p(f_n - f_m, r) \leq \epsilon.$$

Thus $\|f - f_m\|_p \to 0$ as $m \to \infty$. $\qquad\square$

The following lemma is readily obvious, but we will make frequent use of it in what follows.

Lemma 9.1.4. *Let* $\psi \in H^\infty$ *and* $f \in H^p$. *Then* $\psi f \in H^p$.

Proof. Since ψ is bounded on the unit disc, $\|\psi f\|_p \leq \|\psi\|_\infty \cdot \|f\|_p < \infty$. $\qquad\square$

The main theme of the theory of H^p spaces is to clarify its various properties arising from the interplay between analyticity and the behavior of the boundary values on the unit circle. To this end, the Poisson integral plays a crucial role, and it will be the topic of the next section.

---

[89]Note that $M_1(f, R) \leq M_p(f, R)$.

9.2   Poisson Kernel and Boundary Values

Let $f(z)$ be analytic on the open unit disc. Then f can be expanded to a power series:

$$f(z) = \sum_{n=0}^{\infty} a_n z^n.$$

Define $f_r(\theta) := f(re^{i\theta})$. For each fixed $0 < r < 1$, this is a function of θ on the unit circle $\mathbb{T}$, and, furthermore,

$$f_r(\theta) = \sum_{n=0}^{\infty} a_n r^n e^{in\theta} \tag{9.8}$$

holds. That is, the Fourier coefficients of f_r are $a_n r^n$ for $n \geq 0$ and 0 for $n < 0$. To see the relationship between this f and its boundary values on the unit circle, let us for the moment assume that f is analytic not only on the (open) unit disc but also on the boundary.[90] This implies that (9.8) is valid for $r = 1$, and hence we obtain the Fourier expansion of $f_1(\theta) = f(e^{i\theta})$. Let us specifically write $f_*(\theta)$ for this $f_1(\theta)$. Then we have

$$f_*(\theta) = \sum_{n=0}^{\infty} a_n e^{in\theta} \tag{9.9}$$

and

$$a_n = \frac{1}{2\pi} \int_{-\pi}^{\pi} f_*(\theta) e^{-in\theta} d\theta. \tag{9.10}$$

Needless to say, by expansion (9.9), $a_n = 0$ for $n < 0$. Let us express the values of $f_r(\theta) = f(re^{i\theta})$ in terms of $f_*(\theta)$ under this condition. We have

$$\begin{aligned}
f_r(\theta) &= \sum_{n=0}^{\infty} a_n r^n e^{in\theta} \\
&= \sum_{n=-\infty}^{\infty} a_n r^{|n|} e^{in\theta} \\
&= \sum_{n=-\infty}^{\infty} \frac{1}{2\pi} \int_{-\pi}^{\pi} f_*(t) e^{-int} dt \, r^{|n|} e^{in\theta} \\
&= \frac{1}{2\pi} \int_{-\pi}^{\pi} f_*(t) \left(\sum_{n=-\infty}^{\infty} r^{|n|} e^{in(\theta-t)} \right) dt. \tag{9.11}
\end{aligned}$$

The change of order of the integration and the infinite sum here is guaranteed by the fact that the last series converges uniformly on each compact disc of fixed radius r.

---

[90]This means that it is analytic on an open set containing the closed unit disc.

Let us calculate the sum of the series $\sum_{n=-\infty}^{\infty} r^{|n|} e^{in\theta}$. Put $z = re^{i\theta}$. We easily obtain

$$\sum_{n=-\infty}^{\infty} r^{|n|} e^{in\theta} = \sum_{n=1}^{\infty} r^n e^{-in\theta} + \sum_{n=0}^{\infty} r^n e^{in\theta}$$

$$= \sum_{n=0}^{\infty} \bar{z}^n + \sum_{n=0}^{\infty} z^n - 1$$

$$= \mathrm{Re} \left[2 \frac{1}{1-z} - 1 \right] \tag{9.12}$$

$$= \mathrm{Re} \left[\frac{1 + re^{j\theta}}{1 - re^{i\theta}} \right] \tag{9.13}$$

$$= \frac{1 - r^2}{1 - 2r\cos\theta + r^2}. \tag{9.14}$$

The last expression is denoted by $P_r(\theta)$ and is called the *Poisson kernel*. Then (9.11) shows that $f_r(\theta)$ is given by the convolution of the Poisson kernel and the boundary values $f_*(\theta)$ of f on the unit circle:

$$f_r(\theta) = \frac{1}{2\pi} \int_{-\pi}^{\pi} f_*(t) P_r(\theta - t)\,dt. \tag{9.15}$$

The right-hand side integral here is called the *Poisson integral*.

The formula above has been shown under the assumption that f is analytic on the closed unit disc $\{z : |z| \le 1\}$. This assumption can in fact be quite relaxed; for example, one can show (rather easily) that it holds when f is harmonic on the unit disc including the boundary. A function f is called a *harmonic function* if it satisfies Laplace's equation

$$\triangle f := \frac{\partial^2}{\partial x^2} f + \frac{\partial^2}{\partial y^2} f = 0 \tag{9.16}$$

when f is regarded as a function of x, y with substitution $z = x + iy$. Analytic functions are always harmonic due to the Cauchy–Riemann relation, but the converse is not necessarily true, as the counterexample $f(z) = \mathrm{Re}\, z = x$ shows. (We here note that the real part of every analytic function is known to be always harmonic.)

Now the reader is warned that the subsequent discussions will be fairly delicate and technical. While they are mathematically important for the basis of Hardy spaces, it will be rather technical for the scope of this book. We will thus present some fundamental results and only indicate where the difficulties lie. We refer interested readers to [11, 17, 21].

As noted above, if f is harmonic on the unit disc including the boundary, then the values of f in the interior is determined by its boundary values $f_*(\theta) = f(e^{i\theta})$ via the Poisson integral. Actually, even if f is not given as a harmonic function from the outset, it is extensible to the interior as a harmonic function provided that f_* satisfies a certain condition on the circle. The Poisson integral plays a crucial role in showing this.

Theorem 9.2.1. *Let $f_*(\theta)$ be a function defined on the unit circle $\mathbb{T}$ belonging to $L^p(\mathbb{T})$, $1 \le p \le \infty$. Define f on the unit disc $\mathbb{D}$ by*

$$f(re^{i\theta}) = \frac{1}{2\pi} \int_{-\pi}^{\pi} f_*(t) P_r(\theta - t)\,dt. \tag{9.17}$$

180 Chapter 9. Hardy Spaces

Then f is a harmonic function on the unit disc $\mathbb{D}$, and $f_r(\theta) = f(re^{i\theta})$ converges to f_* as $r \to 1$ with respect to the following:

- *the $L^p(\mathbb{T})$-norm when $1 \le p < 1$ and*

- *in the weak* topology[91] of L^∞ when $p = \infty$.*

In other words, the boundary values on $\mathbb{T}$ naturally determine the values of f as a harmonic function on the unit disc, and this boundary value function $f_*(\theta)$ turns out to be the radial limit of the extended $f_r(\theta)$.

The proof is not very difficult but requires some technical preliminaries and is hence omitted. For details, the reader is referred to the existing books on Hardy spaces, e.g., Hoffman [21]. We give only an outline here.

Outline of the Proof of Theorem 9.2.1. The harmonicity of $f(re^{i\theta})$ follows easily by noting the following:

1. $f(re^{i\theta})$ is given as the convolution of f_* and the Poisson kernel.

2. The derivative of f can be computed by differentiating $P_r(\theta - t)$ under the integral sign.

3. The Poisson kernel is a harmonic function as the real part of an analytic function, as shown in (9.12).

4. Then, by Theorem 6.6.7, f is also seen to satisfy (9.16).

It remains to show the convergence as $r \to 1$; this can be ensured by noting that $P_r(\theta)$ converges to Dirac's δ as $r \to 1$. In fact, the following properties hold:

1. $P_r(\theta) \ge 0$.

2. $\frac{1}{2\pi} \int_{-\pi}^{\pi} P_r(\theta) d\theta = 1, \quad 0 \le r < 1$.

3. For $0 < \delta < \pi$,
$$\lim_{r \to 1} \sup_{|\theta| \ge \delta} |P_r(\theta)| = 0.$$

The first statement can be shown via direct calculation. The second property can be seen easily by observing that this integral is the Poisson integral of the identity function 1. The third property follows from the estimate $P_r(\theta) \le (1 - r^2)/(1 - 2r\cos\delta + r^2)$ when $\delta \le |\theta| \le \pi$. Hence, according to Proposition 6.5.3 (p. 121) in Chapter 6, $P_r(\theta) \to \delta$. This implies $f_r(\theta) \to f_*(\theta)$ as $r \to 1$ in the sense of distributions; in fact, it is known that this result also holds in the stronger sense as stated above when f belongs to L^p [21]. This completes the proof of the theorem. $\square$

Summarizing, we have seen that *an L^p function on the unit circle can be extended to the interior as a harmonic function via the Poisson integral.* Let us pose the converse question: *Can a harmonic function on the unit disc be extended to the circle $\mathbb{T}$ in some sense?* The following theorem answers this question.

---

[91] Here we regard $L^\infty = (L^1)'$ according to Example 4.1.8, page 80, Chapter 4.

Theorem 9.2.2. *Let f be a harmonic function given on the unit disc $\mathbb{D}$, and let $f_r(\theta) := f(re^{i\theta})$. The following facts hold:*

- *Let $1 < p \leq \infty$, and suppose that the L^p norm $\|f_r\|_p$ of f_r is bounded (in r). Then there exists a boundary function $f_*(\theta) \in L^p(\mathbb{T})$ such that f can be expressed as the Poisson integral (9.17) of $f_*(\theta)$.*

- *If the L^1 norm of f_r is bounded, then there exists a finite Baire measure $m_*(\theta)$ on the circle such that f can be expressed as its Poisson integral (9.17).*

The proof of this theorem is more involved than that of Theorem 9.2.1. We only indicate its outline: First it is known that a real-valued harmonic function is always given as the real part of some analytic function $g(z) = \sum_{n=0}^{\infty} g_n z^n$. Using this fact, we can expand the real part $u_r(\theta)$ of $f_r(\theta)$ as

$$u_r(\theta) = 2\operatorname{Re} g_0 + \sum_{n=1}^{\infty}(g_n z^n + \overline{g}_n \overline{z}^n) = \sum_{n=-\infty}^{\infty} u_n r^{|n|} e^{i\theta}.$$

Since the same holds true for the imaginary part, f can be expanded as

$$f_r(\theta) = \sum_{n=-\infty}^{\infty} c_n r^{|n|} e^{i\theta} \tag{9.18}$$

with some coefficients c_n. The main question here is under what conditions may such a function be extended to the limit as $r \to 1$. It can be regarded as a problem of Fourier series as to whether such c_n's can be the Fourier coefficients of a function on $\mathbb{T}$, and also which function. The theorem above gives a precise answer: it requires the boundedness of the L^p norm of these functions. The mechanism relies on the fact that when $p > 1$, L^p space is the dual of an L^q space, every bounded set is relatively compact with respect to the weak* topology according to the Bourbaki–Alaoglu theorem, Theorem 4.3.6 (p. 84), and hence we can always choose a convergent subsequence in it.[92] In contrast to this, this argument does not directly apply to L^1, and this is why we have to consider Baire measures on the unit circle. But we will not go into more detail since further discussion would require more elaborate measure theory.

Summarizing the above, there is, roughly, a one-to-one correspondence between L^p on the unit circle and harmonic functions on the unit disc with bounded L^p norm. However, when $p = 1$, i.e., for L^1 functions, the situation is slightly exceptional. Note also that the convergence as $r \to 1$ is in the sense of L^p. In fact, a function such as

$$\exp\left(\frac{z+1}{z-1}\right)$$

is analytic and bounded on the unit disc, but it does not possess a limit as $z \to 1$. Thus we cannot expect $f_r(\theta)$ to possess a limit for every θ as $r \to 1$. On the other hand, such exceptional points are not many, and it is known that $f_r(\theta)$ indeed possesses a limit[93] as

---

[92]Thus the boundedness of the L^p norms plays a role here.

[93]This limit of course agrees with that guaranteed by Theorem 9.2.2.

$r \to 1$ for *almost every* θ. This result is known as Fatou's theorem, but we do not need this result in this book. For details, the reader is referred to Hoffman [21], etc.

Since analytic functions are always harmonic, the discussions above apply equally well to H^p functions. H^p functions are those obtained by taking the limits in L^p guaranteed by Theorem 9.2.2 that are also analytic in the unit disc. The norm is taken as

$$\|f\|_p := \lim_{r \to 1} \|f_r\|_p.$$

This norm is common to H^p and L^p, and it also agrees with the norm of $L^p(\mathbb{T})$ on the unit circle. In this sense, H^p is a subspace of $L^p(\mathbb{T})$ and is also a subspace of harmonic functions on the unit disc with their L^p norms bounded as $r \to 1$. As we have seen in Theorem 9.1.3 (p. 176), H^p is a Banach space and hence is a closed subspace of $L^p(\mathbb{T})$. (However, as we noted above, the case $p = 1$ is an exception.)

It is awkward to distinguish $L^p(\mathbb{T})$ from the space of harmonic functions that is isomorphic to it, and we will hereafter write simply L^p, meaning both.

In the above isomorphic correspondence, H^p not only is a closed subspace of L^p but also allows for a concrete characterization due to its analyticity as follows.

Theorem 9.2.3. *For $1 < p \leq \infty$, space H^p is a closed subspace of L^p, and it consists of those functions with Fourier coefficients corresponding to $\{e^{-in\theta}\}_{n=1}^{\infty}$ that are all 0, that is,*

$$\int_{-\pi}^{\pi} f(\theta)e^{in\theta} d\theta = 0, \quad n = 1, 2, \ldots. \tag{9.19}$$

When $p = 1$, H^1 is the subspace of the space of finite Baire measures on the unit circle, consisting of the L^1 functions $f(\theta)$ satisfying (9.19).

We omit the proof, but let us give a few relevant remarks.

First consider the case $1 < p \leq \infty$. Every function in H^p is analytic in the unit disc and hence admits the expansion

$$f(z) = \sum_{n=0}^{\infty} c_n z^n.$$

This, along with the statement after (9.18), implies that the c_n's give the Fourier coefficients of the boundary function $f_*(\theta)$. Hence we should obtain the Fourier expansion

$$f_*(\theta) = \sum_{n=0}^{\infty} c_n e^{in\theta}.$$

This readily yields (9.19).

The case $p = 1$ is a little singular. It is known that every element of H^1 becomes necessarily absolutely continuous among Baire measures and is orthogonal to $e^{-in\theta}$, $n = 1, 2, \ldots$. This result is due to F. and M. Riesz [11, 21].

The result above takes a particularly simple form when $p = 2$, because L^2 and H^2 are Hilbert spaces. Every $g \in L^2(\mathbb{T})$ can be expanded into Fourier series as

$$g(e^{i\theta}) = \sum_{n=-\infty}^{\infty} g_n e^{in\theta}, \quad \sum_{n=-\infty}^{\infty} |g_n|^2 < \infty. \tag{9.20}$$

As seen above, H^2 functions consist of those g such that $g_n = 0$ for $n < 0$. Also, extending the expression (9.20) to $\mathbb{D}$, we naturally obtain the expression

$$g(z) = \sum_{n=-\infty}^{\infty} g_n z^n, \quad \sum_{n=-\infty}^{\infty} |g_n|^2 < \infty. \tag{9.21}$$

This yields the following theorem.

Theorem 9.2.4. *The family $\{z^n\}_{n=-\infty}^{\infty}$ constitutes an orthonormal basis[94] for L^2. H^2 is the closed subspace spanned by $\{z^n\}_{n=0}^{\infty}$. The orthogonal complement of H^2 in L^2 is denoted by H_-^2. The family $\{z^{-n}\}_{n=1}^{\infty}$ gives an orthonormal basis for H_-^2.*

We thus have the direct sum decomposition $L^2 = H^2 \oplus H_-^2$. We denote the orthogonal projections from L^2 to H^2 and H_-^2 by P_+ and P_-, respectively.

Let us give a few remarks on the inner product in L^2 and the Fourier expansion there. First note that the inner product must take the form

$$(x, y) = \frac{1}{2\pi} \int_{-\pi}^{\pi} x(\theta)\overline{y(\theta)}d\theta \tag{9.22}$$

to make it consistent with the norm of H^2 ((9.6), p. 176). For $\psi \in L^\infty$, define

$$\tilde{\psi}(z) := \overline{\psi(1/\bar{z})}. \tag{9.23}$$

Then

$$(x, \psi y)_{L^2} = (\tilde{\psi}x, y)_{L^2}, \quad (\psi x, y)_{L^2} = (x, \tilde{\psi}y)_{L^2} \tag{9.24}$$

hold for $x, y \in L^2$. They follow readily by substituting these expressions directly into (9.22) by noting that $\bar{z} = z^{-1}$ when $|z| = 1$. We will also apply the definition (9.23) to all L^p functions other than L^∞. For example, $\tilde{\psi} \in H_-^2$ for $\psi \in H^2$.

For g in (9.21), define $\hat{g}$ by

$$\hat{g}(\omega) := \frac{1}{2\pi} \int_0^{2\pi} g(e^{i\theta})e^{-i\omega\theta}d\theta. \tag{9.25}$$

Then its nth Fourier coefficient g_n (i.e., the coefficient of z^n) is given by

$$g_n = \hat{g}(n). \tag{9.26}$$

The following fact will be frequently referred to in section 9.4 on page 189 on shift operators. This is just a special case of Lemma 9.1.4 for $p = 2$, but we state it separately in view of its frequent use.

Lemma 9.2.5. *If $f \in H^\infty$ and $g \in H^2$, then $fg \in H^2$.*

---

[94] see (9.22).

9.3   Canonical Factorization

Let f be a nonzero H^1 function. Such a function can be decomposed into the product of a so-called inner function and an outer function (*inner-outer factorization* or a *canonical factorization*). We will briefly state some fundamental facts.

Since f belongs to H^1, its boundary function f_* exists (almost everywhere) on the unit circle $\mathbb{T}$, and f can be represented by the Poisson integral:

$$f(re^{i\theta}) = \frac{1}{2\pi} \int_{-\pi}^{\pi} f_*(t) P_r(\theta - t) dt.$$

According to (9.13), we can rewrite $P_r(\theta - t)$ as

$$P_r(\theta - t) = \text{Re}\left[\frac{1 + re^{i(\theta - t)}}{1 - re^{i(\theta - t)}}\right] = \text{Re}\left[\frac{e^{it} + re^{i\theta}}{e^{it} - re^{i\theta}}\right]. \tag{9.27}$$

In view of this, define $F(z)$[95] as

$$F(z) := \exp\left[\frac{1}{2\pi} \int_{-\pi}^{\pi} \frac{e^{it} + z}{e^{it} - z} \log|f_*(t)| dt\right]. \tag{9.28}$$

It is known that $\log|f_*(t)|$ is integrable on the unit circle [21]. Hence the above integral is well defined. Since $(e^{it} + z)/(e^{it} - z)$ is analytic with respect to z inside the unit disc, differentiation under the integral yields the analyticity of $F(z)$ inside the unit disc. Furthermore, (9.27) yields

$$\begin{aligned}
|F(re^{i\theta})| &= \exp\left(\text{Re}\left[\frac{1}{2\pi} \int_{-\pi}^{\pi} \frac{e^{it} + re^{i\theta}}{e^{it} - re^{i\theta}} \log|f_*(t)| dt\right]\right) \\
&= \exp\left[\frac{1}{2\pi} \int_{-\pi}^{\pi} \text{Re}\, \frac{e^{it} + re^{i\theta}}{e^{it} - re^{i\theta}} \log|f_*(t)| dt\right] \\
&= \exp\left[\frac{1}{2\pi} \int_{-\pi}^{\pi} P_r(\theta - t) \log|f_*(t)| dt\right] \\
&\leq \frac{1}{2\pi} \int_{-\pi}^{\pi} P_r(\theta - t)|f_*(t)| dt \\
&\leq \frac{1}{2\pi} \int_{-\pi}^{\pi} |f_*(t)| dt.
\end{aligned} \tag{9.29}$$

We have used $\|\phi * \psi\|_1 \leq \|\phi\|_1 \|\psi\|_1$ ((6.32), p. 126) in deriving the last line. The derivation of the fourth line from the third is due to a special case of what is known as Jensen's inequality [11, 47] applied to the measure $d\mu = \frac{1}{2\pi} P_r(\theta - t) dt$ (note that $\int d\mu = 1$, p. 180):

$$\exp\left(\int \phi \, d\mu\right) \leq \int e^\phi \, d\mu. \tag{9.30}$$

---

[95]Roughly speaking, this is based on the idea of taking the log of the modulus of f_* and then taking the exponential in the hope of recovering the same modulus.

Hence

$$\frac{1}{2\pi}\int_{-\pi}^{\pi}|F(re^{it})|dt \le \frac{1}{2\pi}\int_{-\pi}^{\pi}|f_*(t)|dt,$$

and this shows that $F \in H^1$. Now recall that the Poisson kernel ($P_r/2\pi$ to be precise) approaches δ as $r \to 1$ (p. 180). Then take the log of the third line of (9.29) and let $r \to 1$ to obtain $\log|F(e^{i\theta})| = \log|f_*(\theta)|$ almost everywhere on the unit circle. Taking the exponentials of both sides yields $|F(e^{i\theta})| = |f_*(\theta)|$.

Since $F(z)$ is an exponential function, it does not possess any zero in the unit disc. Noticing that $\log|f(z)|$ is a so-called *subharmonic function*, we see that $|f(z)| \le |F(z)|$ holds at every point $z \in \mathbb{D}$.[96] Then we put

$$g(z) := \frac{f(z)}{F(z)}.$$

Obviously $g \in H^\infty$ and $g(e^{i\theta}) = 1$ almost everywhere on the unit circle $\mathbb{T}$. That is, we have decomposed f as $f = gF$. The functions g and F are, respectively, what are known as inner and outer functions. Let us give their precise definitions.

Definition 9.3.1. A function g analytic in the unit disc $\mathbb{D}$ is called an *inner function* if $|g(z)| \le 1$ on $\mathbb{D}$ and $|g(e^{i\theta})| = 1$ almost everywhere on the unit circle $\mathbb{T}$. A function F analytic in the unit disc $\mathbb{D}$ is said to be an *outer function* if there exist a real-valued Lebesgue integrable function $\phi(\theta)$ on $\mathbb{T}$ and a complex number c of modulus 1 such that

$$F(z) = c\exp\left[\frac{1}{2\pi}\int_{-\pi}^{\pi}\frac{e^{it}+z}{e^{it}-z}\phi(t)\,dt\right]. \tag{9.31}$$

The discussion above shows that a nonzero $f \in H^1$ can be decomposed as the product of an inner function and an outer function. That is, we have the following theorem.

Theorem 9.3.2 (Canonical Factorization Theorem 1). *Let f be an H^1 function that is not identically 0. Then f can be decomposed as $f = gF$ with inner function g and outer function F. This decomposition is unique up to multiplication of constants of modulus 1.*

Let us consider as a simple example the rational function $f(z) = (z-1/2)/(z+2)$.

$$\frac{z-\frac{1}{2}}{z+2} = \frac{z-\frac{1}{2}}{1-\frac{1}{2}z}\cdot\frac{1-\frac{1}{2}z}{z+2} =: g(z)F(z)$$

gives a canonical factorization. It is easy to see that $g(z)$ is inner and $F(z)$ does not have a zero in the unit disc. To show that $F(z)$ is indeed outer, we must show that (9.31) is satisfied for some integrable $\phi(\theta)$. Since $F(z)$ is analytic in the unit disc and has no zeros there, it can be written as $F(z) = \exp\psi(z)$ for some ψ. This ψ must also be analytic, and hence $\mathrm{Re}\,\psi$ is a harmonic function. Hence by Theorem 9.2.2 (p. 181), $\mathrm{Re}\,\psi$ can be expressed as a Poisson integral as

$$\mathrm{Re}\,\psi(z) = \frac{1}{2\pi}\int_{-\pi}^{\pi}P_r(\theta-t)\phi(t)\,dt = \frac{1}{2\pi}\int_{-\pi}^{\pi}\left[\mathrm{Re}\,\frac{e^{it}+z}{e^{it}-z}\right]\phi(t)\,dt.$$

---

[96]We omit the details. See, for example, Hoffman [21].

This yields

$$\psi(z) = \frac{1}{2\pi} \int_{-\pi}^{\pi} \frac{e^{it}+z}{e^{it}-z} \phi(t)dt.$$

Since $F(z) = \exp \psi(z)$, this implies that F is indeed outer.

As seen above, for a rational function one can obtain the inner-outer factorization by collecting all zeros (in the unit disc) to make the inner part and then taking the remaining part as the outer part. For the general case, however, the situation is more complex. For example, there exists an inner function that has no zeros. Consider, for example,

$$g(z) = \exp\left(\frac{z+1}{z-1}\right).$$

Since $\mathrm{Re}[(z+1)/(z-1)] = -(1-|z|^2)/|z-1|^2$, this function $(z+1)/(z-1)$ maps the unit disc $\mathbb{D}$ onto the open left half plane and also maps the circle $\mathbb{T}$ to the imaginary axis (thus having modulus 1). Hence g is an inner function defined everywhere except $z = 1$. This function clearly has no zeros.

Such inner functions with no zeros are called *singular inner functions*. In fact, it is known that every inner function can be decomposed into the product of an inner function consisting of zeros in the unit circle, called a *Blaschke product*, and a singular inner function; see Canonical Factorization Theorem 2, Theorem 9.3.6, below.

Let us introduce the notion of Blaschke products. We start with the following lemma on infinite products.

Lemma 9.3.3. *Let $\{\alpha_n\}_{n=1}^{\infty}$ be a sequence of complex numbers. The infinite product $\prod_{n=1}^{\infty} |\alpha_n|$ converges if and only if $\sum_{n=1}^{\infty}(1 - |\alpha_n|) < \infty$.*

We omit the proof since it can be found in many textbooks on analysis (e.g., [3]). The crux lies in the idea of taking the logarithm of $\prod_{n=1}^{\infty} |\alpha_n|$ to change the product into a sum and observing that $\log |\alpha_n|$ is on the same order of infinitesimal as $1 - \alpha_n$ when $1 - \alpha_n$ is small.

The following proposition claims that the zeros of an arbitrary H^p function satisfy the condition of this lemma.

Proposition 9.3.4. *Suppose $f \in H^{\infty}$, and $f(0) \neq 0$. Let $\{\alpha_n\}$ be the zeros of f in the open unit disc $\mathbb{D}$ counting the multiplicities. Then*

$$\sum_{n=1}^{\infty}(1 - |\alpha_n|) < \infty.$$

That is, $\prod_{n=1}^{\infty} |\alpha_n|$ converges.

According to Canonical Factorization Theorem 1, Theorem 9.3.2, the zeros of every $f \in H^p$ are contained in its inner part g, which belongs to $g \in H^{\infty}$. Hence the theorem above applies to any $f \in H^p$ ($p \geq 1$).

Proof. Let us assume $\|f\|_{\infty} \leq 1$ without loss of generality. If there are only finitely many zeros of f, there is nothing to prove. Hence we assume that there are countably infinitely many zeros.

Let $B_n(z)$ be the *finite Blaschke product* defined by

$$B_n(z) := \prod_{k=1}^{n} \frac{z - \alpha_k}{1 - \overline{\alpha}_k z}. \tag{9.32}$$

Clearly B_n is a rational function that is analytic on the closed unit disc $\mathbb{D} \cup \mathbb{T}$ (whose poles $1/|\alpha_k|$ are outside the unit disc). Since $|(e^{i\theta} - \alpha_k)/(1 - \overline{\alpha}e^{i\theta})| = 1$ for every $\theta \in \mathbb{R}$, $|B_n(\theta)| = 1$. That is, B_n is an inner function, and f/B_n belongs to H^∞. Moreover,

$$\frac{|f(e^{i\theta})|}{|B_n(e^{i\theta})|} = |f(e^{i\theta})| \leq 1,$$

and hence by the maximum modulus principle, $|f(z)| \leq |B_n(z)|$ for every z in the disc. In particular,

$$0 < |f(0)| \leq |B_n(0)| = \prod_{k=1}^{n} |\alpha_k|.$$

Since $|\alpha_k| < 1$ for each k, the sequence on the right forms a monotone decreasing sequence that is bounded from below by $|f(0)|$ and hence is convergent as $n \to \infty$. Thus the conclusion follows by Lemma 9.3.3. $\qquad\square$

The following object, obtained by slightly modifying the finite Blaschke product in (9.32),

$$z^p \prod_{n=1}^{\infty} \frac{\overline{\alpha}_n}{|\alpha_n|} \frac{\alpha_n - z}{1 - \overline{\alpha}_n z}, \tag{9.33}$$

is called a *Blaschke product*. The factor z^p corresponds to the zeros at the origin. The factor $\overline{\alpha}_n/|\alpha_n|$ is necessary to guarantee the convergence, because the form (9.32) of finite Blaschke products does not necessarily guarantee convergence. The following lemma is fundamental in this respect.

Lemma 9.3.5. *Let $\{\alpha_n\}$ be a nonzero sequence in $\mathbb{D}$. The Blaschke product*

$$\prod_{n=1}^{\infty} \frac{\overline{\alpha}_n}{|\alpha_n|} \frac{\alpha_n - z}{1 - \overline{\alpha}_n z} \tag{9.34}$$

converges uniformly on every compact subset of $\mathbb{D}$ if and only if

$$\sum_{n=1}^{\infty} (1 - |\alpha_n|) < \infty, \tag{9.35}$$

that is, when $\prod_{n=1}^{\infty} |\alpha_n|$ is convergent. Furthermore, (9.34) gives an inner function that vanishes precisely at $\alpha_1, \alpha_2, \ldots$.

Proof. We prove the sufficiency only. For the necessity, see [21, 47], etc.

Suppose that condition (9.35) holds. Take any compact set $K \subset \mathbb{D}$. Then there can be only finitely many factors of (9.34) that vanish in K. Otherwise, it contradicts (9.35).

Likewise, α_n cannot accumulate at 0. Hence we may assume without loss of generality that there exists $\epsilon > 0$ such that $|\alpha_n| \geq \epsilon$ for all n.

Take any $r < 1$. We prove that the n-partial product of (9.34) converges uniformly on the closed disc $K = \{z : |z| \leq r\}$. Write $f_n(z)$ as

$$f_n(z) := \left(\frac{\overline{\alpha}_n}{|\alpha_n|} \right) \frac{\alpha_n - z}{1 - \overline{\alpha}_n z}.$$

Then

$$1 - f_n(z) = \frac{1}{|\alpha_n|} \left[1 - \frac{|\alpha_n|^2 - \overline{\alpha}_n z}{1 - \overline{\alpha}_n z} \right] + 1 - \frac{1}{|\alpha_n|}$$

$$= \frac{1 - |\alpha_n|}{|\alpha_n|} \left[\frac{1 + |\alpha_n|}{1 - \overline{\alpha}_n z} - 1 \right].$$

Hence, if $|z| \leq r$, then

$$\sum_{n=1}^{\infty} |1 - f_n(z)| \leq \sum_{n=1}^{\infty} \frac{1 - |\alpha_n|}{|\alpha_n|} \left[\frac{2}{1 - r} + 1 \right] \leq \frac{\sum_{n=1}^{\infty}(1 - |\alpha_n|)}{\epsilon} \left[\frac{2}{1 - r} + 1 \right].$$

Thus $\sum |1 - f_n(z)|$ converges uniformly on this disc. Hence the infinite product $\prod_{n=1}^{\infty} f_n(z)$ converges absolutely uniformly on the disc $|z| \leq r$. Since r is arbitrary, this proves the convergence of the Blaschke product.

We omit the proof of the facts that (9.34) is inner and has only $\{\alpha_n\}$ as its zeros; see, for example, [11, 21, 47]. □

According to the Canonical Factorization Theorem, Theorem 9.3.2 (p. 185), every nonzero H^1 function (hence every H^p for $p \geq 1$ also) f can be factorized uniquely as $f = gF$ into the product of an inner function g and an outer function F. Since outer functions do not possess a zero in $\mathbb{D}$, the zeros of f are all zeros of g. Since g is an inner function, it belongs to H^∞ and hence satisfies the hypotheses of Proposition 9.3.4. Form the Blaschke product

$$B(s) := z^p \prod_{n=1}^{\infty} \frac{\overline{\alpha}_n}{|\alpha_n|} \frac{\alpha_n - z}{1 - \overline{\alpha}_n z}$$

consisting of zeros of g, and put $S(z) := g(z)/B(z)$. Then $S(z)$ becomes an inner function without a zero, that is, a *singular inner function*. Summarizing the above, we have the following stronger form of the canonical factorization theorem.

Theorem 9.3.6 (Canonical Factorization Theorem 2). *Let f be an H^1 function that is not identically zero. Then f can be decomposed into the product $f = BSF$, where $B(z)$ is the Blaschke product consisting of the zeros of f, $S(z)$ is a singular inner function, and $F(z) \in H^1$ is an outer function.*

9.4 Shift Operators

As we have seen in Theorem 9.2.4 (p. 183), L^2 is isomorphic, as a Hilbert space, to the space $\ell^2(\mathbb{Z})$ (Example 2.1.17, p. 44) of square summable sequences via the correspondence

$$\mathcal{Z} : \ell^2(\mathbb{Z}) \to L^2 : (\ldots, a_{-m}, \ldots, a_0, a_1, \ldots, a_n, \ldots) \mapsto \sum_{n=-\infty}^{\infty} a_n z^n. \tag{9.36}$$

H^2 is isomorphic to the subspace $\ell^2(\mathbb{Z}_+)$ consisting of those elements such that $a_n = 0$ for $n < 0$. On the other hand, H_-^2 is isomorphic to the subspace $\ell^2(\mathbb{Z}_-)$ consisting of those satisfying $a_n = 0$ for $n \geq 0$. As already noted (p. 183), we denote respectively the orthogonal projection from L^2 to H^2 by P_+ and that from L^2 to H_-^2 by P_-:

$$P_+ \left(\sum_{n=-\infty}^{\infty} a_n z^n \right) = \sum_{n=0}^{\infty} a_n z^n,$$

$$P_- \left(\sum_{n=-\infty}^{\infty} a_n z^n \right) = \sum_{n=-\infty}^{-1} a_n z^n.$$

Define the *shift operator* of space $\ell^2(\mathbb{Z})$ by

$$U : \{a_n\} \mapsto \{a_{n-1}\}. \tag{9.37}$$

Via the correspondence (9.36), this naturally induces a shift operator in L^2. Specifically, this is given by $\mathcal{Z}U\mathcal{Z}^{-1}$. In what follows, identifying L^2 with $\ell^2(\mathbb{Z})$, we will also denote this induced shift operator by U. Observing that U is an operator mapping a_{n-1} to the nth term, we easily see from (9.36) that

$$U : \sum_{n=-\infty}^{\infty} a_n z^n \mapsto \sum_{n=-\infty}^{\infty} a_{n-1} z^n = \sum_{n=-\infty}^{\infty} a_n z^{n+1}.$$

That is, U is simply the multiplication operator by z.

Restrict this U to H^2 and denote it by S. Noticing that if $\sum a_n z^n \in H^2$, then $a_n = 0$ for $n < 0$, we see that the action of S is given by

$$S : \sum_{n=0}^{\infty} a_n z^n \mapsto \sum_{n=0}^{\infty} a_n z^{n+1} = \sum_{n=1}^{\infty} a_{n-1} z^n.$$

An important clue to the properties of linear operators is in characterizing their invariant subspaces. For example, finding a canonical form for a matrix corresponds to decomposing it to the direct sum of its actions on its invariant subspaces. This motivates us to ask to characterize the invariant subspaces of S in H^2.

Obvious ones are those consisting of multiples of a fixed element z^m $(m = 0, 1, \ldots)$:

$$z^m H^2 = \{z^m \phi : \phi \in H^2\}. \tag{9.38}$$

This is clearly S-invariant.[97] It is immediate to see that the set of multiples of an arbitrary (but fixed) polynomial $p(z)$

$$p(z)H^2 = \{p(z)\phi \; : \; \phi \in H^2\} \qquad (9.39)$$

is again S-invariant. The same can be said of an *arbitrary inner* $p(z)$.

The converse of this last statement is also true. It is known as Beurling's theorem.

Theorem 9.4.1 (Beurling). *Let $M \subset H^2$ be an S-invariant subspace. Then there exists an inner function $\psi(z)$ such that $M = \psi H^2$. Such a ψ is unique up to constants of modulus* 1.

We omit the proof. The reader is referred to, for example, Hoffman [21].

9.5   Nehari Approximation and Generalized Interpolation

Various operator representations in Hilbert space arise from the shift operator in H^2. Computation of their norms leads to solutions to varied classical problems in analysis. The generalized interpolation theorem by Sarason [49] gives a fundamental result on the computation of such norms.

Let $V \subset H^2$, $V \neq H^2$ be an arbitrary shift invariant (i.e., S-invariant) subspace. According to Beurling's theorem, Theorem 9.4.1, there exists a unique (up to constants with modulus 1) nonconstant inner function ψ such that $V = \psi H^2$. Denote its orthogonal complement in H^2 by $H(\psi)$, i.e., $H(\psi) = (\psi H^2)^\perp = H^2 \ominus \psi H^2$, where the latter denotes the orthogonal complement of ψH^2 in H^2. Following the usual convention in this context, we adopt the latter notation in what follows. Denote by P_ψ the orthogonal projection from L^2 to $H(\psi)$, and let T be the operator obtained by restricting S to $H(\psi)$ and projecting also to S as $P_\psi S|_{H(\psi)}$, i.e.,

$$T : H(\psi) \to H(\psi) : x \mapsto P_\psi Sx. \qquad (9.40)$$

This T is called the *compressed shift*. Generalizing this, we define for an arbitrary H^∞ function f its *compression* by

$$f(T) : H(\psi) \to H(\psi) : x \mapsto P_\psi M_f x, \qquad (9.41)$$

where M_f denotes the *multiplication operator* in L^2 defined by

$$M_f : L^2 \to L^2 : x \mapsto fx. \qquad (9.42)$$

If an operator A in $H(\psi)$ can be represented as $A = f(T)$ for some $f \in H^\infty$, f is said to *interpolate* A. The idea here is to represent a shift invariant operator in the form of $f(T)$ in $H(\psi)$ for some suitable inner ψ, employing H^2 as a universal model for Hilbert space.

Let us first note that $f(T)$ commutes with T; i.e., the following lemma holds.

Lemma 9.5.1. *For every $f \in H^\infty$, $f(T)T = Tf(T)$.*

Proof. We first show that

$$P_\psi f P_\psi x = P_\psi fx \qquad (9.43)$$

---

[97]i.e., invariant under the action of S.

for every $x \in H^2$. Indeed, write $x = P_\psi x + (x - P_\psi x)$. Then $x - P_\psi x \in H(\psi)^\perp = \psi H^2$. Hence there exists $y \in H^2$ such that $x - P_\psi x = \psi y$. Since $f \in H^\infty$, it follows that $f(x - P_\psi x) = \psi f y \in \psi H^2 = H(\psi)^\perp$ (note that $f y \in H^2$ by Lemma 9.1.4). This yields $P_\psi f(x - P_\psi x) = 0$, and hence (9.43) follows.

Now (9.43) readily implies

$$f(T)Tx = P_\psi f P_\psi zx = P_\psi f zx = P_\psi zf x = P_\psi z P_\psi f x = Tf(T)x. \qquad \square$$

Sarason's theorem gives the converse: It claims that an operator that commutes with the compressed shift is necessarily expressible as $f(T)$ for some $f \in H^\infty$, and its operator norm is equal to the H^∞ norm of f.

Theorem 9.5.2 (Sarason [49]). *Suppose that $A \in \mathcal{L}(H(\psi))$ satisfies $AT = TA$. Then there exists $f \in H^\infty$ such that $A = f(T)$ and $\|A\| = \|f\|_\infty$.*

In what follows, we give a proof of this theorem by way of Nehari's theorem. Nehari's theorem gives the best approximation of an L^∞ function by an H^∞ function, and it is in itself a very interesting and useful result.

Let us start with the definition of Hankel operators. An operator $A : H^2 \to H_-^2$ is a *Hankel operator* if

$$P_- zAp(z) = Azp(z) \qquad (9.44)$$

holds for every polynomial $p(z)$.

Recalling that H^2 and H_-^2 have as their bases $\{z^n\}_{n=0}^\infty$ and $\{z^{-n}\}_{n=1}^\infty$, respectively, we can express A as a matrix in terms of these bases as

$$A = \begin{bmatrix} a_0 & a_1 & a_2 & \cdots \\ a_1 & a_2 & a_3 & \cdots \\ a_2 & a_3 & a_4 & \cdots \\ \vdots & \vdots & \vdots & \ddots \end{bmatrix}$$

according to (9.44). A matrix having this form is called a *Hankel matrix*.

The original problem that Nehari [41] posed is the following: Given a Hankel matrix as above, under what conditions does it give a bounded linear operator on ℓ^2? The solution he gave is as follows.

Theorem 9.5.3 (Nehari [41]). *A necessary and sufficient condition for a Hankel matrix A to give a bounded linear operator on ℓ^2 is that there exists a function $\phi \in L^\infty$ such that its Fourier coefficients (formula (9.26), p. 183) satisfy $\hat{\phi}(n) = a_n$, $n = 0, 1, 2, \ldots$. Under this condition, the operator norm $\|A\|$ of A on ℓ^2 is given by*

$$\|A\| = \inf\{\|\phi\|_\infty : \hat{\phi}(n) = a_n, n = 0, 1, 2, \ldots\}. \qquad (9.45)$$

The theorem above can be reinterpreted as follows: Let us, for the convenience of notation, change the numbering of indices to $z^n \leftrightarrow z^{-n-1}$, $n = 0, 1, \ldots$. Then (9.45) becomes

$$\|A\| = \inf\{\|\phi\|_\infty : \hat{\phi}(-n) = a_{n-1}, n = 1, 2, \ldots\}.$$

Let

$$G(z) := a_1 z^{-1} + a_2 z^{-2} + a_3 z^{-3} + \cdots.$$

Then the set of all such ϕ's constitutes an affine subspace of L^∞:

$$G + H^\infty = \{G + \psi \ : \ \psi \in H^\infty\}.$$

Hence (9.45) means that

$$\|A\| = \inf_{\psi \in H^\infty} \|G - \psi\|_\infty = \mathrm{dist}_{L^\infty}(G, H^\infty) = \|G + H^\infty\|_{L^\infty / H^\infty}. \qquad (9.46)$$

That is, Nehari's theorem gives the best approximation of an L^∞ function by an H^∞ function. Here $\mathrm{dist}_X(f, M)$ denotes the distance of f in the Banach space X to the closed subspace M:

$$\mathrm{dist}_X(f, M) := \inf_{m \in M} \|f - m\|. \qquad (9.47)$$

According to Definition 2.3.3 and (2.26) on page 52, this is the norm of the coset $x + M$ in the quotient space X/M. Hence we see that the norm above is equal to $\|G + H^\infty\|_{L^\infty / H^\infty}$.

We now give a proof of this theorem in the following extended form.

Theorem 9.5.4 (Nehari's theorem). *A continuous linear operator from H^2 to H^2_- is a Hankel operator if and only if there exists an L^∞ function f such that*

$$Ax = \Gamma_f x := P_- M_f x, \quad x \in H^2. \qquad (9.48)$$

Moreover,

$$\|A\| = \mathrm{dist}_{L^\infty}(f, H^\infty) \qquad (9.49)$$

holds.

Remark 9.5.5. If there exists such an f satisfying (9.48), then A clearly becomes a Hankel operator.

Let us first give a dual characterization of space H^∞.

Proposition 9.5.6.

$$H^\infty = (L^1 / H^1_0)' = (H^1_0)^\perp, \qquad (9.50)$$

where H^1_0 denotes the subspace of H^1 consisting of functions that vanish at the origin $z = 0$.

Proof. As was already noted in Theorem 9.2.3 on page 182,

$$H^\infty = \left\{ f \in L^\infty \ : \ \int f(\theta) e^{in\theta} d\theta = 0, \ n = 1, 2, \ldots \right\}.$$

Recall also that $L^\infty = (L^1)'$ (Example 4.1.8, p. 80). Then H^∞ is the annihilator of the subspace of L^1 spanned by $\{e^{in\theta}\}_{n=1}^\infty$ (observe that the term for $n = 0$ is missing). This subspace is precisely H^1_0. Hence, by Theorem 4.4.3 (p. 87) on duality of quotient spaces, (9.50) readily follows. $\square$

The following theorem gives a representation for $H(\psi)$ as well as for the orthogonal projection $P_\psi : H^2 \to H(\psi)$.

Theorem 9.5.7. *The space $H(\psi)$ can be represented as*

$$H(\psi) = \{x \in H^2 : \tilde{\psi}x \in H^2_-\}. \tag{9.51}$$

Define $\pi^\psi : H^2 \to H(\psi)$ by

$$\pi^\psi : H^2 \to H(\psi) : x \mapsto \psi P_- \tilde{\psi}x. \tag{9.52}$$

This gives the orthogonal projection from H^2 to $H(\psi)$, that is, $\pi^\psi = P_\psi$.

Proof. Take $x \in H^2$ such that $\tilde{\psi}x \in H^2_-$. For every $y \in H^2$,

$$(x, \psi y) = (\tilde{\psi}x, y)_{L^2}.$$

Since $\tilde{\psi}x \in H^2_-$, the right-hand side is 0, and this implies $x \perp \psi H^2$. Conversely, if $x \in H(\psi)$, the identity above implies that $\tilde{\psi}x$ must be orthogonal to every $y \in H^2$. Since $H^2_- = L^2 \ominus H^2$, this means $\tilde{\psi}x \in H^2_-$. Hence (9.51) holds.

For the claim on the orthogonal projection P_ψ, let us first show that $\pi^\psi x \in H^2$ for every $x \in H^2$. Decompose $\tilde{\psi}x$ as

$$\tilde{\psi}x = P_+ \tilde{\psi}x + P_- \tilde{\psi}x.$$

Multiply both sides by ψ. Then

$$\pi^\psi x = \psi P_- \tilde{\psi}x = \psi(\tilde{\psi}x - P_+ \tilde{\psi}x) = x - \psi P_+ \tilde{\psi}x. \tag{9.53}$$

Since $x, \psi P_+ \tilde{\psi}x \in H^2$, $\pi^\psi x \in H^2$ follows.

This readily yields $\pi^\psi x \in H(\psi)$. Indeed, $\tilde{\psi}\pi^\psi x = \tilde{\psi}\psi P_- \tilde{\psi}x = P_- \tilde{\psi}x \in H^2_-$. Furthermore, (9.53) implies $x - \pi^\psi x = \psi P_+ \tilde{\psi}x \in \psi H^2$, which in turn yields $(x - \pi^\psi x) \perp H(\psi)$. This means π^ψ is the orthogonal projection and hence $\pi^\psi = P_\psi$. $\qquad\square$

Let us now turn to the proof of Nehari's theorem (p. 192). The proof follows that in [42].

Proof of Nehari's theorem. The sufficiency is clear as seen in Remark 9.5.5. We need only prove the necessity and (9.49). Let A be a Hankel operator. Condition (9.44) implies $P_- z^n A1 = Az^n$, $n \geq 0$. Hence for every polynomial $p(z)$, $P_- p(z)A1 = Ap(z)$. If q is a polynomial with $q(0) = 0$, then $\tilde{q}(z) \in H^2_-$, and

$$(Ap, \tilde{q}) = (P_- pA1, \tilde{q}) = (pA1, \tilde{q}) = \int_{\mathbb{T}} pA1 \cdot qd\theta = \int_{\mathbb{T}} pA1 \cdot qd\theta = \int_{\mathbb{T}} pq \cdot A1d\theta, \tag{9.54}$$

because $(P_+ pA1, \tilde{q}) = 0$. Here $\int_{\mathbb{T}} f d\theta$ denotes $\int_0^{2\pi} f(e^{i\theta})d\theta$. According to Lemma 9.5.8, to be shown below, the set of all products pq with conditions

$$\|p\|_2 \leq 1, \|q\|_2 \leq 1,$$
$$q(0) = 0,$$

and p, q are polynomials

constitutes a dense subset of the unit ball of $H_0^1 := \{g \in H^1 \ : \ g(0) = 0\}$. Hence, by using (9.54), we can compute the operator norm of A as

$$
\begin{aligned}
\|A\| &= \sup_{\|p\|_2 \leq 1} \|Ap\| \quad \text{from Theorem 5.1.1 on page 89} \\
&= \sup_{\|p\|_2, \|q\|_2 \leq 1} |(Ap, \tilde{q})| \quad \text{from Remark 4.1.7 on page 80} \\
&= \sup_{\|p\|_2, \|q\|_2 \leq 1} \left| \int_{\mathbb{T}} pq\, A 1 d\theta \right| \\
&= \sup_{\|g\|_1 \leq 1, g(0)=0} \left| \int_{\mathbb{T}} g A 1 d\theta \right|.
\end{aligned}
$$

Since this last quantity is finite, the linear functional

$$
g \mapsto \int_{\mathbb{T}} g A 1 d\theta
$$

defined for polynomials g with $g(0) = 0$ can be extended to a continuous linear functional over H_0^1. Since the duality here is taken between L^1 and L^∞, there exists an L^∞ function f such that

$$
\int_{\mathbb{T}} g A 1 d\theta = \int_{\mathbb{T}} f g d\theta, \quad g \in H_0^1. \tag{9.55}
$$

This, along with (9.54), yields

$$
\begin{aligned}
(Ap, \tilde{q}) &= (fp, \tilde{q}) \\
&= (P_- fp, \tilde{q}) \quad \text{(because } \tilde{q} \in H_-^2) \\
&= (\Gamma_f p, \tilde{q})
\end{aligned}
$$

for polynomials p, q such that $q(0) = 0$. When this condition is satisfied, p and $\tilde{q}$ consti-tute, respectively, dense subspaces of H^2 and H_-^2, and this identify holds for all elements of H^2 and H_-^2. Hence $A = \Gamma_f$.

Now recall

$$
H^\infty = (H_0^1)^\perp
$$

(Proposition 9.5.6, p. 192) and $L^\infty = (L^1)'$ (Example 4.1.8, p. 80). Then Theorem 4.4.2 on page 85 yields

$$
L^\infty / H^\infty = (L^1)' / (H_0^1)^\perp = (H_0^1)'.
$$

Hence the norm of the functional (9.55) (on H_0^1) is equal to the norm of $f + H^\infty$ in L^∞ / H^∞, namely, the distance from f to H^∞. This proves (9.49).   □

We now give a proof of the following lemma used above.

Lemma 9.5.8. *Every element g in the unit ball of H^1 can be decomposed as the product $g = g_1 g_2$ of some elements g_1, g_2 belonging to the unit ball of H^2.*

Proof. Take the canonical factorization of g:

$$
g = g_i g_o.
$$

Here the inner factor g_i belongs to H^∞, and its absolute value is 1 almost everywhere on the unit circle. On the other hand, the outer factor $g_o \in H^1$ can be expressed as $g_o = e^h$ as an exponential. It now suffices to take $g_1 := g_i e^{h/2}$ and $g_2 := e^{h/2}$. Also, $\|g_1\|_2 = \|g_2\|_2 = \|g\|_1$ readily follow from these definitions.   □

This completes the proof of Nehari's theorem. Let us return to the proof of Sarason's theorem. The following lemma establishes a bridge between Sarason's theorem and Nehari's theorem.

Lemma 9.5.9. *Let A be a continuous linear operator of $H(\psi)$ into itself, and let $A_* := \psi^\sim A P_\psi$, which is a continuous linear operator from H^2 to H^2_-. Then a necessary and sufficient condition for A to commute with the compressed shift T ((9.40) on page 190), i.e., $AT = TA$, is that A_* be a Hankel operator.*

Proof. If $x \in H(\psi)$, then $x \perp \psi H^2$. Then for every $y \in H^2$, $\langle \psi^\sim x, y \rangle = \langle x, \psi y \rangle = 0$. This means that $\psi^\sim x \in H^2_-$. Hence A_* surely maps H^2 into H^2_-.

Suppose $AT = TA$. Since $T = P_\psi z|_{H(\psi)}$, this means that $P_\psi z A = A P_\psi z|_{H(\psi)}$. Composing this with P_ψ, we obtain

$$P_\psi z A P_\psi = A P_\psi z. \tag{9.56}$$

Now we observe that $\psi^\sim P_\psi x = P_- \psi^\sim x$ holds for every $x \in H^2$. Indeed, decompose x as $x = P_\psi x + \psi y$, $y \in H^2$. Then $\psi^\sim x = \psi^\sim P_\psi x + y$, and applying P_- to both sides, we obtain $P_- \psi^\sim x = P_- \psi^\sim P_\psi x = \psi^\sim P_\psi x$ because $\psi^\sim P_\psi x \in H^2_-$, as shown above. It now follows that

$$P_- z A_* = P_- z \psi^\sim A P_\psi = P_- \psi^\sim z A P_\psi$$
$$= \psi^\sim P_\psi z A P_\psi$$
$$= \psi^\sim A P_\psi z \quad \text{from (9.56)}$$
$$= A_* z.$$

That is, A_* is a Hankel operator.

Conversely, if A_* is a Hankel operator, reversing the procedure above, we obtain (9.56), and this readily implies that $TA = AT$, i.e., that A commutes with the compressed shift T.   □

We are now ready to complete the proof of Sarason's theorem (p. 191). This proof is also based on that in [42].

Proof of Sarason's theorem. Suppose that A commutes with T. Then Lemma 9.5.9 and Nehari's theorem, Theorem 9.5.4 (p. 192), imply that there exists $f \in L^\infty$ such that $\psi^\sim A P_\psi = \Gamma_f$. In particular, for $v \in \psi H^2$, $P_- f v = \psi^\sim A P_\psi v = 0$ (note that $P_\psi v = 0$). In other words, for every $u \in H^2$, $f \psi u \in H^2$. Note that $f\psi = f\psi 1 \in H^2$. This means that $f\psi$ is analytic in the unit disc. Since its boundary function must be bounded because of $f\psi \in L^\infty$, this implies that $f\psi$ must belong to H^∞. Hence f can be written as $f = \psi^\sim h$, $h \in H^\infty$. Then for $g \in H(\psi)$, we have

$$\psi^\sim A g = \psi^\sim A P_\psi g = \Gamma_f g = P_- f g = P_- \psi^\sim h g \quad \text{(note that } g = P_\psi g).$$

Multiplying by ψ on both sides, we obtain

$$Ag = \psi P_- \tilde{\psi} hg = P_\psi hg = h(T)g. \tag{9.57}$$

Here we have used $P_\psi = \psi P_- \tilde{\psi}$ (Theorem 9.5.7). Moreover, if we replace h by any element $h + \psi \chi$ ($\chi \in H^\infty$) in $h + \psi H^\infty$, we obtain the same result because $P_\psi \psi \chi g = \psi P_- \tilde{\psi} \psi \chi g = \psi P_- \chi g = 0$. Now since $\Gamma_f = \tilde{\psi} A P_\psi$, ψ is inner, and P_ψ is a projection, we obtain $\|\Gamma_f\| = \|A P_\psi\| = \|A\|$. Hence it follows from Nehari's theorem, Theorem 9.5.4 (or from (9.49)), that

$$\begin{aligned}
\|A\| = \|\Gamma_f\| &= \mathrm{dist}_{L^\infty}(f, H^\infty) \\
&= \inf\{\|\tilde{\psi} h + u\|_{L^\infty} : u \in H^\infty\} \\
&= \inf\{\|h + \psi u\|_\infty : u \in H^\infty\}.
\end{aligned}$$

If we show the existence of an H^∞ function ϕ that attains the infimum of the last row, then that is what is claimed in Sarason's theorem.

Suppose that the H^∞ norm of the sequence $\{\phi_n = h + \psi u_n\}$ converges to the infimum $\|A\|$. Then u_n is a sequence of functions analytic in the unit disc and bounded there. Hence this sequence forms what is known as a normal family in complex analysis, and by Montel's theorem,[98] it contains a subsequence that is convergent uniformly on every compact subset in the unit disc, and hence the limit is analytic in $\mathbb{D}$ and also bounded there. Let u be the limit. Then by the continuity of norms we clearly have $\|A\| = \|h + \psi u\|_\infty$. Putting $\phi := h + \psi u$ readily yields $A = \phi(T)$ and $\|A\| = \|\phi\|_\infty$, and this completes the proof. $\square$

9.6   Application of Sarason's Theorem

Sarason's theorem has a wide variety of applications. We here show its applications to the Nevanlinna–Pick interpolation problem and to the proof of the Carathéodory–Fejér theorem.

9.6.1   Nevanlinna–Pick Interpolation Theorem

Consider the following problem.

The Nevanlinna–Pick interpolation problem. Given distinct points $a_1, \ldots, a_n \in \mathbb{D}$ and $b_1, \ldots, b_n \in \mathbb{C}$, find conditions under which there exists a function $f \in H^\infty$ such that

1. interpolation conditions $f(a_i) = b_i$, $i = 1, \ldots, n$, are satisfied, and

2. $\|f\|_\infty \leq 1$,

and if such an $f \in H^\infty$ exists, construct one from the data a_i's and b_i's.

In order that the constraint $\|f\|_\infty \leq 1$ be satisfied, it is clearly necessary that $|b_i| \leq 1$. But this alone does not guarantee the existence of a solution. We need an extra condition involving the so-called Pick matrix.

Regarding the actual construction of an interpolant f, there exists an inductive algorithmic solution, known as Nevanlinna's algorithm. But we here give a very elegant, albeit not constructive, solution based on Sarason's theorem for an existence condition for f. The proof below follows those given in [49, 57].

---

[98] See standard textbooks on complex analysis, e.g., [3].

Put

$$\psi(z) := \prod_{i=1}^{n} \left(\frac{z - a_i}{1 - \overline{a}_i z} \right), \tag{9.58}$$

and recall $H(\psi) = H^2 \ominus \psi H^2$.

We start with the following lemma.

Lemma 9.6.1. *For ψ defined above, $H(\psi)$ is an n-dimensional vector space and possesses the basis*

$$g_i(z) := \frac{1}{1 - \overline{a}_i z}, \quad i = 1, \ldots, n. \tag{9.59}$$

Proof. Clearly ψH^2 is the set of all $q \in H^2$ that satisfy $q(a_i) = 0$, $i = 1, \ldots, n$. Cauchy's integral formula yields for $q \in H^2$

$$(q, g_i) = \frac{1}{2\pi} \int_{-\pi}^{\pi} \frac{q(e^{i\theta})}{1 - a_i e^{-i\theta}} d\theta = \frac{1}{2\pi} \int_{-\pi}^{\pi} \frac{e^{i\theta} q(e^{i\theta})}{e^{i\theta} - a_i} d\theta = \frac{1}{2\pi i} \int_{\mathbb{T}} \frac{q(z)}{z - a_i} dz = q(a_i). \tag{9.60}$$

Hence q belongs to ψH^2 if and only if $(q, g_i) = 0$, $i = 1, \ldots, n$. This means $\mathrm{span}[g_1, \ldots, g_n]$ is the orthogonal complement of ψH^2, i.e., is equal to $H(\psi)$.

It remains only to prove the linear independence of $\{g_1, \ldots, g_n\}$. Assume the contrary; that is, we suppose $g_1 \in \mathrm{span}[g_2, \ldots, g_n]$, without loss of generality. Let ψ_2 denote the Blaschke product consisting of $\{g_2, \ldots, g_n\}$. Then we should have $H^2 \ominus \psi H^2 = H(\psi) = H^2 \ominus \psi_2 H^2$, since $g_1 \in \mathrm{span}[g_2, \ldots, g_n]$. Hence $\psi H^2 = \psi_2 H^2$. Then every function q vanishing at $\{a_2, \ldots, a_n\}$ should also vanish at $a_1, \ldots, a_n$. Since $a_1, \ldots, a_n$ are mutually distinct, this is absurd. $\qquad\square$

The following lemma shows that g_i is an eigenvector of the adjoint T^* of the compressed shift T.

Lemma 9.6.2. *Under the notation above, $T^* g_i = \overline{a}_i g_i$, $i = 1, \ldots, n$.*

Proof. Take an arbitrary point ζ in $\mathbb{T}$. It is easy to check that $T^* g(\zeta) = (g(\zeta) - g(0))/\zeta$. Then

$$T^* g(\zeta) = \frac{g(\zeta) - g(0)}{\zeta} = \frac{\frac{1}{1 - \overline{a}_i \zeta} - 1}{\zeta} = \frac{\overline{a}_i}{1 - \overline{a}_i \zeta} = \overline{a}_i g_i. \qquad\square$$

Recall that we say an H^∞ function f interpolates a given operator A if $A = f(T)$. Why this is referred to as interpolation will be clear from the following proposition.

Proposition 9.6.3. *Define an operator $A : H(\psi) \to H(\psi)$ by $A^* g_i = \overline{b}_i g_i$, $i = 1, \ldots, n$. A necessary and sufficient condition for $f \in H^\infty$ to interpolate A is that $f(a_i) = b_i$, $i = 1, \ldots, n$.*

First note that defining A^* in this way makes A^* always commute with T^*, and then A commutes with T. Hence, by Sarason's theorem, there always exists $f \in H^\infty$ such that $A = f(T)$. This proposition claims that this abstract interpolation condition is indeed the same as the interpolation condition at $a_1, \ldots, a_n$.

Proof. Let $A = f(T) = \pi^\psi M_f$. Then there exist $q_i \in H^2$ such that $Ag_i = fg_i - \psi q_i$, $i = 1,\ldots,n$. Hence

$$(Ag_i)(a_i) = f(a_i)g_i(a_i). \tag{9.61}$$

On the other hand, the definition of A^* and (9.60) yield

$$(Ag_i)(a_i) = (Ag_i, g_i) = \left(g_i, A^* g_i\right) = \left(g_i, \overline{b}_i g_i\right) = b_i g_i(a_i).$$

This along with (9.61) implies $f(a_i) = b_i$. (Note here that $g_i(a_i) \neq 0$ by the definition (9.59) of g_i.)

Conversely, suppose $f(a_i) = b_i$, $i = 1,\ldots,n$, and we show that $f(T) = A$. Since $\{g_1,\ldots,g_n\}$ forms a basis, it is enough to show $f(T)^* g_i = \overline{b}_i g_i$.

Similarly to the above, $f(T)g_i = fg_i - \psi q_i$, $i = 1,\ldots,n$, for some $q_i \in H^2$. Then for every $1 \leq i, j \leq n$, we have

$$
\begin{aligned}
\left(f(T)g_i, g_j\right) &= \left(g_i, f(T)^* g_j\right) = \left(fg_i - \psi q_i, g_j\right) \\
&= \left(fg_i, g_j\right) \quad \text{(because } g_j \perp \psi H^2) \\
&= f(a_j)g_i(a_j) \quad \text{(by (9.60))} \\
&= b_j g_j(a_i) = b_j \overline{g_i(a_j)} \quad \text{(because } g_j(a_i) = \overline{g_i(a_j)}) \\
&= \left(g_i, \overline{b}_i g_j\right) \quad \text{(again by (9.60))} \\
&= \left(g_i, A^* g_j\right) = \left(Ag_i, g_j\right).
\end{aligned}
$$

Hence $f(T)g_i = Ag_i$, $i = 1,\ldots,n$. Since $\{g_1,\ldots,g_n\}$ is a basis, $A = f(T)$. $\square$

We are now ready to prove the Nevanlinna–Pick interpolation theorem.

Theorem 9.6.4 (Nevanlinna–Pick interpolation theorem). *Let $a_1,\ldots,a_n \in \mathbb{D}$, $b_1,\ldots,b_n \in \mathbb{C}$ be given. A necessary and sufficient condition for $f \in H^\infty$ to exist with the interpolation condition $f(a_i) = b_i$, $i = 1,\ldots,n$, $\|f\|_\infty \leq 1$ is that the Pick matrix*

$$\left[\frac{1 - \overline{b}_i b_j}{1 - \overline{a}_i a_j}\right]_{i,j=1,\ldots,n} \tag{9.62}$$

be nonnegative definite.

Proof. According to Proposition 9.6.3, $f \in H^\infty$ satisfies the interpolation condition if and only if the operator A defined in this proposition satisfies $A = f(T)$. According to Sarason's theorem, Theorem 9.5.2 (p. 191), there exists f satisfying this interpolation condition and $\|A\| = \|f\|_\infty$. Hence for an f to exist such that $\|f\|_\infty \leq 1$, it is necessary and sufficient that $\|A\| \leq 1$. Since $\|A\| = \|A^*\|$ (Proposition 5.2.2, p. 91), this is equivalent to $\|A^*\| \leq 1$. Hence it is enough to find a condition under which

$$\left(A^* g, A^* g\right) \leq (g, g) \tag{9.63}$$

holds for every $g \in H(\psi)$. Since by Lemma 9.6.1 $g \in H(\psi)$ admits a representation in terms of the basis $g_1,\ldots,g_n$ as

$$g = \alpha_1 g_1 + \cdots + \alpha_n g_n,$$

we must have

$$(g,g) = \sum_{1 \le i,j \le n} \alpha_i \overline{\alpha}_j \left(g_i, g_j \right) = \sum_{1 \le i,j \le n} \alpha_i \overline{\alpha}_j g_i(a_j) = \sum_{1 \le i,j \le n} \alpha_i \overline{\alpha}_j \frac{1}{1 - \overline{a}_i a_j}.$$

On the other hand, $A^* g = \alpha_1 \overline{b}_1 g_1 + \cdots + \alpha_n \overline{b}_n g_n$ by the definition of A^* (Proposition 9.6.3), and hence

$$\left(A^* g, A^* g \right) = \sum_{1 \le i,j \le n} \alpha_i \overline{\alpha}_j b_i \overline{b}_j \frac{1}{1 - \overline{a}_i a_j}.$$

Thus (9.63) holds if and only if

$$\sum_{1 \le i,j \le n} \alpha_i \overline{\alpha}_j b_i \overline{b}_j \frac{1}{1 - \overline{a}_i a_j} \le \sum_{1 \le i,j \le n} \alpha_i \overline{\alpha}_j \frac{1}{1 - \overline{a}_i a_j}$$

holds for every $[\alpha_1, \ldots, \alpha_n]^T$, i.e.,

$$\sum_{1 \le i,j \le n} \alpha_i \overline{\alpha}_j \frac{1 - b_i \overline{b}_j}{1 - \overline{a}_i a_j} \ge 0.$$

Since $[\alpha_1, \ldots, \alpha_n]^T$ is arbitrary, this means that the Pick matrix (9.62) is nonnegative definite. $\square$

9.6.2 The Carathéodory–Fejér Theorem

The Carathéodory–Fejér theorem gives a condition under which for given complex numbers $c_0, \ldots, c_n$ there exists a power series $f = \sum_{k=0}^{\infty} \hat{f}(k) z^k$ such that its first $n+1$ coefficients $\hat{f}(0), \ldots, \hat{f}(n)$ agree with $c_0, \ldots, c_n$ and its H^∞ norm is less than or equal to 1. (For an expression of coefficients $\hat{f}(k)$, see (9.26) on page 183.) We here derive this result by way of Sarason's theorem.

The problem considered here is the following.

The Carathéodory–Fejér interpolation problem. Let $(c_1, \ldots, c_n)$ be arbitrary complex numbers. Find a condition under which there exists an $f \in H^\infty$ such that

1. the interpolation condition $f = \sum_{k=0}^{\infty} \hat{f}(k) z^k$, $\hat{f}(k) = c_k$, $k = 0, \ldots, n$, is satisfied, and

2. $\| f \|_\infty \le 1$.

Theorem 9.6.5 (Carathéodory–Fejér). *A necessary and sufficient condition for a solution f to exist for the Carathéodory–Fejér interpolation problem is that the norm (maximal singular value) of the matrix*

$$A = \begin{bmatrix} c_0 & 0 & \cdots & 0 \\ c_1 & c_0 & \ddots & \vdots \\ \vdots & \ddots & \ddots & 0 \\ c_n & \cdots & c_1 & c_0 \end{bmatrix} \tag{9.64}$$

be less than or equal to 1.

Proof. To reduce the problem to Sarason's theorem, let us put $\psi := z^{n+1}$. Then $H(\psi) = H^2 \ominus \psi H^2$ is clearly the set of polynomials of degree at most n. In addition, P_ψ becomes the truncation operator

$$P_\psi : \sum_{k=0}^{\infty} a_k z^k \mapsto \sum_{k=0}^{n} a_k z^k.$$

Furthermore, since S is the multiplication by z, the action of the compressed shift ((9.40), p. 190) $T = P_\psi S|_{H(\psi)}$ on z^k is given by

$$T z^k = z^{k+1}, \quad k = 0, 1, \dots, n-1, \tag{9.65}$$
$$T z^n = 0. \tag{9.66}$$

Since $H(\psi)$ is an $n + 1$-dimensional subspace with basis $1, z, \dots, z^n$, a linear mapping $A : H(\psi) \to H(\psi)$ is completely determined by specifying how it maps the basis elements (Lemma 1.2.11, p. 16). Among such linear operators, those commuting with T, i.e., satisfying $TA = AT$, are of interest here.

Put

$$A1 = \sum_{k=0}^{n} c_k z^k.$$

Then by (9.65) and (9.66),

$$Az = AT1 = TA1 = \sum_{k=0}^{n-1} c_k z^{k+1}.$$

Note that the sum extends only to $n - 1$ because $T z^n = 0$. Similarly,

$$Az^j = \sum_{k=0}^{n-j} c_k z^{k+j}, \quad j = 0, \dots, n.$$

Expressing these in terms of a matrix, we obtain

$$A[1, \dots, z^n] = [1, \dots, z^n] \begin{bmatrix} c_0 & 0 & \cdots & 0 \\ c_1 & c_0 & \ddots & \vdots \\ \vdots & \ddots & \ddots & 0 \\ c_n & \cdots & c_1 & c_0 \end{bmatrix}.$$

Then Sarason's theorem 9.5.2 (p. 191) asserts that there exists an H^∞ function f such that $A = f(T)$ and $\|f\|_\infty = \|A\|$. (Here $\|A\|$ denotes the operator norm of A in $H(\psi)$.) Put $f := \sum_{k=0}^{\infty} \hat{f}(k) z^k$ and write down this condition: $A = f(T)$. Then we obtain $f(T)1 = P_\psi M_f 1 = \sum_{k=0}^{n} \hat{f}(k) z^k$, $f(T)z = P_\psi M_f z = \sum_{k=0}^{n-1} \hat{f}(k) z^{k+1}$, etc. Hence, as in the representation for A, we obtain

$$f(T)[1, \dots, z^n] = [1, \dots, z^n] \begin{bmatrix} \hat{f}(0) & 0 & \cdots & 0 \\ \hat{f}(1) & \hat{f}(0) & \ddots & \vdots \\ \vdots & \ddots & \ddots & 0 \\ \hat{f}(n) & \cdots & \hat{f}(1) & \hat{f}(0) \end{bmatrix}.$$

Therefore, $A = f(T)$ if and only if $\hat{f}(k) = c_k, k = 0, \ldots, n$. This is nothing but the required interpolation condition. Hence a necessary and sufficient condition for f with $\|f\|_\infty \leq 1$ to exist is that A in (9.64) satisfy $\|A\| \leq 1$. Since $H(\psi)$ is finite-dimensional with ℓ^2 norm, the operator norm here coincides with the maximal singular value (Problem 11 of Chapter 5, p. 105). $\qquad\square$

9.7 Nehari's Theorem—Supplements

There are still various other consequences of Nehari's theorem. One of them deals with the characterization of $\inf_{q \in H^\infty} \|f - q\|$ for a given $f \in L^\infty$. This plays a key role in H^∞ control theory, and we state the result. For a proof, the reader is referred to, for example, [67, 68]. We will discuss H^∞ control theory further in the next chapter.

Theorem 9.7.1. *Let* $f \in L^\infty$ *and* $\|\Gamma_f\| = \|f - q\|$, $q \in H^\infty$. *In addition, suppose that there exists* $v \in H^2$ *such that* $\|\Gamma_f v\| = \|\Gamma_f\| \|v\|$. *Then*

$$(f - q)v = \Gamma_f v. \qquad (9.67)$$

In fact, any H^2 function is nonzero almost everywhere in $\mathbb{T}$, and hence both sides of (9.67) can be divided by v. Hence the optimal solution q is uniquely determined by

$$q = f - \frac{\Gamma_f v}{v}. \qquad (9.68)$$

The theorem above assumes the existence of v such that $\|\Gamma_f v\| = \|\Gamma_f\| \|v\|$, i.e., a maximizing vector v that attains $\|\Gamma_f\|$. This does not necessarily exist, but when Γ_f is compact, it does. This is as seen in Problem 12 in Chapter 5.

Chapter 10

Applications to Systems and Control

Modern control and system theory is built upon a solid mathematical basis. Advanced mathematical concepts developed in this book are indispensable for further in-depth study of elaborate theory of systems and control. We here present some fundamentals of this theory.

10.1 Linear Systems and Control

Many physical systems are described by differential equations, ordinary or partial. We often wish to *control* such systems, natural or artificial. In other words, given a system, we want it to behave in such a way that is in some sense desirable for us. Normally, a system may contain three types of variables: an *input variable* that drives the system, an *output variable* that can either be observed by some devices or affect the external environment, and a *state variable* that describes the internal behavior of the system which is not necessarily observable from the external environment. Summarizing, a differential equation description may take the following form:

$$\frac{dx}{dt}(t) = f(x(t), u(t)), \tag{10.1}$$

$$y(t) = g(x(t)), \tag{10.2}$$

where $x(t) \in \mathbb{R}^n$, $u(t) \in \mathbb{R}^m$, and $y(t) \in \mathbb{R}^p$ denote, respectively, the state, input, and output variables. These two equations have the structure that

1. starting with an *initial state* x_0 at some time t_0, and for a given input $u(t)$, $t \geq 0$, (10.1) describes how the system state $x(t)$ evolves in time, and

2. the output $y(t)$ is determined by the present value of the state $x(t)$ according to (10.2).

The fundamental objective of *control* is to design or synthesize $u(t)$ so as to make the above system behave desirably. Here the control input $u(t)$ is something we can maneuver to control the system state $x(t)$, and the output $y(t)$ is then controlled accordingly. In general, it is necessary to make $x(t)$ behave nicely rather than merely controlling the behavior of $y(t)$ alone, as there can be some hidden behavior in $x(t)$ that does not explicitly appear in that of $y(t)$.

While most systems are nonlinear as above, or maybe even more complex as to be partial differential or integral equations, their treatment is often difficult. In such cases, one often resorts to a more simplified approximation. This amounts to linearizing the differential equation along a given "reference" trajectory, by taking the error (variation) from this prespecified reference trajectory.

We thus consider the simplified ordinary linear differential system equations as follows:

$$\frac{dx}{dt}(t) = Ax(t) + Bu(t), \tag{10.3}$$

$$y(t) = Cx(t). \tag{10.4}$$

Here $x(t) \in \mathbb{R}^n$, $u(t) \in \mathbb{R}^m$, and $y(t) \in \mathbb{R}^p$ are the state, input, and output vectors, as above, and matrices $A \in \mathbb{R}^{n \times n}$, $B \in \mathbb{R}^{n \times m}$, and $C \in \mathbb{R}^{p \times n}$ are constant matrices. When necessary, we regard an initial state x_0 as fixed at time t_0 as $x(t_0) = x_0$. For every such $x_0 \in \mathbb{R}^n$, (10.3) yields a unique solution once u is specified.

The interpretation of these equations is as follows: There is a "system" Σ described by (10.3), (10.4), and this system accepts input $u(t)$, and the state $x(t)$ changes according to the *state transition equation* (10.3) with initial state $x(t_0) = x_0$. The state variable $x(t)$ simultaneously produces an output $y(t)$ according to the *output equation* (10.4). Since the correspondences $[x, u] \mapsto dx/dt$ and $x \mapsto y$ are linear in (10.3), (10.4), we call system Σ a *linear system*. It is also *finite-dimensional*, since the space (called the *state space* of Σ) $\mathbb{R}^n$ to which $x(t)$ belongs is a finite-dimensional vector space over $\mathbb{R}$. For brevity of notation, we will denote the system Σ by the triple (A, B, C).

Example 10.1.1. Consider Figure 10.1. The pendulum is placed upside down and is supported by a free joint at the bottom. Ignoring the friction at the joint, and also that between the cart and the ground, its equation of motion is given as follows:

$$(M + m)\ddot{x} + ml(\ddot{\theta}\cos\theta - \dot{\theta}^2\sin\theta) = u, \tag{10.5}$$

$$\ddot{x}\cos\theta + l\ddot{\theta} - g\sin\theta = 0. \tag{10.6}$$

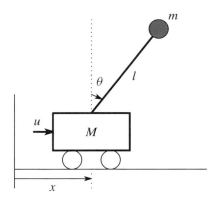

Figure 10.1. *Inverted pendulum*

Set $[x_1, x_2, x_3, x_4]^T := [x, \dot{x}, \theta, \dot{\theta}]^T$. Then the above equations can be rewritten as

$$
\frac{d}{dt}\begin{bmatrix} x_1 \\ x_2 \\ x_3 \\ x_4 \end{bmatrix} = \begin{bmatrix} x_2 \\ \dfrac{mlx_4^2 \sin x_3 - mg \sin x_3 \cos x_3}{M + m \sin^2 x_3} \\ x_4 \\ \dfrac{g}{l} \sin x_3 - \dfrac{m \sin x_3 \cos x_3 (lx_4^2 - g \cos x_3)}{(M + m \sin^2 x_3)l} \end{bmatrix} + \begin{bmatrix} 0 \\ \dfrac{1}{M + m \sin^2 x_3} \\ 0 \\ -\dfrac{\cos x_3}{(M + m \sin^2 x_3)l} \end{bmatrix} u,
$$

which is in the form of (10.1)–(10.2).

If we assume that θ and $\dot{\theta}$ are sufficiently small, the following linearized equations can be obtained from (10.5)–(10.6).

$$
\begin{aligned}
(M + m)\ddot{x} + ml\ddot{\theta} &= u, \\
\ddot{x} + l\ddot{\theta} - g\theta &= 0.
\end{aligned}
\tag{10.7}
$$

This can further be rewritten in state space form (i.e., that not involving higher-order derivatives) as

$$
\frac{d}{dt}\begin{bmatrix} x_1 \\ x_2 \\ x_3 \\ x_4 \end{bmatrix} = \begin{bmatrix} 0 & 1 & 0 & 0 \\ 0 & 0 & -\dfrac{mg}{M} & 0 \\ 0 & 0 & 0 & 1 \\ 0 & 0 & \dfrac{(M+m)g}{Ml} & 0 \end{bmatrix} \begin{bmatrix} x_1 \\ x_2 \\ x_3 \\ x_4 \end{bmatrix} + \begin{bmatrix} 0 \\ \dfrac{1}{M} \\ 0 \\ -\dfrac{1}{Ml} \end{bmatrix} u
$$

$$
:= Ax + Bu,
\tag{10.8}
$$

which is in the form of (10.3).

10.2 Control and Feedback

Let us examine Example 10.1.1 a little further. The eigenvalues of the A matrix in (10.8) are $0, 0, \pm\omega$, where $\omega = \sqrt{(M+m)g/Ml}$. The fundamental solution for $u = 0$ is given by $e^{At}x_0$ with initial condition $x(0) = x_0$. Clearly the positive eigenvalue ω and also the double root 0 make the total system *unstable*; i.e., $e^{At}x_0$ does not approach zero (i.e., the equilibrium state) as $t \to \infty$. That is, if the initial state were nonzero, the solution $x(t)$ would not remain in a neighborhood of the origin, and the pendulum would fall. While this conclusion is based on the linearized model (10.8) or (10.7), the conclusion easily carries over to the nonlinear model (10.5) and (10.6) as well.

Now consider the following problem: *is it possible to keep the system state $x(t)$ in a neighborhood of the origin (or even approach the origin)?* As it is, without the action of the input u, this is clearly not possible, but it can be with a suitable choice of an input u.

How do we do this? A rough idea is to choose $u(t)$ such that when the pendulum is falling to the right, we apply an $u(t)$ to move the cart to the right to "catch up" and do the opposite when it is falling to the left. This is only a rough scenario, and it need not work without a more elaborate choice of u.

There is another serious problem: even if we happen to find a suitable control input u, that depends on the initial state x_0. If we start from a different initial state x, then we need to find a different u for this x. This appears to be an impossibly complicated problem: for each initial state x, we need to find a suitable input function u corresponding to x.

Note that this is very different in nature from the seemingly easy appearance of the differential equations (10.3), (10.4). The objective here is not merely to solve them. Indeed, it is almost trivial to give a concrete solution,

$$x(t) = e^{At}x_0 + \int_0^t e^{A(t-\tau)}Bu(\tau)d\tau; \tag{10.9}$$

this is far from our objective to *design* the controller here to make the total system behave desirably.

The notion of *feedback* is particularly useful. It uses the measured output signal $y(t)$ and modifies the input signal $u(t)$ accordingly. A simple example is a *linear feedback*. This gives a control input u as a linear function of the output y or the state x. The former is called an *output feedback*, while the latter is called a *state feedback*. Consider the simplest case where $u(t) = Kx(t)$, where K is a constant matrix.[99] Substituting this into (10.3), we obtain a new differential equation,

$$\frac{dx}{dt}(t) = (A + BK)x(t), \tag{10.10}$$

and this changes the system dynamics completely. Part of the problem in control can be reduced to designing a suitable K to make the system behave desirably. But what do we mean by "desirable?" One requirement is stability. The linear system (10.3) is said to be *(asymptotically) stable* if $e^{At}x \to 0$ as $t \to \infty$ for every initial state x. It is *unstable* if it is not stable. System (10.3) is stable if and only if the real parts of all eigenvalues of A are negative. The inverted pendulum in Example 10.1.1 is clearly unstable. Note that the input term B does not enter into this definition. The role of feedback as in (10.10) is to change the dynamics A as $A \mapsto A + BK$ with some K. How can we achieve stability then?

Lord Maxwell's paper [34] is probably the first tractable theoretical account on stability of control systems. There he derived a condition for stability of a third-order linear differential equation obtained from an equation of a centrifugal governor. Needless to say, centrifugal governors played a crucial role in Watt's steam engine invention, and its stability was obviously of central importance. The problem is how one can obtain a condition on coefficients without solving for the characteristic roots of the system matrix A.

If Σ is not stable, we may want to modify the system (for example, by feedback) to make it stable. This is called a stabilization problem. In other words, we want to study properties of A, B, and C or combinations thereof to find out how Σ can or cannot be stabilized. Solving (10.3), (10.4) is only a small part of it, or may not even be necessary, to arrive at a solution.

However, before asking more elaborate system questions, let us ask a fundamental question: *Can Σ be controlled at all?* This is no doubt one of the most basic questions that can be raised if one wants to *control Σ.*

---

[99]In general, we should consider output feedback, and the feedback may also depend on the past values of y—not just the present value $y(t)$; but this simple example case gives rise to a generic construction in conjunction with a *state estimator* often known as the *Kalman filter* [24] or an *observer*.

Somewhat surprisingly, it was relatively recently, with the advent of so-called modern control theory in the 1950s (see [25]), that such a question was properly addressed in the literature. Part of the reason was that classical control theory did not employ the "state space" model (10.3), (10.4) but described its direct input to output relationship via so-called *transfer functions*. We will come back to this issue later, but let us proceed to formulate the notion of *controllability*.

10.3   Controllability and Observability

Controllability and observability are two fundamental notions in modern system and control theory. The former deals with the capability of how control inputs can affect the system, while the latter is concerned with how we can identify the internal behavior of the system by observing the output. Interestingly, for finite-dimensional linear systems such as (10.3)–(10.4), they are completely dual to each other.

Let us start our discussion with controllability.

10.3.1   Controllability

Let $x_0, x_1 \in \mathbb{R}^n$ be two arbitrary elements in the state space $\mathbb{R}^n$. We say that system Σ is *controllable* if there exists a control input u and a time interval $[t_0, t_1)$ such that u steers the initial state x_0 given at time t_0 to the final state x_1 at time t_1 along the trajectory given by (10.3). Since (10.3) is clearly invariant under time shifts, we may take $t_0 = 0$. Formalizing this and using (10.9), we arrive at the following definition.

Definition 10.3.1. System Σ defined by (10.3)–(10.4) is said to be *controllable* if for every pair $x_0, x_1 \in \mathbb{R}^n$, there exists an input function u given on some interval $[0, t_1)$ such that

$$x_1 = e^{At_1} x_0 + \int_0^{t_1} e^{A(t_1 - \tau)} Bu(\tau) d\tau. \tag{10.11}$$

If this holds for $x_0, x_1 \in \mathbb{R}^n$, we also say that x_0 is *controllable* to x_1 in time t_1. See Figure 10.2.

A related notion is *reachability*. We say that x_1 is *reachable* from 0 if (10.11) holds for $x_0 = 0$.[100] System Σ is said to be *reachable* if every x is reachable from 0.

Note here that for a fixed pair (x_0, x_1), t_1 and u need not be unique, and hence the trajectory connecting these two states is not unique, either (Figure 10.2).

Reachability as above is clearly a special case of controllability. However, the two notions are actually equivalent in the present context. We state this as a lemma.

Lemma 10.3.2. *System Σ as above is controllable if and only if it is reachable.*

Proof. We need only prove the sufficiency.

Suppose that Σ is reachable. We want to show that for every pair $x_0, x_1 \in \mathbb{R}^n$, x_0 is controllable to x_1 (or x_1 can be reached from x_0) by some action of an input.

---

[100]We also say that x_1 is reachable from 0 in time t_1.

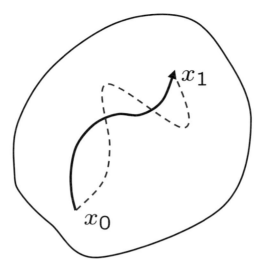

Figure 10.2. *Control steering x_0 to x_1*

First, there exist T and u such that

$$x_1 = \int_0^{T_1} e^{A(T_1-\tau)} Bw(\tau) d\tau. \qquad (10.12)$$

Since every state must be reachable from 0, we see that $-e^{AT_1}x_0$ is also reachable from 0. That is, there exist u and T_0 such that

$$-e^{AT_1}x_0 = \int_0^{T_0} e^{A(T_0-\tau)} Bu(\tau) d\tau. \qquad (10.13)$$

Let $T := \max\{T_0, T_1\}$. Note that we may replace both T_0 and T_1 above by T as follows:

$$-e^{AT}x_0 = \int_0^{T} e^{A(T-\tau)} Bu(\tau) d\tau$$

and

$$x_1 = \int_0^{T} e^{A(T-\tau)} Bw(\tau) d\tau.$$

If $T > T_0$, then set u to zero on $[T - T_0, T)$ to obtain

$$-e^{AT}x_0 = e^{A(T-T_0)} \int_0^{T_0} e^{A(T_0-\tau)} Bu(\tau) d\tau = \int_0^{T} e^{A(T-\tau)} Bu(\tau) d\tau \qquad (10.14)$$

because u is zero on $[T - T_0, T)$. Similarly, if $T > T_1$, set

$$\overline{w} := \begin{cases} 0, & 0 \le t < T - T_1, \\ w(t - T + T_1), & T - T_1 \le t < T, \end{cases}$$

to obtain

$$x_1 = \int_0^T e^{A(T-\tau)} B\overline{w}(\tau)d\tau. \tag{10.15}$$

Defining $v(t) := u(t) + \overline{w}(t)$ for $0 \le t < T$ then easily yields

$$e^{AT}x_0 + \int_0^T e^{A(T-\tau)} Bv(\tau)d\tau = e^{AT}x_0 + \int_0^T e^{A(T-\tau)} Bu(\tau)d\tau + \int_0^T e^{A(T-\tau)} B\overline{w}(\tau)d\tau = x_1,$$

by (10.14) and (10.15), as desired. $\qquad\square$

If the state space is finite-dimensional, reachability implies that the system is reachable in a *uniformly bounded time*.

Proposition 10.3.3. *Suppose that system Σ defined by (10.3) is reachable. Then there exists $T > 0$ such that for every $x \in \mathbb{R}^n$ there exists u such that*

$$x = \int_0^T e^{A(T-\tau)} u(\tau)d\tau.$$

Sketch of proof. Consider a basis $x_1, \dots, x_n$ for $\mathbb{R}^n$, and take suitable $u_1, \dots, u_n$ that gives

$$x_i = \int_0^{T_i} e^{A(T-\tau)} u_i(\tau)d\tau.$$

Take $T := \max\{T_1, \dots, T_n\}$, and modify u_i suitably as in the proof of Lemma 10.3.2 to obtain

$$x_i = \int_0^T e^{A(T-\tau)} u_i(\tau)d\tau.$$

Express x as a linear combination of x_i, and also take u as a linear combination of u_i accordingly. This u drives 0 to x in time T. $\qquad\square$

We now give some criteria for reachability. Given (10.3), consider the mapping

$$\mathcal{R}_T : L^2[0,T] \to \mathbb{R}^n : u \mapsto \int_0^T e^{At} Bu(t)dt. \tag{10.16}$$

It is easy to see that $x \in \mathbb{R}^n$ is reachable from zero in time T if and only if $x \in \operatorname{im} \mathcal{R}_T$.

Now let $\mathcal{R}_T^* : \mathbb{R}^n \to L^2[0,T]$ be the adjoint operator of $\mathcal{R}_T$. According to Example 5.5.2, the adjoint operator $\mathcal{R}_T^*$ is given by

$$\mathcal{R}_T^* : \mathbb{R}^n \to L^2[0,T] : x \mapsto B^* e^{A^* t} x.$$

Now consider the composed operator $M_T := \mathcal{R}_T \mathcal{R}_T^* : \mathbb{R}^n \to \mathbb{R}^n$. We first show the following lemma.

Lemma 10.3.4. *The kernel of M_T consists of the elements that are not reachable from zero in time T, except 0 itself.*

Proof. Let $x \in \ker M_T$, i.e., $M_T x = 0$. Then

$$x^* M_T x = \int_0^T x^* e^{At} B B^* e^{A^* t} x \, dt = 0.$$

Hence

$$\int_0^T \left\| x^* e^{At} B \right\|^2 dt = 0,$$

where $\|\cdot\|$ is the Euclidean norm of $\mathbb{R}^n$. Since the integrand is continuous, it follows that $x^* e^{At} B \equiv 0$ on $[0, T]$.

Now suppose that there exists $u \in L^2[0, T]$ such that

$$x = \int_0^T e^{A(T-\tau)} B u(\tau) d\tau.$$

Multiply x^* from the left to obtain

$$x^* x = \int_0^T x^* e^{A(T-\tau)} B u(\tau) d\tau = \int_0^T x^* e^{At} B u(T - t) dt = 0,$$

and hence $x = 0$.   □

We are now ready to give the following theorem.

Theorem 10.3.5. *The following conditions are equivalent:*

1. *System Σ defined by* (10.3) *is reachable.*

2. *M_T is of full rank, i.e., it is invertible.*

3.
$$\mathrm{rank} \left[B, AB, \dots, A^{n-1} B \right] = n, \tag{10.17}$$

   where n is the dimension of the state space, i.e., the size of A.

Proof. Suppose Σ is reachable. Then by Lemma 10.3.4, $\ker M_T = 0$. Since M_T is a square matrix, this means that M_T must be of full rank.

Conversely, suppose that M_T is invertible. Then Theorem 3.3.1 (p. 75) implies that

$$u(t) := \mathcal{R}_T^* M_T^{-1} x$$

gives a solution to $\mathcal{R}_T u = x$; that is, x is reachable from 0 in time T by input $u \in L^2[0, T]$. (Moreover, this u is the input with minimum norm, according to Theorem 3.3.1.) Hence property 2 implies reachability.

Now suppose that condition 2 fails; i.e., there exists a nonzero $x \in \ker M_T$. Then, as in the proof of Lemma 10.3.4, $x^* e^{At} B \equiv 0$ on $[0, T]$. Differentiating successively and evaluating at $t = 0$, we see that

$$x^* \left[B, AB, \dots, A^{n-1} B \right] = 0. \tag{10.18}$$

Since $x \neq 0$, $[B, AB, \ldots, A^{n-1}B]$ cannot have full rank. Hence condition 3 implies condition 2.

Conversely, suppose that (10.17) fails; i.e., there exists a nonzero vector x such that (10.18) holds. Note that by the Cayley–Hamilton theorem

$$A^n = -\alpha_1 A^{n-1} - \cdots - \alpha_n I. \tag{10.19}$$

Hence we have

$$x^* A^k B = 0, \quad k = 0, 1, 2, \ldots.$$

This implies that $x^* e^{At} B$ is identically zero since it is an analytic function and its Taylor coefficients are all zero. Hence $B^* e^{A^* t} x \equiv 0$, which readily implies $x \in \ker M_T$. $\square$

Remark 10.3.6. Note that M_T is always a nonnegative definite matrix. Hence the above condition 2 is equivalent to M_T being positive definite.

A remarkable consequence of controllability is that it enables the property known as *pole-shifting*. We state this as a theorem without giving a detailed proof.

Theorem 10.3.7 (pole-shifting/pole-assignment theorem). *Let Σ be the system defined by (10.3), and suppose that it is controllable. Take any complex numbers $\lambda_1, \ldots, \lambda_n$. Then there exists an $n \times m$ matrix K such that the characteristic polynomial $\chi(s) = \det(sI - A + BK)$ has zeros $\lambda_1, \ldots, \lambda_n$.*

For a complete proof, we refer the reader to standard textbooks such as [19, 23, 56]. We here give a simple proof for the sufficiency for the special case of single inputs, i.e., $m = 1$ in (10.3).

Proof of sufficiency for the case $m = 1$. In this case, B is a column vector, and we write b for B. By Theorem 10.3.5,

$$\{b, Ab, \ldots, A^{n-1}b\}$$

is linearly independent and hence forms a basis for $\mathbb{R}^n$. Let $\det(sI - A) = s^n + \alpha_1 s^{n-1} + \cdots + \alpha_n$ be the characteristic polynomial of A. Consider the set of vectors

$$e_1 = A^{n-1}b + \alpha_1 A^{n-2}b + \cdots + \alpha_n b$$

$$\vdots$$

$$e_{n-1} = Ab + \alpha_1 b,$$

$$e_n = b.$$

This is a triangular linear combination of $\{b, Ab, \ldots, A^{n-1}b\}$ and hence forms a basis. With respect to this basis, it is easy to see that A and b take the form

$$A = \begin{bmatrix} 0 & 1 & 0 & \cdots & 0 \\ 0 & 0 & 1 & \cdots & 0 \\ & & \cdots & \cdots & 0 \\ 0 & 0 & 0 & \cdots & 1 \\ -\alpha_n & -\alpha_{n-1} & -\alpha_{n-2} & \cdots & -\alpha_1 \end{bmatrix}, \quad b = \begin{bmatrix} 0 \\ 0 \\ \vdots \\ 0 \\ 1 \end{bmatrix}. \tag{10.20}$$

Let $\chi(s) = s^n + \beta_1 s^{n-1} + \cdots + \beta_n$. To make $\xi(s)$ have roots $\lambda_1, \ldots, \lambda_n$, it is sufficient that any such $\xi(s)$ can be realized as $\det(sI - A + bK)$. Just choose

$$K := [\beta_n - \alpha_n, \beta_{n-1} - \alpha_{n-1}, \ldots, \beta_1 - \alpha_1].$$

Then $A - bK$ takes the companion form

$$A = \begin{bmatrix} 0 & 1 & 0 & \cdots & 0 \\ 0 & 0 & 1 & \cdots & 0 \\ & & \cdots & \cdots & 0 \\ 0 & 0 & 0 & \cdots & 1 \\ -\beta_n & -\beta_{n-1} & -\beta_{n-2} & \cdots & -\beta_1 \end{bmatrix}.$$

Hence $\det(sI - A + bK) = \chi(s) = s^n + \beta_1 s^{n-1} + \cdots + \beta_n$.   □

Remark 10.3.8. Naturally, when the λ_i's are complex, K will also be a complex matrix. In order that K be real, these λ_i's should satisfy the condition that it be symmetric against the real axis; i.e., if λ belongs to this set, then $\bar{\lambda}$ should also belong to it.

10.3.2 Observability

A completely dual notion is *observability*. For simplicity, we confine ourselves to system (10.3)–(10.4).

Conceptually, observability means the determinability of the initial state of the system under the assumption of output observation with suitable application of an input. Let $x_0 \in \mathbb{R}^n$ be an initial state of system (10.3)–(10.4). We say that x_0 is *indistinguishable* from 0 if the output derived from x_0 satisfies

$$Ce^{At}x_0 = 0 \tag{10.21}$$

for all $t \geq 0$. That is, there is no way to distinguish x_0 from 0 by observing its output. Note that the input term can play no role in view of the linearity in u entering into the solution (10.9). We say that x_0 is *distinguishable* if it is not indistinguishable from 0, i.e., if there exists $T \geq 0$ such that

$$Ce^{AT}x_0 \neq 0. \tag{10.22}$$

This leads to the following definition.

Definition 10.3.9. *The system* (10.3)–(10.4) *is said to be* observable *if every state $x_0 \in \mathbb{R}^n$ is distinguishable from* 0.

Some remarks are in order. First, in principle, there is no uniform upper bound for T in (10.22) for different x_0 even if the system is observable. However, when the system is finite-dimensional, as in the present situation, there is indeed an upper bound for T.

We first note the following lemma.

Lemma 10.3.10. *Suppose that the initial state $x \in \mathbb{R}^n$ satisfies $Ce^{At}x \equiv 0$ on $[0, T]$ for some $T > 0$. Then $CA^k x = 0$ for every nonnegative integer k, and $Ce^{At}x \equiv 0$ for all $t \geq 0$.*

Proof. Evaluate $d^k(Ce^{At}x)/dt^k$ at 0 to obtain

$$Cx = CAx = \cdots = 0.$$

Since $Ce^{At}x$ is a real analytic function in t, it readily follows that it is identically zero on the real axis.   □

Fix $T > 0$, and consider the mapping

$$\mathcal{O}_T : \mathbb{R}^n \to L^2[0,T] : x \mapsto Ce^{At}x. \tag{10.23}$$

In view of Lemma 10.3.10 above, system (10.3)–(10.4) is observable if and only if (10.23) is an injective mapping for some $T > 0$ (indeed, for any $T > 0$).

The adjoint operator of $\mathcal{O}_T$ is given by

$$\mathcal{O}_T^* : L^2[0,T] \to \mathbb{R}^n : u \mapsto \int_0^T e^{A^*t} C^* u(t)dt$$

according to Example 5.5.3 on page 97.

Consider the composed operator

$$N_T := \mathcal{O}_T^* \mathcal{O}_T : \mathbb{R}^n \to \mathbb{R}^n.$$

We have the following theorem, which is dual to Theorem 10.3.5.

Theorem 10.3.11. *The following conditions are equivalent:*

1. *System Σ defined by (10.3) and (10.4) is observable.*

2. *N_T is of full rank for every $T > 0$, i.e., it is invertible.*

3.
$$\text{rank} \begin{bmatrix} C \\ CA \\ \vdots \\ CA^{n-1} \end{bmatrix} = n, \tag{10.24}$$

   where n is the dimension of the state space, i.e., the size of A.

Proof. Suppose Σ is observable and also that there exists $T > 0$ such that $N_T x = 0$ for some x. Then

$$x^* N_T x = \int_0^T x^* e^{A^*t} C^* C e^{At} x dt = 0.$$

Hence

$$\int_0^T \left\| C e^{At} x \right\|^2 dt = 0.$$

This readily implies (cf. the proof of Lemma 10.3.4) $Ce^{At}x \equiv 0$ on $[0,T]$, which in turn yields $Ce^{At}x \equiv 0$ for all $t \geq 0$ by Lemma 10.3.10. Since Σ is observable, x must be zero. Hence N_T is of full rank.

Now suppose that N_T is of full rank, i.e., it is invertible. Suppose also that (10.24) fails. Then by the Cayley–Hamilton theorem (see (10.19)), $CA^k x = 0$ for every nonnegative integer k for some nonzero x (cf. the proof of Theorem 10.3.5). This implies $Ce^{At}x \equiv 0$ for every $t \geq 0$, and hence $N_T x = 0$. This is a contradiction; i.e., observability should hold.

Suppose now that Σ is not observable. Then there exists a nonzero x such that $Ce^{At}x \equiv 0$ for every $t \geq 0$. Lemma 10.3.10 readily yields $CA^k x = 0$ for every nonnegative integer k, which contradicts (10.24).

Finally, suppose that (10.24) fails. There exists a nonzero x such that $CA^k x = 0$ for $k = 0, \ldots, n-1$. By the Cayley–Hamilton theorem, this implies $CA^k x = 0$ for all $k \geq 0$. It follows that $Ce^{At}x \equiv 0$, $t \geq 0$, contradicting observability. $\square$

10.3.3   Duality between Reachability and Observability

Theorems 10.3.5 and 10.3.11 are in clear duality. In this subsection, we will further establish a more direct relationship.

Let Σ be given by (10.3) and (10.4). Its *dual system* denoted Σ^* is defined by

$$\Sigma^* : \begin{cases} \frac{dz}{dt}(t) & = & A^* z(t) + C^* v(t), \\ w(t) & = & B^* z(t), \end{cases} \tag{10.25}$$

where $*$ denotes complex conjugation, and $z(t) \in \mathbb{R}^n$, $v(t) \in \mathbb{R}^p$, $w(t) \in \mathbb{R}^m$.

The following theorem is immediate from Theorems 10.3.5 and 10.3.11, particularly (10.17) and (10.24).

Theorem 10.3.12. *System Σ given by (10.3) and (10.4) is reachable if and only if Σ^* (10.25) is observable; Σ is observable if and only if Σ^* is reachable.*

We now give a more intrinsic interpretation of this fact. First suppose for simplicity that Σ is stable. This is by no means necessary but makes the subsequent treatment easier. Define the mappings $\mathcal{R}$ and $\mathcal{O}$ as follows:

$$\mathcal{R} : (L^2(-\infty, 0])^m \to \mathbb{R}^n : u \mapsto \int_{-\infty}^0 e^{-At} Bu(t)dt, \tag{10.26}$$

$$\mathcal{O} : \mathbb{R}^n \to (L^2[0, \infty))^p : x \mapsto Ce^{At}x. \tag{10.27}$$

Dually, we also define

$$\overline{\mathcal{R}} : (L^2(-\infty, 0])^p \to \mathbb{R}^n : v \mapsto \int_{-\infty}^0 e^{-A^*t} C^* v(t)dt, \tag{10.28}$$

$$\overline{\mathcal{O}} : \mathbb{R}^n \to (L^2[0, \infty))^m : z \mapsto B^* e^{A^*t} z. \tag{10.29}$$

The following lemma is then readily obvious.

Lemma 10.3.13. *System Σ (10.3)–(10.4) is reachable if and only if $\mathcal{R}$ is surjective and is observable if and only if $\mathcal{O}$ is injective. Similarly, the dual system Σ^* is reachable if and only if $\overline{\mathcal{R}}$ is surjective and is observable if and only if $\overline{\mathcal{O}}$ is injective.*

Now define the following bilinear form between $(L^2(-\infty, 0])^\ell$ and $(L^2[0, \infty))^\ell$:

$$(L^2(-\infty, 0])^\ell \times (L^2[0, \infty))^\ell : (\omega, \gamma) \mapsto \langle \omega, \gamma \rangle := \int_{-\infty}^0 \gamma^*(-t)\omega(t)dt = \int_0^\infty \gamma^*(t)\omega(-t)dt, \tag{10.30}$$

where ℓ is either m or p. With respect to this duality, it is easily seen that $((L^2(-\infty,0])^\ell)' = (L^2[0,\infty))^\ell$ and $((L^2[0,\infty))^\ell)' = (L^2(-\infty,0])^\ell$. The following theorem gives a duality between reachability and observability.

Proposition 10.3.14. *With respect to the duality* (10.30) *above, the following duality relations hold:*

$$\mathcal{R}^* = \overline{\mathcal{O}}, \quad \mathcal{O}^* = \overline{\mathcal{R}}. \tag{10.31}$$

Proof. Apply Example 5.5.2 (p. 96) in Chapter 5 to (10.28) with $K(t) := e^{-At}B$, $a = -\infty$, $b = 0$.[101] By reversing the time direction as demanded by (10.30), we obtain $K^*(t) = B^*e^{A^*t}$, and hence $\mathcal{R}^* = \overline{\mathcal{O}}$. The same is true for $\mathcal{O}^*$. □

Hence if Σ is reachable, $\mathcal{R}$ is surjective, and by Proposition 5.2.3 in Chapter 5, $\mathcal{R}^* = \overline{\mathcal{O}}$ must be injective, and hence Σ^* is observable. The same is true for $\mathcal{O}$. This readily yields Theorem 10.3.12 and establishes a complete duality between reachability and observability in a function space setting.

10.4 Input/Output Correspondence

The system description (10.3)–(10.4) gives the following flows of signals: inputs $\mapsto$ states and states $\mapsto$ outputs.

There is also a contrasting viewpoint that directly describes the correspondence from inputs to outputs. In so-called classical control theory, one often deals with such a correspondence.

Let an input u be applied to (10.3) on the interval $[0,t)$ with initial state x at time 0. Then we have

$$x(t) = e^{At}x + \int_0^t e^{A(t-\tau)}Bu(\tau)d\tau,$$

and hence

$$y(t) = Ce^{At}x + \int_0^t Ce^{A(t-\tau)}Bu(\tau)d\tau. \tag{10.32}$$

We have a family of input to output correspondences

$$u \mapsto \left\{ Ce^{At}x + \int_0^t Ce^{A(t-\tau)}Bu(\tau)d\tau : x \in \mathbb{R}^n \right\} \tag{10.33}$$

parametrized by state x.

Hence, strictly speaking, the correspondence from inputs to outputs is not a mapping; that is, an input gives rise to a family of output functions parametrized by state x, which is unknown. To bypass this difficulty, in the approach of classical control theory, one assumes that x is zero in the above and considers the correspondence

$$u \mapsto \int_0^t Ce^{A(t-\tau)}Bu(\tau)d\tau. \tag{10.34}$$

---

[101] with a slight abuse of the result because in Example 5.5.2 $[a,b]$ is a bounded interval, but this is not essential.

This is perhaps an acceptable assumption when the underlying system is a priori known to be stable, so that an effect due to the initial state will soon decay to zero. However, strictly, or philosophically, speaking, this is rather unsatisfactory, and one should attack the correspondence (10.33) right in front. This is the viewpoint of *behavioral system theory* proposed by Willems and is studied extensively in the literature; see [44].

However, we will not pursue this relatively modern approach here and will content ourselves with the classical viewpoint, setting the initial state $x = 0$, mainly for technical convenience.

Another rationale for assuming the initial state is zero is the following: Almost every linear system arises as a linearized approximation of a nonlinear system. This linearization is performed around an equilibrium trajectory, and the state variable $x(t)$ represents a deviation from such a reference trajectory. In this sense it is rather natural to assume that the initial state $x(0)$ is zero. See Figure 10.3.

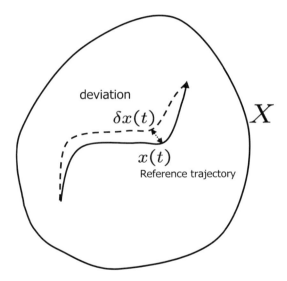

Figure 10.3. *Linearization around a reference trajectory*

Under this assumption, (10.32) takes the form

$$y(t) = \int_0^t Ce^{A(t-\tau)}Bu(\tau)d\tau = \int_0^t g(t-\tau)u(\tau)d\tau, \qquad (10.35)$$

where $g(t) := Ce^{A(t-\tau)}B$. In other words, $y = g * u$, and g is called the *impulse response* of system Σ. This is because $g = g * \delta$; that is, g is obtained as the response against the impulse δ. The Laplace transform $\mathcal{L}[g]$ is called the *transfer function (matrix)* of Σ. These notions are already encountered under a more specialized situation in Chapter 6, section 6.7 (p. 131).

10.5 Realization

In the previous sections, we have given two different types of representations for linear systems: one via the differential equation description (10.3)–(10.4), and the other via the input/output correspondence $y = g * u$ (under the assumption that the initial state is zero). Let us now denote the system Σ described by (10.3)–(10.4) by the triple (A, B, C). The dimension of the state is implicitly specified by the size of A.

Given the triple (A, B, C), the impulse response g is given by $Ce^{At}B$, $t \geq 0$, and the transfer function by $C(sI - A)^{-1}B$. The *realization problem* asks the following converse question.

The realization problem. Given a $p \times m$ matrix function g with support in $[0, \infty)$, find a system $\Sigma = (A, B, C)$ such that $g(t) = Ce^{At}B$ for all $t \geq 0$. The obtained system $\Sigma = (A, B, C)$ is called a *realization* of g.

If necessary, assume sufficient regularity for g. For $t < 0$, we set $g(t) = 0$, and hence it will usually have a discontinuity at $t = 0$, as shown in Figure 10.4. Naturally, not every such g can be realized by a finite-dimensional system as above. For example, $g(t) = e^{t^2}$ is clearly seen to be nonrealizable by any (A, B, C) because it grows much faster than any exponential order.

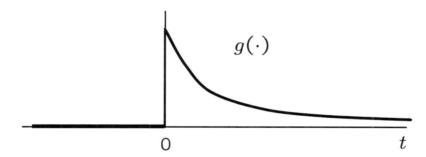

Figure 10.4. *Impulse response*

Even if g is realizable by some (A, B, C), the realization is not necessarily unique. For example, consider

$$g(t) := \begin{cases} e^{-t}, & t \geq 0, \\ 0, & t < 0. \end{cases}$$

Then both

$$A = -1, \quad B = 1, \quad C = 1 \tag{10.36}$$

and

$$A = \begin{bmatrix} 0 & 1 \\ -1 & -2 \end{bmatrix}, \quad B = \begin{bmatrix} 0 \\ 1 \end{bmatrix}, \quad C = \begin{bmatrix} 1 & 1 \end{bmatrix} \tag{10.37}$$

satisfy $Ce^{At}B = e^{-t}$, but their state space dimensions are not equal.

Remark 10.5.1. As seen above, there exist more than one realization for a given impulse response, and their dimension is not uniquely determined. A realization Σ is said to be *minimal* if it possesses the least dimension among all realizations for a given impulse response. It is known that a minimal realization is essentially uniquely determined (up to basis change in the state space) for a given impulse response [23].

In general, reasonably regular (smooth) g always admits a realization, but not necessarily with a finite-dimensional state space. We say that a linear system Σ is *finite-dimensional* if the state space (in which $x(t)$ resides) is finite-dimensional.

For example, consider the input/output relation

$$u(t) \mapsto u(t-1), \tag{10.38}$$

which represents a unit time delay. In order to represent this relation, it is clear that we need to store a complete function piece defined on a unit time interval, and a function space consisting of all such functions. For example, if we consider L^2 inputs u, this space will become $L^2[0,1]$, which is clearly not finite-dimensional.

The following criterion is well known for its finite-dimensional realizability.

Theorem 10.5.2. *Let g be an impulse response. A necessary and sufficient condition for g to admit a finite-dimensional realization is that the associated transfer function matrix $\hat{g}(s)$ be a strictly proper rational function of s.*

Proof. For simplicity, we give a proof for a single-input/single-output system, i.e., the case $m = p = 1$.

Necessity. Let $g(t) = Ce^{At}B$ for some A, B, C. Then $\hat{g}(s) = C(sI - A)^{-1}B$, and this is clearly rational in s. Rewriting $C(sI - A)^{-1}B$ in terms of Cramer's rule, we immediately see that this is strictly proper.

Sufficiency. Conversely, suppose that $\hat{g}(s) = b(s)/a(s)$, where $\deg b < \deg a$. Without loss of generality, we can take a to be a monic polynomial; i.e., its highest order coefficient is 1. Write $a(s) = s^n + a_1 s^{n-1} + \cdots + a_n$, and $b(s) = b_1 s^{n-1} + \cdots + b_n$. Then the triple

$$A := \begin{bmatrix} 0 & 1 & 0 & \cdots \\ 0 & 0 & 1 & \cdots \\ \vdots & & & \ddots \\ 0 & \cdots & & 1 \\ -a_n & -a_{n-1} & \cdots & -a_n \end{bmatrix}, \quad B := \begin{bmatrix} 0 \\ 0 \\ \vdots \\ 0 \\ 1 \end{bmatrix}, \quad C := [b_n, b_{n-1}, \cdots, b_n]$$

is easily seen to satisfy $C(sI - A)^{-1}B = b(s)/a(s)$. This readily yields $g(t) = Ce^{At}B$. $\square$

Exercise 10.5.3. Generalize the above to the matrix case.

10.5.1   Transfer Functions and Steady-State Response

Let us return to formula (10.35) for the input/output relations. In terms of the Laplace transform, this means (Theorem 8.1.2)

$$\hat{y} = \hat{g}\hat{u}. \tag{10.39}$$

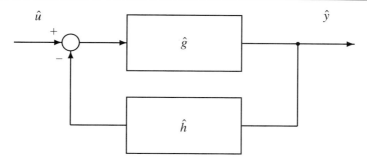

Figure 10.5. *Feedback control system*

Figure 10.6. *Input/output relationship with transfer function $\hat{g}$*

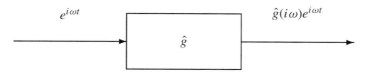

Figure 10.7. *Steady-state response against $e^{i\omega t}$*

An advantage of this representation is that it can place the whole framework of control systems into an algebraic structure. For example, the *feedback* structure

$$\hat{y} = \hat{g}\hat{e}, \quad \hat{e} = \hat{u} - \hat{h}\hat{y}$$

can be schematically represented by the "block diagram" Figure 10.5.

A remarkable feature of this representation is that it yields a concise expression of the "steady-state response" against sinusoidal inputs (see Figure 10.7). Consider Figure 10.6, and suppose that (i) the system is "stable" and (ii) an input $e^{i\omega t}u_0$ is applied to the system. The following proposition states that the system output becomes asymptotically equal to a multiple of this sinusoid, with the same frequency ω, and no other frequencies are present.

Proposition 10.5.4. *Consider the linear system with a strictly proper rational transfer function $\hat{g}$ as in Figure* 10.6. *Suppose that $\hat{g}$ has its poles only in the open left half complex plane $\mathbb{C}_- = \{s : \operatorname{Re} s < 0\}$. Apply an input $u(t) = H(t)e^{i\omega t}u_0$ to this system. Then the output $y(t)$ asymptotically approaches $\hat{g}(i\omega)e^{i\omega t}u_0$ as $t \to \infty$.*

Proof. For simplicity, assume $m = p = 1$ and $u_0 = 1$. Let us first suppose that the initial state x_0 at $t = 0$ is zero. Then the response y against u must satisfy $\hat{y}(s) = \hat{g}\hat{u}(s) = \hat{g}(s)/(s - i\omega)$. Since $\hat{g}$ is analytic at $s = i\omega$, we can expand it as

$$\hat{g}(s) = \hat{g}(i\omega) + (s - i\omega)\hat{g}_1(s),$$

where $\hat{g}_1(s)$ is analytic at $i\omega$ and has no poles in the closed right half complex plane $\overline{\mathbb{C}_+} = \{s : \operatorname{Re} s \geq 0\}$, because neither does $\hat{g}$. It follows that

$$\frac{\hat{g}(s)}{s - i\omega} = \frac{\hat{g}(i\omega)}{s - i\omega} + \hat{g}_1(s).$$

Taking the inverse Laplace transform of the terms on the right, we see that the first term becomes $\hat{g}(i\omega)e^{i\omega t}$, and the second term approaches zero as $t \to \infty$ because of the absence of closed right half plane poles. Hence the system response asymptotically approaches $\hat{g}(i\omega)e^{i\omega t}$.

When there is a nonzero initial state, simply observe that the corresponding response approaches zero as $t \to \infty$ due to the nonexistence of closed right half plane poles of $\hat{g}$. $\square$

The response $\hat{g}(i\omega)e^{i\omega t}$ is called the *steady-state response* against $e^{i\omega t}$.

This proposition shows the following fact: *For a linear stable system described by a transfer function $\hat{g}$, the steady-state response against a sinusoid $e^{i\omega t}$ is proportional to it, obtained by multiplying $\hat{g}(i\omega)$.* Rewriting $\hat{g}(i\omega)e^{i\omega t}$, we obtain

$$\hat{g}(i\omega)e^{i\omega t} = |\hat{g}(i\omega)|e^{i\omega t + \psi}, \quad \psi = \angle \hat{g}(i\omega), \tag{10.40}$$

where $\angle \hat{g}(i\omega)$ denotes the argument (phase angle) of $\hat{g}(i\omega)$. This means that in steady state, the amplitude is multiplied by $|\hat{g}(i\omega)|$, and the phase angle is shifted by $\angle \hat{g}(i\omega)$. The mapping

$$\mathbb{R} \to \mathbb{C} : \omega \mapsto \hat{g}(i\omega) \tag{10.41}$$

is called the *frequency response* of the system.

In general, when an input is applied to a system, it can be decomposed into frequency components via the Fourier transform, and the resulting output is a composite (i.e., the sum or integral) of respective responses at all frequencies. The plot exhibiting the magnitude $|\hat{g}(i\omega)|$ and the phase $\angle \hat{g}(i\omega)$ for each frequency is very useful in analyzing the steady-state response of $\hat{g}$. The plot that shows the curves of $|\hat{g}(i\omega)|$ and $\angle \hat{g}(i\omega)$ against the frequency axis is called the *Bode plot*, named after Hendrick Bode, who devised this scheme. In particular, the plot exhibiting the *gain* $|\hat{g}(i\omega)|$ is called the *Bode magnitude plot*, while that for phase $\angle \hat{g}(i\omega)$ is called the *Bode phase plot*. To show the gain, its dB value (i.e., $20\log_{10}|\hat{g}(i\omega)|$) is used. This is often convenient for evaluating a composite effect when two systems are connected in a cascade way.[102] Figure 10.8 shows the Bode magnitude and phase plots of $\hat{g}(s) = 100/(s^2 + 2s + 100)$. This example shows that this system shows a highly sensitive response at $\omega = 10$ [rad/sec].

Observe also that the maximum of the Bode magnitude plot gives the H^∞ norm of $\hat{g}$.

10.6 H^∞ Control

We are now at the point of discussing some basics of H^∞ control.

Consider the control system depicted in Figure 10.9. Here $\hat{P}(s)$ is the plant to be controlled, and $\hat{C}(s)$ is the controller to be designed. Such a control system is usually called

---

[102]In this case, the transfer function is the product of the two component transfer functions, and the gain becomes the product at each frequency; hence the dB value can be obtained as the sum.

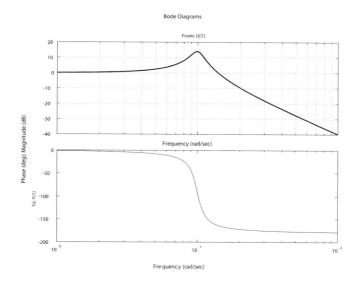

Figure 10.8. *Bode plots for* $100/(s^2 + 2s + 100)$

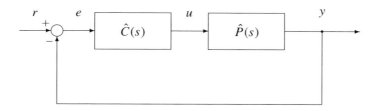

Figure 10.9. *Unity feedback system*

a *unity feedback system* in the sense that the feedback loop has the unity gain. The feedback signal is subtracted from the input, and the controller $\hat{C}(s)$ is driven by the error e. In this sense the *closed-loop system* has a *negative feedback*. The original idea of feedback, from the time of Watt, was to *reduce* the sensitivity of the control system against various fluctuations inherent to the system and operation conditions. Hence the error from the reference is to be measured, and this leads to the negative feedback. This structure is fairly standard. If one needs to incorporate a different gain in the feedback loop, that can be accomplished by suitably scaling the output y.

The objective here is to design $\hat{C}(s)$ so as to make the characteristics of this *closed-loop system* desirable with respect to some performance index.

Observe that

$$\hat{y} = \hat{P}\hat{u}, \quad \hat{u} = \hat{C}\hat{e}, \quad \hat{e} = \hat{r} - \hat{y}.$$

Eliminating $\hat{u}$ and $\hat{y}$, we obtain the transfer function $\hat{S}(s) = (1 + \hat{P}(s)\hat{C}(s))^{-1}$ from the *reference input* r to the error e. This $\hat{S}$ is called the *sensitivity function*. An objective here is to make this function small. This will accomplish small errors against references, even under modeling errors and fluctuations in the plant.

Unfortunately, this objective cannot be accomplished uniformly over all frequencies (Bode's theorem [8]). If the gain S becomes lower at a certain frequency range, it becomes higher somewhere else. On the other hand, making the sensitivity function uniformly small is not necessarily always required. There is usually a relatively important frequency range where $\hat{S}(i\omega)$ has to be small, while in other ranges this is not necessarily so. In view of this, one usually takes a weighting function $\hat{W}(s)$ (usually taken to be a proper and stable rational function) and then attempts to minimize $\hat{W}(i\omega)\hat{S}(i\omega)$.

A classical criterion is to take the H^2 norm as the performance index:

$$J = \inf_{\hat{C}(s):\text{stabilizing}} \|\hat{W}(s)\hat{S}(s)\|_2, \qquad (10.42)$$

where $\hat{C}$ runs over all controllers that make the closed-loop (i.e., the feedback) system stable. This is a minimization problem in the Hilbert space H^2, and the projection theorem as described in Theorem 3.2.7 (Chapter 3) can be invoked to derive its solution.

While this H^2 minimization has played some historically important roles in classical optimal control theory, it gives only a limited estimate of the closed-loop performance. For example, if there is a sharp peak in $|\hat{W}(i\omega)\hat{S}(i\omega)|$ at a certain frequency, it is not necessarily reflected upon its H^2 norm as a large value, because the H^2 norm measures an integrated (i.e., averaged) performance over all frequency range.

Around 1980, George Zames introduced the H^∞ performance index [69]

$$\gamma_{\text{opt}} = \inf_{\hat{C}(s):\text{stabilizing}} \|\hat{W}(s)\hat{S}(s)\|_\infty \qquad (10.43)$$

and asserted that it gives a very natural performance criterion for many control objectives. We can safely say that this is the origin of H^∞ control theory.

The performance index (10.43) attempts to minimize the maximum (supremum) of the Bode magnitude plot. Hence, if this index can be made small, it guarantees the worst-case response to be within a tolerable range and hence is desirable. On the other hand, since H^∞ is not a Hilbert space, the projection theorem, a key to many optimization problems, is not available to us here. Naturally, at the time, there was doubt as to whether this problem is solvable in sufficient generality as to be applicable to general control systems.

However, contrary to such concerns, a general solution was found in subsequent developments in a few years [15, 70]. In due course, deep connections with Sarason's generalized interpolation theory (Theorem 9.5.2, p. 191), Nehari's theorem (Theorem 9.5.4, p. 192), and Nevanlinna–Pick interpolation theory (Theorem 9.6.4, p. 198) were found and clarified.

This is one of the cornerstone accomplishments in control theory in the past 30 years, and it has become standard. It is now largely applied in many control problems.

We here give solutions to an elementary H^∞ control problem. Our objective is to find the optimal value

$$\inf_{\hat{C}(s):\text{stabilizing}} \|\hat{W}(s)(1 + \hat{P}(s)\hat{C}(s))^{-1}\|_\infty \qquad (10.44)$$

and find a controller $\hat{C}$ accomplishing this value.

10.6.1 Preliminaries for the H^∞ Solution

For brevity of exposition, let us assume that the plant $\hat{P}(s)$ is stable, i.e., it has poles only in the open left-half complex plane $\mathbb{C}_- := \{s : \operatorname{Re} s < 0\}$. This assumption is not at all necessary or desirable,[103] and it is indeed removable but substantially simplifies our exposition here. Also, H^∞ design for stable plants, for improving performance, is also quite meaningful.

A problem in (10.44) above is that the design parameter $\hat{C}$ enters into $\hat{W}(s)(1 + \hat{P}(s)\hat{C}(s))^{-1}$ nonlinearly. This makes the problem quite complicated. To bypass this problem, introduce a new parameter $\hat{Q}(s) = (1 + \hat{P}(s)\hat{C}(s))^{-1}\hat{C}(s)$. Now observe the following:

- $\hat{S}(s) = 1 - \hat{P}(s)\hat{Q}(s)$;

- the closed-loop system is stable (i.e., $\hat{S} \in H^\infty$) if and only if $Q(s) \in H^\infty$;

- once we find $\hat{Q}$, then $\hat{C}$ can be determined by $\hat{C}(s) = (1 - \hat{P}(s)\hat{Q}(s))^{-1}\hat{Q}(s)$.

Then the stability of the closed-loop system is reduced to that of $\hat{Q}$, and, moreover, the new parameter $\hat{Q}$ enters into the transfer function $\hat{S}$ in an affine way.

Our problem is thus reduced to finding the following:

$$\inf_{\hat{Q} \in H^\infty} \|\hat{W} - \hat{W}\hat{P}\hat{Q}\|_\infty. \tag{10.45}$$

Let us further execute the inner-outer factorization for $\hat{W}\hat{P} \in H^\infty$ as $\hat{W}(s)\hat{P}(s) = v_i(s)v_o(s)$, where v_i, v_o are inner and outer parts of $\hat{W}\hat{P}$, respectively. Since outer functions do not possess zeros in $\mathbb{C}_+$, introduce yet a new parameter $\tilde{Q} := \hat{Q}v_o$, and rewrite (10.45) as

$$\inf_{\tilde{Q} \in H^\infty} \|\hat{W} - m\tilde{Q}\|_\infty \tag{10.46}$$

with $m := v_i(s)$. In what follows we deal with the *sensitivity minimization problem*[104] in this form.

In the present context, one may ask when the closed-loop system in Figure 10.9 is stable. Suppose, for the moment, C is also stable. An easy sufficient condition is then

$$\|PC\|_\infty < 1. \tag{10.47}$$

This is because $I + PC$ is boundedly invertible due to the Neumann series expansion (5.7). This is a special case of the so-called *small-gain theorem*, e.g., [8, 73]. While this criterion is rather restrictive in the present context, it will become more useful in a general robustness analysis; see, e.g., [8, 73]. Of course, a general solution to H^∞ control does not presuppose such a restriction.

Remark 10.6.1. The sensitivity minimization problem as defined above is a very special case of the H^∞ control problem. For actual control syntheses, a more refined and elaborate problem must be considered. For example, we often need to consider the so-called *complementary sensitivity function* $\hat{T}(s) := 1 - \hat{S}(s)$ to reduce the effect of uncertainty in the model $\hat{P}$, e.g., parameter fluctuations, unmodeled dynamics, etc.

---

[103]An objective of control system design is to stabilize an unstable system.

[104]Also called the *model matching problem*.

Since $\hat{S}(s) + \hat{T}(s) \equiv 1$, however, it is impossible to attenuate these two measures simultaneously. In many practical cases, $\hat{S}$ is required to be small in a low-frequency range, while $\hat{T}(s)$ needs to be small mainly in a high-frequency range. This is because the tracking performance is more important in low frequency (typically against the step inputs (see Remark 10.6.2 below) for which $\omega = 0$), while plant uncertainties are more dominant in high frequency. To accommodate this constraint, we often employ weighting functions $\hat{W}_1, \hat{W}_2$ and attempt to minimize

$$\left\| \begin{bmatrix} \hat{W}_1 S & \hat{W}_2 T \end{bmatrix} \right\|_\infty$$

instead of just one sensitivity function. This problem is called the *mixed sensitivity problem*. See, e.g., [8, 14, 73] for more general and advanced materials.

Remark 10.6.2. The step input here means a constant multiple of the Heaviside unit step function $H(t)$. In many applications, a control command is given in the form of tracking such a signal. For example, when we ride on an elevator, we push the button of a target floor. This action is transferred to the system as a proper voltage value that should be tracked, and it is representable as a constant multiple of the unit step function. The Laplace transform of the unit step function is $1/s$, and hence it corresponds to $\omega = 0$ in the frequency response, i.e., the direct current component. In some other cases, one may wish to reject disturbances arising from AC power supply, which is often represented as a sinusoid at $\omega = 60[\text{Hz}]$. These simple examples show why we are interested in tracking/rejection properties in control systems.

On the other hand, the complementary sensitivity function $T = (1 + PC)^{-1}PC$ gives the correspondence $r \mapsto y$ in Figure 10.9. This has much uncertainty at high frequencies, particularly when identified from experimental data.

10.7 Solution to the Sensitivity Minimization Problem

We here give two different types of solutions—one via the Nevanlinna–Pick interpolation and the other via Nehari's theorem.

However, before going into detail, let us note the difference of the treatment of Hardy spaces given in Chapter 9 and those needed in this section. While Hardy spaces in Chapter 9 are given on $\mathbb{D}$, we here deal with those defined on half planes in $\mathbb{C}$. Typically, H^2 is the space consisting of functions f analytic on the open right half plane $\mathbb{C}_+ := \{s : \operatorname{Re} s > 0\}$ and satisfying

$$\sup_{x>0} \int_{-\infty}^{\infty} |f(x+iy)|^2 dy < \infty.$$

The H^2 norm of $f \in H^2$ is defined by

$$\|f\|_2 := \sup_{x>0} \left\{ \frac{1}{2\pi} \int_{-\infty}^{\infty} |f(x+iy)|^2 dy \right\}^{1/2}. \tag{10.48}$$

Similarly, the space H_-^2 is the one consisting of functions f analytic on the open left half plane $\mathbb{C}_- := \{s : \operatorname{Re} s < 0\}$ with

$$\sup_{x<0} \int_{-\infty}^{\infty} |f(x+iy)|^2 dy < \infty.$$

Its norm is defined by

$$\|f\|_2 := \sup_{x<0} \left\{ \frac{1}{2\pi} \int_{-\infty}^{\infty} |f(x+iy)|^2 dy \right\}^{1/2}. \tag{10.49}$$

$L^2(-i\infty, i\infty)$ is the space of functions on the imaginary axis such that

$$\int_{-\infty}^{\infty} |f(i\omega)|^2 d\omega < \infty$$

with norm

$$\|f\|_2 := \left\{ \frac{1}{2\pi} \int_{-\infty}^{\infty} |f(i\omega)|^2 d\omega \right\}^{1/2}. \tag{10.50}$$

While H^2 and H^2_- are not given a priori on the imaginary axis, they can be extended there.

Theorem 10.7.1. *Let $f \in H^2$. Then for almost every ω the nontangential limit*

$$\tilde{f}(i\omega) := \lim_{x \to 0} f(x+i\omega)$$

exists and belongs to $L^2(-i\infty, i\infty)$. The mapping $H^2 \ni f \mapsto \tilde{f} \in L^2(-i\infty, i\infty)$ is linear and norm-preserving. The same is true for H^2_-.

Furthermore, the Fourier transform gives norm-preserving isomorphisms between $L^2[0, \infty)$ and H^2, and between $L^2(-\infty, 0]$ and H^2_-.

Theorem 10.7.2. *The Fourier transform $\mathcal{F}$ gives a norm-preserving isomorphism between $L^2(-\infty, \infty)$ and $L^2(-i\infty, i\infty)$. The Fourier transform $\mathcal{F}$ maps $L^2[0, \infty)$ onto H^2 and $L^2(-\infty, 0]$ onto H^2_-.*

The first half is nothing but the Plancherel–Parseval theorem, Theorem 7.5.5 (p. 159). The latter half follows from the Paley–Wiener theorem; see, e.g., Hoffman [21, 47]. Identifying H^2 (H^2_-) with its boundary values on the imaginary axis by Theorem 10.7.1, this statement indeed shows that H^2 is precisely the image of $L^2[0, \infty)$ under the Laplace transform, i.e., $H^2 = \mathcal{L}L^2[0, \infty)$. Likewise, $H^2_- = \mathcal{L}L^2(-\infty, 0]$. These facts give a complete analogue of Theorem 9.2.4. We omit the proof; see [21, 47, 66], etc.

In this context, we also note the following, which was somewhat implicit in our motivation of H^∞ optimization.

Proposition 10.7.3. *Let $f \in H^\infty$. Then $M_f H^2 \subset H^2$ and*

$$\|M_f\| = \|f\|_\infty, \tag{10.51}$$

where M_f denotes the multiplication operator by f.

The first half is nothing but Lemma 9.2.5. For the second half, $\|M_f\| \le \|f\|_\infty$ is almost trivial, but the converse is a little technical, and we again omit its proof. See, e.g., [67]. This proposition shows that if we measure the size of input and output signals by the H^2 (or L^2) norm, then the H^∞ norm of the transfer function gives the "gain" of the system, that is, how the inputs are magnified or attenuated by the system.

10.7.1 Solution via the Nevanlinna–Pick Interpolation

The problem (10.46) is equivalent to finding the least γ_{opt} among all γ's that satisfy

$$\|\hat{W} - m\hat{Q}\|_\infty \le \gamma, \tag{10.52}$$

subject to $\hat{Q} \in H^\infty$. We give a solution based on the Nevanlinna–Pick interpolation, following the treatment in [8]. Let us first characterize the *suboptimal* inequality (10.52).

Take any $\gamma > 0$, and set

$$G := \frac{1}{\gamma}(\hat{W} - m\hat{Q}).$$

If $\hat{Q} \in H^\infty$, then G belongs to H^∞, but not necessarily conversely. We need some interpolation conditions.

Let $\{s_1, \ldots, s_n\}$ be the zeros of $m(s)$ in $\mathbb{C}_+$. For simplicity, we assume that these s_i's are all distinct. Since m is inner, there is no zero on the imaginary axis, and hence $\mathrm{Re}\, s_i > 0$, $i = 1, \ldots, n$. This readily means that the interpolation condition

$$G(s_i) = \frac{1}{\gamma}\hat{W}(s_i), \quad i = 1, \ldots, n,$$

must be satisfied. Conversely, if $G \in H^\infty$ satisfies this condition, then

$$\hat{Q} := m^{-1}(\hat{W} - \gamma G)$$

clearly belongs to H^∞. All unstable poles s_i arising from m^{-1} are cancelled by the numerator $\hat{W} - \gamma G$. Hence the optimum γ_{opt} is given by the minimum of all γ such that

$$G(s_i) = \frac{1}{\gamma}W(s_i), \quad i = 1, \ldots, n,$$
$$\|G\|_\infty \le 1$$

is satisfied for some H^∞ function G. This is nothing but the Nevanlinna–Pick interpolation problem with respect to the interpolation data

$$
\begin{array}{ccc}
s_1, & \cdots & s_n \\
\updownarrow & \cdots & \updownarrow \\
W(s_1)/\gamma, & \cdots & W(s_n)/\gamma \, .
\end{array}
$$

We know from Theorem 9.6.4 (p. 198) that γ satisfies the condition above if and only if the Pick matrix P satisfies

$$P = (p_{ij}) = \left(\frac{1 - W(s_i)\overline{W(s_j)}/\gamma^2}{s_i + \overline{s}_j} \right) \ge 0.$$

The optimum γ_{opt} is the minimum of such γ. The corresponding $\hat{Q}$ is given by

$$\hat{Q} := \frac{W - \gamma_{\text{opt}}G}{m}.$$

10.7.2 Solution via Nehari's Theorem

We here give another solution via Nehari's theorem. (The following exposition mainly follows that of [14].)

Multiplying $m\tilde{}(s) = m^{-1}(s)$ on both sides of (10.46), we obtain

$$\gamma_{\text{opt}} = \inf_{\hat{Q} \in H^{\infty}} \|m\tilde{}\hat{W} - \hat{Q}\|_{\infty}. \tag{10.53}$$

However, note here that $m\tilde{}W \notin H^{\infty}$, and hence the norm cannot be taken in the sense of H^{∞}. It should be understood in the sense of $L^{\infty}(-i\infty, i\infty)$.

Decompose $m\tilde{}\hat{W}$ as

$$m\tilde{}\hat{W} = W_1 + W_2,$$

such that $W_1 \in H^{\infty}$, W_2 is strictly proper and has all its poles in the open right-half complex plane.

Define the Hankel operator Γ_{W_2} associated with W_2 as

$$\Gamma_{W_2} : H^2 \to H_-^2 : x \mapsto P_- W_2 x, \tag{10.54}$$

where P_- is the canonical projection: $L^2(-i\infty, i\infty) \to H_-^2$.

Since W_2 clearly belongs to $L^{\infty}(-i\infty, i\infty)$, this gives a Hankel operator according to Nehari's theorem, Theorem 9.5.4 (p. 192), and this implies that

$$\gamma_{\text{opt}} = \|\Gamma_{W_2}\|.$$

An actual computation of γ_{opt} may be done as follows: Let (A, B, C) be a minimal realization of W_2 (note that W_2 is strictly proper; i.e., the degree of the numerator is less than that of the denominator). Hence $W_2(s) = C(sI - A)^{-1}B$. Taking the inverse Laplace transform (via the bilateral Laplace transform), we have

$$W_2(t) = \begin{cases} -Ce^{At}B, & t < 0, \\ 0, & t \geq 0. \end{cases}$$

Note that the time direction is reversed. Using this and via the inverse Laplace transform, we can consider the Hankel operator over the time domain as a correspondence $L^2[0, \infty) \to L^2(-\infty, 0]$ rather than from H^2 to H_-^2. Since the Laplace transform gives an isometry between $L^2[0, \infty)$ and H^2 (and also between $L^2(-\infty, 0]$ and H_-^2) by Theorem 10.7.2 (and the text following that), respectively, this change of domains does not affect the norm of the Hankel operator. Since multiplication in the Laplace transformed domain translates to convolution in the time domain, the correspondence $u \mapsto y$ in the time domain is given by

$$y(t) = \int_0^{\infty} W_2(t - \tau)u(\tau)d\tau = -Ce^{At}\int_0^{\infty} e^{-A\tau}Bu(\tau)d\tau, \quad t < 0.$$

Introduce two operators

$$\Psi_c : L^2[0, \infty) \to \mathbb{C}^n : \Psi_c u := -\int_0^{\infty} e^{-A\tau}Bu(\tau)d\tau,$$

$$\Psi_o : \mathbb{C}^n \to L^2(-\infty, 0] : (\Psi_o x)(t) := Ce^{At}x, \quad t < 0.$$

Note here that the time direction is again reversed from that discussed in section 10.3. Then $\Gamma_{W_2} = \Psi_o \Psi_c$ (using the same notation Γ_{W_2} for the transformed Hankel operator). Since Γ_{W_2}

is decomposed as the composition of two (finite-rank) operators Ψ_c and Ψ_o, it is a compact operator. Hence its norm is given by its maximal singular value (Chapter 5, Problem 12). To this end, consider the eigenvalue problem

$$\Psi_c^* \Psi_o^* \Psi_o \Psi_c u = \lambda u. \tag{10.55}$$

Since the operator $\Psi_o \Psi_c$ must be nonzero, it suffices to consider nonzero eigenvalues λ. Then by Lemma 10.7.4 below, we can interchange the order of operators to obtain

$$\Psi_c \Psi_c^* \Psi_o^* \Psi_o x = \lambda x. \tag{10.56}$$

Moreover,

$$\Psi_c \Psi_c^* = \int_0^\infty e^{-At} B B^T e^{A^T t} dt,$$

$$\Psi_o^* \Psi_o = \int_0^\infty e^{-A^T t} C^T C e^{At} dt$$

hold true.

Observe that (10.56) is a finite-dimensional eigenvalue problem, in contrast to (10.55). It is also known that matrices

$$L_c := \Psi_c \Psi_c^*, \quad L_o := \Psi_o^* \Psi_o$$

are unique solutions of the following Lyapunov equations:

$$A L_c + L_c A^T = B B^T,$$
$$A^T L_o + L_o A = C^T C.$$

Solving these equations and finding the maximal eigenvalue of $L_c L_o$, we arrive at the solution.

It remains to show the following lemma. (See also Problems 2 and 8, Chapter 5, for the case $X = Y$.)

Lemma 10.7.4. *Let X, Y be normed linear spaces, and let $A \in \mathcal{L}(X, Y)$, $B \in \mathcal{L}(Y, X)$. Then a nonzero $\lambda \in \mathbb{C}$ belongs to the resolvent set $\rho(AB)$ if and only if $\lambda \in \rho(BA)$. As a consequence, $\sigma(AB) \setminus \{0\} = \sigma(BA) \setminus \{0\}$.*

Proof. Note first that if $\lambda \neq 0$, then

$$(\lambda I - T)^{-1} = \lambda^{-1} (I - T/\lambda)^{-1}.$$

Hence we may assume without loss of generality that $\lambda = 1$.

Suppose $1 \in \rho(AB)$. Consider[105] $T := I + B(I - AB)^{-1} A$, which is clearly a bounded operator. Then

$$
\begin{aligned}
T(I - BA) &= I - BA + B(I - AB)^{-1} A - B(I - AB)^{-1} ABA \\
&= I - BA + B(I - AB)^{-1}(I - AB)A \\
&= I - BA + BA = I.
\end{aligned}
$$

---

[105]This may appear ad hoc. For a supporting intuition, consider the Neumann series expansion (5.7), page 92, $(I - BA)^{-1} = I + BA + BABA + \cdots + (BA)^n + \cdots$, which is equal to $I + B(I + AB + ABAB + \cdots)A = I + B(I - AB)^{-1} A$.

Similarly, $(I - BA)T = I$, and hence $T = (I - BA)^{-1}$. This means $1 \in \rho(BA)$.

Since $\sigma(AB) = \mathbb{C} \setminus \rho(AB)$, the last assertion for the spectra immediately follows. $\quad\square$

10.8 General Solution for Distributed Parameter Systems

The solutions given above depend crucially on the rationality of $m(s)$. When the given plant has spatially dependent parameters, i.e., when it is a *distributed parameter system*, the plant becomes infinite-dimensional, and these solutions do not directly carry over. There is much research that attempts similar ideas. However, since many of these attempts end up with infinite-dimensional Lyapunov equations or so-called Riccati equations, obtaining practically computable solutions is nontrivial.

In contrast to these approaches, there is an alternative studied by Bercovici, Foias, Özbay, Tannenbaum, Zames, and others (see, e.g., [13]) that makes use of the rationality of W, and this leads to an interesting rank condition in spite of underlying infinite-dimensionality. Since one has freedom in choosing the weighting function W, this is a very plausible choice.

Let $m(s)$ be an arbitrary (real) inner function that is not necessarily rational. Then space $m H^2$ is a right-shift invariant closed subspace of H^2 which is indeed invariant under multiplication by an arbitrary H^∞ function (see Theorem 9.4.1, p. 190). As seen in Chapter 9 (however, see Remark 10.8.1 below), its orthogonal complement is $H(m) = H^2 \ominus m H^2$. For each $t \geq t$, let $M_{e^{-ts}}$ denote the multiplication operator induced by e^{-ts}, i.e.,

$$M_{e^{-ts}} : H^2 \to H^2 : \phi(s) \mapsto e^{-ts}\phi(s).$$

Since this occurs in the Laplace transformed domain, it corresponds to the right shift operator by t in L^2. let $T(t)$ denote its compression to $H(m)$:

$$T(t) = \pi^m M_{e^{-ts}}|_{H(m)},$$

where $\pi^m : H^2 \to H(m)$ is the canonical projection. More generally, let W_2 be an arbitrary function in H^∞, and we define its *compression* $W_2(T)$ to $H(m)$ by

$$W_2(T) := \pi^m M_{W_2}|_{H(m)}. \tag{10.57}$$

Remark 10.8.1. While in Chapter 9 H^p spaces are defined on $\mathbb{D}$, we deal here with those spaces defined on $\mathbb{C}_+$, etc. Via the bilinear transform $s \leftrightarrow (z-1)/(z+1)$, the results for $\mathbb{D}$ naturally carry over to results over H^p spaces on $\mathbb{C}_+$.

We state Sarason's theorem (Theorem 9.5.2, p. 191) in the following modified form, which is more suitable to the present context. The difference here is that we take $T(t)$ instead of T as in Chapter 9 (p. 191).

Theorem 10.8.2 (Sarason [49]). *Let $F : H(m) \to H(m)$ be a bounded linear operator that commutes with $T(t)$ for every $t \geq 0$. Then there exists $f \in H^\infty$ such that $F = f(T)$ and $\|F\| = \|f\|_\infty$.*

10.8.1 Computation of Optimal Sensitivity

Let us now return to the computation of the optimal sensitivity:

$$\gamma_{\text{opt}} = \inf_{Q \in H^\infty} \| W - m Q \|_\infty.$$

Let $F := W(T)$. According to Sarason's theorem, there exists $f_{\text{opt}} \in H^\infty$ such that $f_{\text{opt}}(T) = W(T)$ and $\| W(T) \| = \| f \|_\infty$. The action of W and f must agree on $H(m)$, and hence

- $f_{\text{opt}} = W - m Q_{\text{opt}}$, and

- $\gamma_{\text{opt}} = \| W(T) \|$.

Hence computation of the optimum γ_{opt} is reduced to that of $\| W(T) \|$. For a large class of systems, it is known that this is equivalent to computing the maximal singular value of $W(T)$, and, moreover, its computation is reducible to computing the rank of a certain matrix [13].

10.8.2 Zhou–Khargonekar Formula

We here present a beautiful formula, first proven by Zhou and Khargonekar [74] for $m(s) = e^{-Ls}$, and later extended to general inner functions by Lypuchuk, Smith, and Tannenbaum [30, 55]. Later some simplified proofs were given in [16, 62]. Let (A, B, C) be a minimal realization of $W(s)$. Define the Hamiltonian matrix H_γ by

$$H_\gamma := \begin{bmatrix} A & B B^T / \gamma \\ -C^T C / \gamma & -A^T \end{bmatrix}.$$

Then $\gamma > 0$ is a singular value of the Hankel operator $\Gamma_{m^\sim W}$ if and only if

$$\det m^\sim (H_\gamma)|_{22} = 0, \tag{10.58}$$

where $m^\sim(s) = m^{-1}(s) = m(-s)$ and $(\cdot)|_{22}$ denotes the $(2,2)$-block when partitioned conformally with H_γ. In the case of $m(s) = e^{-Ls}$, (10.58) takes the form

$$\det \exp \begin{bmatrix} A & B B^T / \gamma \\ -C^T C / \gamma & -A^T \end{bmatrix} L|_{22} = 0.$$

Needless to say, the maximal singular value is the largest γ among those satisfying the above condition. (For brevity we omit some extra conditions; see, e.g., [62].)

10.9   Supplementary Remarks

We have given a brief overview of systems and control theory, with emphasis on the frequency domain treatment of the sensitivity optimization of H^∞ control. Except some extensions to nonlinear or infinite-dimensional systems, H^∞ control theory has become fairly complete and been used in many important applications.

Control theory can provide solutions to many practical problems. It is no exaggeration to say that almost no system, artificial or natural, can work without control, and control theory also raises many interesting mathematical problems, whose resolution can lead to fruitful new directions. It is hoped that the brief overview here motivates the reader to look more closely into the subject.

For a more general treatment of H^∞ control, the reader is referred to [9, 13, 73]. For a functional analytic approach to a more general H^∞ problem [13], one needs the commutant lifting theorem, which is a generalization of Sarason's theorem treated here. For this, the reader is referred to [40].

For generalizations to distributed parameter systems, see, e.g., [4, 13] and also [27] for some recent developments. While there are many standard textbooks in systems and control, we list [23] as a classical textbook and also [19, 56] for a more recent treatment. The book by Young [67] is also useful in grasping H^∞ control from a Hilbert space point of view.

Appendix A

Some Background on Sets, Mappings, and Topology

A.1 Sets and Mappings

Let us briefly review and summarize the fundamentals on sets and mappings.

We will content ourselves with the naive understanding that a set is a collection of objects. Rigorously speaking, this vague "definition" is unsatisfactory because it can lead to a contradiction if one proceeds to enlarge this class of objects too much, for example, by considering the "set of all sets." But we do not deal with such overly big objects.

A member of a set is called an *element* of the set. For vector spaces, it may be called a *vector* or a *point*. When we wish to say that a property holds for every element of a set X, we may often use the expression "for an *arbitrary* element of X," meaning "for every (any) element of X." This expression does not seem to be so common in daily life as it is in mathematics and perhaps requires a little care.

Given a proposition $P(x)$ on an element x, we denote by

$$\{x : P(x)\}$$

the set of all x such that $P(x)$ holds true. In particular, when we explicitly show that x is an element in X and consider the above as a subset of X, then we write

$$\{x \in X : P(x)\}.$$

For example, if $X = \mathbb{R}$ and $P(x)$ is given by $|x| \leq 1$, then

$$\{x \in \mathbb{R} : |x| \leq 1\}$$

is the interval $[-1, 1]$.

The union $\cup$, intersection $\cap$, and inclusion $\subset$ are assumed to be already familiar to the reader. We just note that $A \subset B$ does not exclude the case $A = B$. For a subset A of X, its *complement* A^c is defined as

$$A^c := \{x \in X : x \notin A\}.$$

A set consisting of finitely many elements is called a *finite set*; otherwise it is called an *infinite set*. The size of an infinite set, called the *cardinal number* or *cardinality*, i.e., the number of elements in the set, is beyond the scope of this book and is hence omitted. We only note that if the elements of the set can be shown with ordering as $A = \{x_1, x_2, x_3, \ldots\}$, the set A is called a *countable set*, or possesses a countable number of elements. The sets $\mathbb{N}$ of natural numbers, $\mathbb{Z}$ of integers, $\mathbb{Q}$ of rational numbers are countable sets.

Let X, Y be sets. If for every x in X there is always associated one (and only one) element y in Y, this correspondence is called a *mapping* or simply a *map* (it may be called a *function* depending on the context; the distinction is ambiguous) and is denoted as

$$f : X \to Y, \text{ or } X \xrightarrow{f} Y.$$

The set X is called the *domain*, and Y the *codomain*, of f. When we wish to explicitly include in this notation the correspondence between x and y under f, we write it (in this book) as

$$f : X \to Y : x \mapsto y$$

using the notation $\mapsto$. For example,

$$\sin : \mathbb{R} \to \mathbb{R} : x \mapsto \sin(x).$$

We may also write simply $x \mapsto y$ when showing the domain and codomain is not necessary. Of course, the notation $y = f(x)$ is also used. We also use the following expressions: f maps (takes, sends) x to y, or the action of f on x is y.

Strictly speaking, $y = f(x)$ shows that the *image* of the element x under f is $f(x)$. In other words, representing a mapping (function) as $f(x)$ (showing its dependence on x explicitly) is somewhat imprecise. To be more rigorous, one should write either f or $f(\cdot)$ without reference to a particular element x.

Let $f : X \to Y$ be a mapping, and A a subset of X. The set

$$f(A) := \{f(a) \in Y : a \in A\}$$

of all elements of Y mapped by f from A is called the *image* of A under f. When $A = X$, it is called the *image*, or the *range*, of f and is denoted by im f or $R(f)$. Here the symbol ":=" means that the right-hand side defines the left-hand one. Given two mappings $f : X \to Y$ and $g : Y \to Z$, we define their *composition* (or *composed mapping*) as $g \circ f : X \to Z$: $(g \circ f)(x) := g(f(x))$.

When $Y = f(X)$, i.e., any element of Y is an image of some element of X under f, we say that the mapping f is an *onto mapping*, a *surjection, or a surjective mapping*. On the other hand, if $f(x_1) = f(x_2)$ always implies $x_1 = x_2$, i.e., f sends different elements to different elements, we say that f is a *one-to-one mapping* or an *injection, injective mapping*. A mapping that is both surjective and injective simultaneously is called a *bijection, bijective mapping* or a *one-to-one correspondence*. For a bijection f, as clearly seen from Figure A.1, there exists its *inverse mapping* f^{-1} that maps every $f(x) \in Y$ to $x \in X$, i.e., $f^{-1} : Y \to X : f(x) \mapsto x$. Clearly $f^{-1} \circ f = I_X$, $f \circ f^{-1} = I_Y$. Here I_X denotes the *identity mapping* on X, that is, the mapping that sends every x to x itself.

For sets X, Y, their (*direct*) *product*) $X \times Y$ means the set of ordered pairs (x, y), $x \in X$, $y \in Y$:

$$X \times Y := \{(x, y) : x \in X, \ y \in Y\}. \tag{A.1}$$

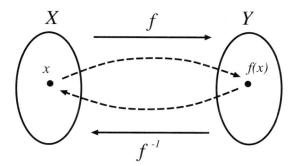

Figure A.1. *Inverse mapping*

Since this concept arose originally from the analytical geometry and coordinate system due to Descartes, it is also commonly referred to as the *Cartesian product*. We can likewise consider the product of an arbitrary number (can be infinite) of sets. In particular, the n product of the same set X is denoted as X^n. Spaces such as $\mathbb{R}^n$ and $\mathbb{C}^n$ are nothing but the n products of $\mathbb{R}$ and $\mathbb{C}$, respectively.

A.2   Reals, Upper Bounds, etc.

The construction of *real numbers* (or, simply, *reals*) is a basis for analysis, but it is often treated very lightly. It would be an interesting experience to read [6] by Dedekind himself.

The crux of the construction of the reals lies, in short, in the completion of the set of rational numbers $\mathbb{Q}$ while maintaining its algebraic structures represented by the four fundamental rules of arithmetic (i.e., addition, subtraction, multiplication, and division), and, further, the order structure and the concept of closeness (topology) determined by the absolute value, etc. In other words, we wish to enlarge the set $\mathbb{Q}$ so that we can freely take limits. As also pointed out in the main text, the source of difficulty lies in the fact that we can use no other concepts than rationals in constructing reals. Unfortunately, we often forget this trivial fact, since we are so used to expressions such as $\sqrt{2}$. One tends to implicitly assume that an object such as $\sqrt{2}$ already exists before proving its existence. Just the fact that one is familiar with an expression such as $\sqrt{2}$ cannot be a proof of existence, of course.

Consider for example the definition of $\sqrt{2}$ using the Dedekind cut (see Figure A.2). It is well known and rather easy to prove that there is no rational number whose square

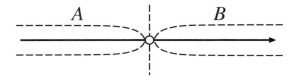

Figure A.2. *Dedekind cut*

is 2. This means that if we draw a "line" with rationals only, there should be a "hole" corresponding to this "number." This hole should be what we desire to define as $\sqrt{2}$. The difficulty here is how we can define, e.g., $\sqrt{2}$, only with rationals, without illegitimately importing the concept of reals which is what we want to define.

The idea of Dedekind is roughly as follows: If there is a "hole," then the rationals should be separated into two disjoint parts: one subset A should consist of all rationals that are less than $\sqrt{2}$, and the other set B should consist of those that are greater than $\sqrt{2}$. Of course, we have to express these two sets A and B without using the expression $\sqrt{2}$. Once this is done, the "hole" (i.e., $\sqrt{2}$) can be identified with the *pair* (A, B). One may think this is trivial, but it is based on the beautiful change of views from the hole itself to the opposite objects (i.e., the subsets separated by the hole). The difference is that while it is difficult to pin down the hole itself, it is possible to describe these sets A and B using rationals only. In the present case, A and B are given by

$$A := \{a \in \mathbb{Q} : a \leq 0 \text{ or } a > 0 \text{ and } a^2 < 2\},$$
$$B := \{b \in \mathbb{Q} : b > 0 \text{ and } b^2 > 2\}.$$

Since every element of A is smaller than any element of B, this pair separates the whole set of rationals into two disjoint parts. Dedekind called such a pair a *cut*. The remaining technical problems include formal procedures of defining the four fundamental rules of arithmetic, absolute values, etc., so that we can safely deal with pairs like (A, B) as bona fide *numbers*. But we will not delve into this further. The totality of all such pairs is called the real numbers.

A possible doubt is that we may arrive at a further extension by considering cuts of real numbers. But it can be shown that we will not obtain any further extension and will remain in the realm of real numbers. Thus any limit of reals remains in the reals, and this leads to the completeness of reals.

A remarkable consequence of the completeness can be seen, for example, by the following fact:

A monotone increasing (nondecreasing) sequence that is bounded above is convergent.

In fact, if we deal only with rationals, there are many (in fact, infinitely many) "holes" in the sense described by the Dedekind cut, and, as we saw in the example of $\sqrt{2}$, the limit does not exist in $\mathbb{Q}$. To circumvent this situation, we had to construct the reals. The statement above asserts that for reals, there are no such "holes."

In the same vein, for a subset M of the reals, its *supremum* or *least upper bound*

$$\sup M = \sup\{x \in \mathbb{R} : x \in M\}$$

always exists within reals, provided that M is bounded above. It is indeed known that this statement is equivalent to the completeness of the reals. The same can be said of an *infimum* or a *greatest lower bound*. A least upper bound is not necessarily a maximum. Hence $\sup M$ does not necessarily belong to the set M. Because of this, some may find it difficult to handle such concepts. It is worthwhile to note the following fundamental properties of the supremum and infimum of a sequence $\{a_n\}$.

Lemma A.2.1. *For the supremum ρ and the infimum σ of a real sequence $\{a_n\}$, the following properties hold:*

- *For every n, $\sigma \leq a_n \leq \rho$.*

- *For every $\epsilon > 0$, there exists n such that $\rho - \epsilon < a_n$.*

- *For every $\epsilon > 0$, there exists n such that $a_n < \sigma + \epsilon$.*

Sketch of proof. The first statement is obvious since the supremum is an upper bound, and the infimum is a lower bound. The second property follows from the fact that any number $\rho - \epsilon$ smaller than the supremum ρ is no longer an upper bound; i.e., ρ is the smallest upper bound. The same is true for the infimum. □

Remark A.2.2. It may be cumbersome to always say "provided that M is bounded above (below)." To avoid this repetition, we can introduce the notion of the *extended reals*. Let us write, formally,

$$\mathbb{R}^e := \mathbb{R} \cup \{+\infty\} \cup \{-\infty\}. \tag{A.2}$$

We said "formally" because $+\infty, -\infty$ are not numbers. But, by introducing this concept, one can write, for example,

$$\sup M = +\infty$$

for a subset M that is not bounded above. The same is true for inf M. This simplifies notation for other functions such as dim, lim, and $\int$ by allowing notation such as

$$\lim_{n \to \infty} a_n = \infty.$$

Upper and Lower Limits

Related concepts are *upper limits* and *lower limits*. Let us give some details on the upper limit of a real sequence $\{a_n\}$. Perhaps the main reason for confusion on upper limits is that they are represented by a single symbol $\limsup_{n \to \infty} a_n$, while they do indeed involve two limiting processes sup and lim; this is often overlooked due to the nature of this single symbol. A precise definition is given by

$$\limsup_{n \to \infty} a_n = \lim_{n \to \infty} \left(\sup_{v \geq n} a_v \right). \tag{A.3}$$

That is, first form another sequence defined by $b_n := \sup_{v \geq n} a_v$ and then take its limit, and this is the upper limit of $\{a_n\}$. As we increase n, b_n is defined by taking a supremum in a smaller range, and hence it is monotone decreasing (nonincreasing, to be exact). Hence $\{b_n\}$ always possesses a limit as $n \to \infty$, provided that we allow $\pm\infty$ as its limit, due to the completeness of the reals. This is how we define the upper limit of the sequence $\{a_n\}$.

In fact, this upper limit can be shown to be the largest accumulation point.[106] While this may be convenient for some cases, it is probably better to return to the definition (A.3) above when there is likely to be a confusion. Needless to say, the same arguments hold for lower limits.

---

[106]A point a is an *accumulation point* of $\{a_n\}$ if every neighborhood a contains a_n for some n.

A.3   Topological Spaces

We have given in section 2.1 (p. 39) some fundamentals on the *topology* of normed linear spaces. The topology of a normed space X is, in short, the concept of "closeness" specified by introducing the *distance $d(x, y)$* between two points x, y by

$$d(x, y) := \|x - y\| \tag{A.4}$$

via the norm of X. By introducing the notion of closeness, we have seen that we can define such concepts as continuity of mappings and open and closed sets (subsection 2.1.4, p. 47). For example, the continuity of a mapping $f : X \to Y$ between normed linear spaces X and Y: The mapping f is *continuous* at $x \in X$ if all points sufficiently close to x should be mapped closely to $f(x)$; in other words, for every neighborhood V_y of $y = f(x)$ there exists a neighborhood V_x such that its image $f(V_x)$ is contained in V_y. This can be rephrased as the property that for every neighborhood V_y of y its inverse image $f^{-1}(V_y)$ is again a neighborhood of x. When f is continuous at every point, it is called a *continuous mapping*.

In fact, to discuss such properties, the space X does not have to be a normed space. It is sufficient that a nonnegative real-valued function *distance $d : X \times X \to \mathbb{R}$* is defined on X satisfying the properties as follows:

1. $d(x, y) \geq 0$, and $d(x, y) = 0 \Leftrightarrow x = y$,

2. $d(x, y) = d(y, x)$,

3. $d(x, z) \leq d(x, y) + d(y, z)$ (triangle inequality)

for all $x, y, z \in X$. When these conditions are satisfied, the pair (X, d) is called a *metric space*. As we saw in Chapter 2, the distance induced by (A.4) from the norm in a normed space X satisfies the above axioms for a distance, and X becomes a metric space. Indeed, almost all topological properties we discussed in Chapter 2 carry over to metric spaces. One can suitably generalize the notion of a Cauchy sequence to metric spaces and then discuss completeness. The Baire category theorem also holds for complete metric spaces.

Let us now further generalize and discuss the notion of topology in a more general context. Let us recall how the concept of neighborhoods was defined in Chapter 2. A neighborhood of a point x was defined (p. 40) as a set containing an open ϵ neighborhood $B(x, \epsilon) = \{y : \|y - x\| < \epsilon\}$. We can generalize the notion of neighborhoods as follows: A subset $V \ni x$ is a neighborhood of x if it contains an open set containing x. Advancing this idea, one can introduce a topology by using only the notion of open sets, and many mathematics books indeed take this standpoint. Roughly, it goes as follows: one first starts out by specifying what should qualify as open sets and then defines other topological concepts such as closed sets, neighborhoods, continuity, and convergence.

In specifying a family of open sets, the following requirements must be satisfied:

O1. The whole space X and the empty set $\emptyset$ are open.

O2. The intersection of finitely many open sets is open.

O3. The union of open sets is open.

In other words, given a family of subsets $\mathcal{T}$ satisfying the conditions above, we say that a topology is introduced to X. That the notion of open sets defined in Chapter 2 satisfies this condition is as shown in subsection 2.1.4 on page 47.

Let us now review how other notions such as that of closed sets can be defined using the concept of open sets as introduced above.

Let $(X, \mathcal{T})$ denote a topological space; i.e., $\mathcal{T}$ is a family of open sets that satisfies the conditions above. A subset F of X is said to be a *closed set* if its complement F^c is an open set, that is, $F^c \in \mathcal{T}$. For an arbitrary subset A of X, its *closure* $\overline{A}$ is defined to be the smallest closed set containing A. This is indeed equal to

$$\overline{A} = \bigcap \{F \ : \ A \subset F, \ F \text{ is closed}\}. \tag{A.5}$$

Let us note that X is always closed as the complement of the empty set, and hence the right-hand side above is not empty.

A point belonging to the closure $\overline{A}$ is called a *closure* (or *adherent*) *point* of A. A subset A is said to be *dense* in X if $X = \overline{A}$. If this holds, then every point of X is a closure point of A. A subset V is a *neighborhood* of x if V contains an open set $O \in \mathcal{T}$ that contains x. A topological space X is said to be *separable* if there exists a dense subset A that has at most countable cardinality. If a Hilbert space X possesses a complete orthonormal system $\{e_n\}_{n=1}^{\infty}$, then X is easily seen to be separable. Most Hilbert or Banach spaces appearing in practice are known to be separable, and hence we consider only separable spaces in this book.

A mapping $f : X \to Y$ between topological spaces X, Y is said to be *continuous* if for every open set O_Y in Y its inverse image $f^{-1}(O_Y)$ is an open set in X. It should not be difficult to check that this definition coincides with the one given in terms of the concept of neighborhoods.

A.4 Product Topological Spaces

Given topological spaces $(X, \mathcal{T}_X)$ and $(Y, \mathcal{T}_Y)$, we can introduce in a natural and canonical way a topology to their product space $X \times Y$. For open sets $O_1 \in \mathcal{T}_X$ in X and $O_2 \in \mathcal{T}_Y$ in Y, we first require that the product set $O_1 \times O_2$ be an open set in $X \times Y$. But taking all such sets does not satisfy the axioms O1–O3 as defined in the previous section. We thus require that

- the intersection of arbitrary but finitely many of such sets and

- the union of an arbitrary family of such sets

also be open sets in $X \times Y$. Then it is easy to ensure that the collection of all such sets satisfies the axioms O1–O3. Hence this procedure induces a topology on $X \times Y$. The product space equipped with this topology is called a *product topological space*.

When $(X, \|\cdot\|_X)$ and $(Y, \|\cdot\|_Y)$ are normed linear spaces, this topology is known to be equal to the one induced by the norm[107]

$$\|(x, y)\| := \|x\|_X + \|y\|_Y . \tag{A.6}$$

---

[107]Its proof, although not difficult, is omitted here since it requires some preliminaries.

A.5   Compact Sets

Compactness plays a role in various aspects of analysis.

There are several equivalent definitions. The most general is the one given in terms of covering, but we here give a more elementary one. While the definition works for more general cases, we content ourselves with a definition for normed linear spaces that can be given by sequences.

Definition A.5.1. Let M be a subset of a normed linear space X. M is said to be *compact* if for every infinite sequence $\{x_n\}_{n=1}^{\infty}$ in M there always exists a convergent subsequence of it.

The definition above follows a standard wording in such a case. A warning is, however, in order. When we say that a (sub)sequence is convergent, it means that the limit itself should belong to the set M. (When we use such an expression, we do not consider the "outside" of the set considered.) Confusion can arise when M is embedded in a larger space X, and a subsequence can have a limit in X but not in M, for example, $M := \{x \in \mathbb{Q} : 0 \leq x \leq 1\}$, and $X = \mathbb{R}$. The above requirement really says that there exists a subsequence that converges to a point in M. While this paraphrasing is certainly less confusing, the reader may encounter the expression in the above definition more often.

Hence a compact set is necessarily a closed set. It should also be easy to see that a compact set is always bounded in the sense already defined.[108]

Exercise A.5.2. Prove the statement above.

Let us then ask the following question: Is every bounded and closed set compact? In fact, this is valid in finite-dimensional space $\mathbb{R}^n$. This fact is known as the Bolzano–Weierstrass theorem. To summarize, for finite-dimensional Banach spaces, a compact set is another name for a bounded and closed set.

Of course, if compact sets are just bounded and closed sets, there is certainly not much value to such a different name. It is when the dimensionality becomes infinite where these two notions depart from each other.

To exemplify this, let us state the following fact.

Example A.5.3. Let X be an infinite-dimensional Banach space. Then the unit ball B_1 of X is never compact.

Let us show this in the case of Hilbert spaces. Let $\{e_n\}_{n=1}^{\infty}$ be an orthonormal system in a Hilbert space X. Since X is infinite-dimensional, this always exists; see the Schmidt orthogonalization given in Problem 1 of Chapter 3. Observe that

$$\|e_i - e_j\|^2 = 2$$

for every pair $i \neq j$. Clearly no subsequence of this sequence can be convergent. Since $\{e_n\}_{n=1}^{\infty}$ are elements of B_1, this shows that B_1 is not compact.

---

[108]cf. Definition 2.1.8, page 42.

A.6   Norms and Seminorms

Let X be a vector space. A nonnegative valued function $\|\cdot\|$ on X is called a *seminorm* if it satisfies for all $x, y \in X$ and $\alpha \in \mathbb{R}$

$$\|\alpha x\| = |\alpha|\, \|x\|, \tag{A.7}$$

$$\|x + y\| \leq \|x\| + \|y\|, \tag{A.8}$$

that is, the conditions omitting the requirement that $\|x\| = 0 \Rightarrow x = 0$ in the definition of norms.

Recall that the topology of a normed linear space is always defined by a single norm. This is actually a rather special case of topological vector spaces, and there are a number of topological spaces of interest in practice that do not satisfy the conditions above—for example, the space $C(-\infty, \infty)$ of all continuous functions on the real line, the spaces $\mathcal{D}$, $\mathcal{S}$ of test functions, or various spaces of distributions introduced in Chapter 6. The topology of such a space is defined in terms of many (in general, infinitely many) seminorms. For the general theory, we must refer the reader to more advanced treatments of topological vector spaces such as Köthe [29], Schaefer [51], Treves [58], Yamanaka [64], etc. We here briefly explain how a topology can be introduced from a family of seminorms.

Let $\|\cdot\|_\lambda$, $\lambda \in \Lambda$, be a family of seminorms given on a vector space X. For every finite subset $\Sigma \subset \Lambda$ and $\epsilon > 0$, define

$$V(\Sigma, \epsilon) := \{x \in X : \|x\|_\lambda < \epsilon \quad \forall \lambda \in \Sigma\}. \tag{A.9}$$

A subset V is defined to be a neighborhood of the origin 0 if there exist Σ and $\epsilon > 0$ such that $V(\Sigma, \epsilon) \subset V$. A neighborhood of a point x can be defined by translation; i.e., V is a neighborhood of x if and only if its translate $V - x$ is a neighborhood of the origin. We can then show that the totality of all neighborhoods as defined satisfies the axiom of neighborhoods. A vector space equipped with the topology by a family of seminorms in the way as given above is called a *locally convex topological vector space*.[109] Now consider the condition

for every $x \neq 0$, there exists a seminorm $\|\cdot\|_\lambda$, $\lambda \in \Lambda$, such that $\|x\|_\lambda \neq 0$

in place of

$$\|x\| = 0 \quad \Rightarrow \quad x = 0$$

required for normed linear spaces. When this condition is satisfied, the topological vector space X is called a *Hausdorff space* or is said to satisfy the *Hausdorff separation principle*. Most spaces appearing in applications are Hausdorff spaces.

A.7   Proof of the Hahn–Banach Theorem

For topological vector spaces such as normed linear spaces, a major role is played by the theory of duality, i.e., the study of various properties arising from the pair of the base spaces and the respective duals. In such a study, it is implicitly assumed that the dual space is sufficiently rich; that is, it contains sufficiently many elements to make such a study meaningful. Actually it is known that there exists a topological vector space with its

---

[109] Since the fundamental neighborhoods defined by (A.9) are convex, this is named as such.

dual consisting only of 0, and the assumption above is not trivial at all.[110] Fortunately, for normed linear spaces or, more generally, locally convex spaces where topology is defined via seminorms, such a pathology does not occur and it is known that their dual spaces contain sufficiently many elements. What guarantees this fact is the Hahn–Banach theorem to be stated and proven here. It claims that every continuous[111] linear functional given on an arbitrary subspace can be extended to the whole space maintaining continuity.

The importance of the Hahn–Banach theorem lies not only in guaranteeing the richness of dual spaces. It is also very important for applications. One of them is that it guarantees the existence of a *supporting hyperplane* in optimization problems. For this, the reader is referred to, for example, [31].

We give a proof in a way applicable to locally convex spaces but also assume that the space is separable. This greatly simplifies the argument. The reader familiar with the axiom of choice or Zorn's lemma will find no difficulty in generalizing the proof to nonseparable spaces.

Theorem A.7.1 (Hahn–Banach extension theorem). *Let X be a normed linear space, M its subspace, and $\|\cdot\|$ a continuous seminorm on X. Let f be a linear functional defined on M satisfying the following inequality:[112]*

$$|f(x)| \leq \|x\| \quad \forall x \in M. \tag{A.10}$$

Then f can be extended to the whole space X maintaining inequality (A.10) *to hold on the whole space.*[113]

Before going into the proof, let us prepare the notion of a *real vector subspace* of a complex vector space. A subset N of a vector space X is said to be a *real vector subspace* if N is closed under addition and scalar multiplication by real numbers; that is, $x, y \in N$ and $\alpha, \beta \in \mathbb{R}$ should imply $\alpha x + \beta y \in N$. Note that this places no conditions on scalar multiplication by complex numbers. Every vector subspace is necessarily a real vector subspace, but not conversely.[114] Of course, if X is a vector space over $\mathbb{R}$, the two concepts clearly coincide. Let N be a real vector subspace. A functional ψ defined on N is called a *real linear functional* or a *real linear form* if ψ is real-valued and satisfies $\psi(\alpha x + \beta y) = \alpha \psi(x) + \beta \psi(y)$ for $x, y \in N$ and $\alpha, \beta \in \mathbb{R}$.

Proof. This theorem does not require X to be a normed linear space, and it indeed holds for arbitrary vector space. We have assumed X to be a normed space only for the need of using separability, to avoid the usage of Zorn's lemma or the transcendental induction. The continuity of $\|\cdot\|$ is not necessary either, and this is assumed also for simplicity of the arguments. For Zorn's lemma, see [10, 20, 33]. For the proof of the general case, see, for example, Köthe [29], Schaefer [51], Treves [58], and Yamanaka [64].

---

[110]Needless to say, the algebraic dual space, without a constraint on continuity, always contains sufficiently many elements, but it is possible that none of these elements except 0 are continuous.

[111]with respect to the subspace topology

[112]In this proof, we will write $f(x)$ in place of $\langle f, x \rangle$.

[113]Since $\|\cdot\|$ is continuous, we can conclude that f is continuous as in Proposition 2.1.11.

[114]For example, consider the set $\mathbb{C}$ of complex numbers as a vector space over $\mathbb{C}$ itself. The set $\mathbb{R}$ of reals is a real vector subspace, but it is clearly not closed under a scalar multiplication with a complex number and hence is not a vector subspace of $\mathbb{C}$ in this case. On the other hand, if we regard $\mathbb{C}$ as a vector space over $\mathbb{R}$, then $\mathbb{R}$ becomes a subspace.

Step 1. First we consider $g(x) := \operatorname{Re} f(x)$, and we show that this g can be extended to the whole space X. This functional g is clearly a real linear functional and satisfies

$$|g(x)| \leq \|x\|, \quad x \in M. \tag{A.11}$$

If $M = X$, there is nothing to prove. Assume $M \neq X$. Then there exists x_0 in X that does not belong to M. Consider the real linear subspace M_1 spanned by M and x_0; that is, $M_1 := \{m + \alpha x_0 : m \in M, \alpha \in \mathbb{R}\}$. We first show that g can be extended to M_1 while maintaining inequality (A.11).

Take any element $m + \alpha x_0 \in M_1$. If g can be extended to M_1 as a real linear functional, then we must have $g(m + \alpha x_0) = g(m) + \alpha g(x_0)$. It suffices to show that (A.11) remains valid by taking a real number $g(x_0)$ appropriately.

Take any $m_1, m_2 \in M$. By the triangle inequality (A.8), we have

$$g(m_1) + g(m_2) = g(m_1 + m_2) \leq \|m_1 + m_2\| \leq \|m_1 - x_0\| + \|m_1 + x_0\|.$$

Hence

$$g(m_1) - \|m_1 - x_0\| \leq \|m_2 + x_0\| - g(m_2).$$

This implies

$$\sup_{m \in M} \left[g(m) - \|m - x_0\| \right] \leq \inf_{m \in M} \left[\|m + x_0\| - g(m) \right].$$

Take a real number c that is in between the two numbers of both sides of the above, and define $g(x_0) := c$. The real linearity of the extended g is obvious. Let us confirm the estimate (A.11). When $\alpha > 0$, we have

$$g(m + \alpha x_0) = g(m) + \alpha c \leq g(m) + \alpha \left\{ \left\| \frac{m}{\alpha} + x_0 \right\| - g\left(\frac{m}{\alpha} \right) \right\} = \|m + \alpha x_0\|.$$

On the other hand, for $\alpha < 0$, we have

$$g(m + \alpha x_0) = g(m) + \alpha c \leq g(m) + \alpha \left\{ g\left(-\frac{m}{\alpha} \right) - \left\| -\frac{m}{\alpha} - x_0 \right\| \right\}$$

$$= (-\alpha) \left\| -\frac{m}{\alpha} - x_0 \right\| = \|m + \alpha x_0\|.$$

Hence for every $x \in M_1$, $g(x) \leq \|x\|$ is satisfied.

Step 2. With the preparation above, we now extend g to the whole space. For this we assume that X is separable.[115]

Let $\{x_1, \ldots, x_n, \ldots\}$ be a dense subset of X. We may assume, by removing redundant elements, that this is real-linearly independent and independent on M. Then by Step 1, we can inductively extend g over to $\operatorname{span}\{M, x_1\}$, $\operatorname{span}\{M, x_1, x_2\}$, ..., $\operatorname{span}\{M, x_1, \ldots, x_n\}$, ..., and so on. Hence we can extend g as a continuous liner functional to a dense subspace S, spanned by $\{x_1, \ldots, x_n, \ldots\}$ (since the seminorm $\|\cdot\|$ is continuous, g is also continuous by (A.11)). Hence by continuity, g can be extended to the whole space X as a continuous real linear functional. Indeed, for every element x in X, we can take a sequence $\{s_n\}$ in S that converges to x and define $g(x) := \lim_{n \to \infty} g(s_n)$. The extended g is clearly real linear, and since $|g(s_n)| \leq \|s_n\|$, taking limits on both sides yields

$$|g(x)| \leq \|x\|, \quad x \in X.$$

---

[115]As noted before, this is a technical assumption and can be removed.

Step 3. Let us now define

$$F(x) := g(x) - i g(ix)$$

using the extended real linear functional $g(x)$. We show that this F gives a linear extension of the given f. The additivity of F is obvious. For the scalar multiplication, we have

$$
\begin{aligned}
F((\alpha + i\beta)x) &= g(\alpha x + \beta(ix)) - ig(-\beta x + \alpha(ix)) \\
&= \alpha g(x) + \beta g(ix) - i[-\beta g(x) + \alpha g(ix)] \\
&= (\alpha + i\beta)(g(x) - i g(ix)) \\
&= (\alpha + i\beta)F(x),
\end{aligned}
$$

and hence F is linear. For $x \in M$, $g(x) = \operatorname{Re} f(x)$. Hence $F(x) = \operatorname{Re} f(x) - i \operatorname{Re} f(ix) = \operatorname{Re} f(x) - i \operatorname{Re}(if(x)) = \operatorname{Re} f(x) + i \operatorname{Im} f(x)$, and this shows that F is an extension of f.

Finally, since $F(x)$ can be written as $F(x) = |F(x)|e^{i\theta}$ for some θ, we have

$$|F(x)| = e^{-i\theta} F(x) = F(e^{-i\theta}x) = g(e^{-i\theta}x) \le \left\| e^{-i\theta}x \right\| = \|x\|,$$

because $F(e^{-i\theta})$ is real. Hence F preserves inequality (A.10). $\qquad\square$

This completes the proof of the Hahn–Banach theorem. Let us now give a few corollaries.

The following corollaries guarantee that the dual space of a normed linear space contains sufficiently many elements.

Corollary A.7.2. *Let X be a nontrivial normed linear space and x_0 any element of X. Then there exists a continuous linear functional $F \in X'$ with norm 1 such that $F(x_0) = \|x_0\|$.*

Proof. First consider the case $x_0 \ne 0$. Take the subspace $M := \{\alpha x_0 : \alpha \in \mathbb{K}\}$ generated by x_0, and define a functional f by $f(\alpha x_0) = \alpha \|x_0\|$. Clearly f satisfies

$$|f(x)| \le \|x\|, \quad x \in M.$$

Hence by the Hahn–Banach theorem f can be extended to $F \in X'$ with the above inequality holding for all $x \in X$. Clearly $F(x_0) = f(x_0) = \|x_0\|$.

On the other hand, if $x_0 = 0$, then any $F \in X'$ satisfies the requirements. Also from the first half of this proof, we see $X' \ne \{0\}$, and hence such an F exists. $\qquad\square$

This further yields the following corollary.

Corollary A.7.3. *Let X be a normed linear space and $x \in X$. If $f(x) = 0$ for every $f \in X'$, then $x = 0$.*

Proof. If $x \ne 0$, then by Corollary A.7.2 above, there would exist $F \in X'$ such that $F(x) = \|x\| \ne 0$. This is a contradiction, and hence $x = 0$. $\qquad\square$

The following corollary is often used in the discussion of normed linear spaces.

Corollary A.7.4. *Let X be a normed linear space, and M its subspace. Every element g in M' can be extended to a continuous linear functional G on the whole space without changing its norm.*

Proof. Define $p(x) := \|g\| \, \|x\|$, $x \in X$. This p is clearly a continuous seminorm on X. It is also clear that $g(y) \le p(y)$, $y \in M$. Hence by the Hahn–Banach theorem, Theorem A.7.1, there exists an extension G of g to X such that $|G(x)| \le p(x)$. Then $|G(x)| \le p(x) = \|g\| \, \|x\|$ readily implies $\|G\| \le \|g\|$. On the other hand, since $G = g$ on M, $\|G\| \ge \|g\|$ should follow. Hence $\|G\| = \|g\|$. $\qquad\qquad\qquad\qquad\qquad\qquad\qquad\qquad\qquad\qquad\qquad\qquad\qquad\qquad\square$

A.8   The Hölder–Minkowski Inequalities

We state Hölder's and Minkowski's inequalities that appear frequently in ℓ^p and L^p spaces. When $p = 2$, Hölder's inequality reduces to Schwarz's inequality, and Minkowski's inequality is nothing but the triangle inequality in Hilbert space ℓ^2, and they are proved in Chapter 3. The proofs for general p are more involved and technical, so we omit them here and state the results only. For the proofs, we refer the reader to textbooks on functional analysis—for example, [20, 31].

Proposition A.8.1 (Hölder's inequality). *Let p, q be positive integers satisfying $1 \le p, q \le \infty$ and $1/p + 1/q = 1$. Let $x = \{x_n\}_{n=1}^{\infty} \in \ell^p$ and $y = \{y_n\}_{n=1}^{\infty} \in \ell^q$. Then*

$$\sum_{n=1}^{\infty} |x_n y_n| \le \|x\|_p \, \|y\|_q, \tag{A.12}$$

where the equality holds only when

$$\left(\frac{|x_n|}{\|x\|_p} \right)^{1/q} = \left(\frac{|y_n|}{\|y\|_q} \right)^{1/p} \tag{A.13}$$

is satisfied for each n.

Theorem A.8.2 (Minkowski's inequality). *Suppose $x, y \in \ell^p$, $1 \le p \le \infty$. Then $x + y$ also belongs to ℓ^p, and*

$$\left\{ \sum_{k=1}^{\infty} |x_k + y_k|^p \right\}^{1/p} \le \left\{ \sum_{k=1}^{\infty} |x_k|^p \right\}^{1/p} + \left\{ \sum_{k=1}^{\infty} |y_k|^p \right\}^{1/p}. \tag{A.14}$$

That is, $\|x + y\|_p \le \|x\|_p + \|y\|_p$. In addition, when $p < \infty$, the equality holds if and only if there exist constants k_1, k_2 such that $k_1 x = k_2 y$.

Let us also give the corresponding inequalities in the integral forms.

Proposition A.8.3 (Hölder's inequality–integral form). *Let $1/p + 1/q = 1$ and $p, q > 1$, and suppose that $x \in L^p(a, b)$, $y \in L^q(a, b)$. Then*

$$\int_a^b |x(t)y(t)| dt \le \|x\|_p \, \|y\|_q. \tag{A.15}$$

The equality holds when and only when

$$\left(\frac{|x(t)|}{\|x\|_p}\right)^{1/q} = \left(\frac{|y(t)|}{\|y\|_q}\right)^{1/p} \tag{A.16}$$

holds almost everywhere in (a,b).

Theorem A.8.4 (Minkowski's inequality–integral form). *If $x, y \in L^p(a,b)$ for $1 \le p \le \infty$, then $x + y$ belongs to $L^p(a,b)$, and*

$$\left\{\int_a^b |x(t) + y(t)|^p dt\right\}^{1/p} \le \left\{\int_a^b |x(t)|^p dt\right\}^{1/p} + \left\{\int_a^b |y(t)|^p dt\right\}^{1/p}. \tag{A.17}$$

That is, $\|x + y\|_p \le \|x\|_p + \|y\|_p$.

Using Hölder's inequality, we can show the following inclusion relationship between L^p spaces.

Lemma A.8.5. *Suppose that a and b, $a < b$, are finite, and let $0 < p < q \le \infty$. Then $L^q[a,b] \subset L^p[a,b]$, and for every $f \in L^q[a,b]$,*

$$\left\{\frac{1}{b-a}\int_a^b |f(t)|^p dt\right\}^{1/p} \le \begin{cases} \left\{\dfrac{1}{b-a}\displaystyle\int_a^b |f(t)|^q dt\right\}^{1/q}, & q < \infty, \\[4mm] \|f\|_\infty, & q = \infty. \end{cases} \tag{A.18}$$

Proof. It suffices to show (A.18). First consider the case $q < \infty$. Let $r := q/p > 1$. Defining $r' := q/(q-p)$, we have $1/r + 1/r' = 1$. Then apply Hölder's inequality (A.15) to $x(t) = |f(t)|^p$, $y(t) = 1$ to obtain

$$\frac{1}{b-a}\int_a^b |f(t)|^p dt \le \frac{1}{b-a}\left\{\int_a^b \left(|f(t)|^p\right)^r dt\right\}^{1/r} \left\{\int_a^b 1^{r'} dt\right\}^{1/r'}$$

$$\le \frac{1}{b-a}\left\{\int_a^b \left(|f(t)|^p\right)^{q/p} dt\right\}^{p/q} (b-a)^{(q-p)/q}$$

$$= \left(\frac{1}{b-a}\right)^{p/q}\left\{\int_a^b |f(t)|^q dt\right\}^{p/q}.$$

Taking the pth roots of both sides, we obtain

$$\left\{\frac{1}{b-a}\int_a^b |f(t)|^p dt\right\}^{1/p} \le \left\{\frac{1}{b-a}\int_a^b |f(t)|^q dt\right\}^{1/q}.$$

On the other hand, when $q = \infty$, we easily obtain

$$\int_a^b |f(t)|^p dt \le \int_a^b \|f\|_\infty^p dt = (b-a)\|f\|_\infty^p,$$

and this readily yields the second estimate in (A.18). $\qquad\square$

Appendix B

Table of Laplace Transforms

$f(t)$	$\mathcal{L}[f]$
$f_1(t) + f_2(t)$	$\mathcal{L}[f_1](s) + \mathcal{L}[f_2](s)$
$f(at)$	$\frac{1}{|a|}\mathcal{L}[f]\left(\frac{s}{a}\right)$
$f(t-a)$	$e^{-as}\mathcal{L}[f](s)$
$e^{-at}f(t)$	$\mathcal{L}[f](s+a)$
$f(t)\cos\omega t$	$\frac{1}{2}[\mathcal{L}[f](s+i\omega) + \mathcal{L}[f](s-i\omega)]$
$f_1 * f_2$	$\mathcal{L}[f_1](s)\mathcal{L}[f_2](s)$
f'	$s\mathcal{L}[f](s)$
$H(t)$	$\dfrac{1}{s}$
$H(t)e^{-at}$	$\dfrac{1}{s+a}$
$tH(t)$	$\dfrac{1}{s^2}$
δ	1
δ_a	e^{-as}
δ'	s
$H(t)\cos\omega t$	$\dfrac{s}{s^2+\omega^2}$
$H(t)\sin\omega t$	$\dfrac{\omega}{s^2+\omega^2}$
$H(t)te^{-at}$	$\dfrac{1}{(s+a)^2}$
$H(t)e^{-at}\cos\omega t$	$\dfrac{s+a}{(s+a)^2+\omega^2}$
$H(t)e^{-at}\sin\omega t$	$\dfrac{\omega}{(s+a)^2+\omega^2}$

Appendix C

Solutions

C.1 Chapter 1

Exercise 1.1.12. Omitted.

Exercise 1.1.17. It is obvious that every polynomial of degree n can be written as a linear combination of $1, \ldots t^n$. To show that they are linearly independent (although this is not necessary for showing the finite-dimensionality of P_n), one needs only to show that $\sum_{k=0}^{n} \alpha_k t^k \equiv 0$ implies $\alpha_k = 0$ for all k. To see this, differentiate both sides successively, and set $t = 0$ at each step.

Exercise 1.1.21. Imitate the solution above.

Exercise 1.2.2. Combining (1.17) and (1.18) easily leads to (1.19). Conversely, put $\alpha = \beta = 1$ in (1.19) to obtain (1.17), and also put $\beta = 0$ to obtain (1.18).

Exercise 1.2.3. $g(y) = x \Rightarrow f(x) = y \Rightarrow \alpha y = f(\alpha x) \Rightarrow \alpha x = g(\alpha y)$, by definition. On the other hand, $\alpha x = \alpha g(y)$.

Exercise 1.2.5. Since $K(t,s)$ is uniformly continuous on $[0,1] \times [0,1]$, for any $\epsilon > 0$ there exists $\delta > 0$ such that $|t - t'| < \delta \Rightarrow |K(t,s) - K(t',s)| < \epsilon$ for all $s \in [0,1]$. Also, there exists $M > 0$ such that $|K(t,s)| \leq M$ on $[0,1] \times [0,1]$. Hence if $|t - t'| < \delta$, then $|(A\phi)(t) - (A\phi)(t')| \leq |\int_0^{t'} |K(t,s) - K(t',s)||\phi(s)|ds| + |\int_{t'}^{t} |K(t,s)||\phi(s)|ds| \leq \epsilon \sup_{0 \leq t \leq 1} |\phi(t)| + M\delta \sup_{0 \leq t \leq 1} |\phi(t)|$. Thus $(A\phi)(t)$ is a continuous function.

Exercise 1.2.12. Combining $f[e_1, \ldots, e_m] = [v_1, \ldots, v_n]F$ with $g[v_1, \ldots, v_n] = [w_1, \ldots, w_p]G$, we directly obtain $g \circ f[e_1, \ldots, e_m] = [w_1, \ldots, w_p]GF$.

Exercise 1.3.3. Consider $f^{-1}(N)$. Suppose $x, y \in N$, $\alpha, \beta \in \mathbb{K}$. Then $f(x), f(y) \in N$. Since N is a subspace, $\alpha f(x) + \beta f(y) \in N$. On the other hand, the linearity of f yields $\alpha f(x) + \beta f(y) = f(\alpha x + \beta y)$. Hence by Exercise 1.2.2, $f^{-1}(N)$ is proved to be a subspace. The same is true for images.

Exercise 1.3.4. We show $\ker f = \{0\} \Rightarrow f$. If $f(x) = f(y)$, then $f(x - y) = 0$. Hence $x - y \in \ker f = \{0\} \Rightarrow x = y$. The converse also holds, but it is obvious. The same is true for the surjectivity of f.

Exercise 1.3.6. 1. Obvious. 2. Consider the union of the x and y axes in the 2-dimensional x-y plane. 3. If U is a subspace containing $V \cup W$, then U must contain $V + W$ defined here. On the other hand, since it is easily seen that $V + W$ itself is a subspace, this is the smallest (with respect to the inclusion) subspace containing $V \cup W$.

Exercise 1.3.7. By a similar argument as above, it suffices to show that V is indeed a subspace. But this is easily seen from the definition.

Exercise 1.3.11. Suppose $w = \sum_{j=1}^{n} v_j = \sum_{j=1}^{n} w_j$, $v_j, w_j \in V_j$. Then

$$v_1 - w_1 = \sum_{j=2}^{n} (w_j - v_j).$$

The left-hand side belongs to V_1 while the right-hand side is an element of $\sum_{j \neq 1} V_j$. Then by condition (1.48), both must be 0. Hence $v_1 = w_1$. The same is true for other v_j's.

Exercise 1.3.17. For example, $\pi(x + y) = [x + y] = x + y + M = (x + M) + (y + M) = [x] + [y] = \pi x + \pi y$.

Exercise 1.3.20. Take $[v_1, \ldots, v_r]$ and $[w_1, \ldots, w_s]$ as bases for M and N, respectively, and then show that $[v_1, \ldots, v_r, w_1, \ldots, w_s]$ form a basis for $M \oplus N$. This is easily proven by noting that every element of $M \oplus N$ is uniquely expressible as $m + n$, $m \in M$, $n \in N$.

Exercise 1.4.2. Readily obvious from the definition.

Exercise 1.4.4. Omitted.

Exercise 1.4.7. For example, $\langle x, \alpha \varphi + \beta \psi \rangle = \alpha \langle x, \varphi \rangle + \beta \langle x, \psi \rangle$.

Exercise 1.4.11.

$$y^* \circ \mathcal{A}(\alpha x + \beta y) = y^*(\alpha \mathcal{A}x + \beta \mathcal{A}y) = \alpha y^* \circ \mathcal{A}x + \beta y^* \circ \mathcal{A}y.$$

Exercise 1.4.12. For every $x \in X$,

$$\mathcal{A}^*(\alpha_1 y_1^* + \alpha_2 y_2^*)(x) = (\alpha_1 y_1^* + \alpha_2 y_2^*)(\mathcal{A}x) = \alpha_1 y_1^*(\mathcal{A}x) + \alpha_2 y_2^*(\mathcal{A}x)$$
$$= \alpha_1 \mathcal{A}^* y_1^* + \alpha_2 \mathcal{A}^* y_2^*.$$

Exercise 1.4.16. Let A be surjective. Suppose $A^* y^* = 0$ for some $y^* \in Y^*$. This means that $\langle x, A^* y^* \rangle = 0$ for every x. This yields $\langle Ax, y^* \rangle = 0$. Since A is surjective, it follows $\langle y, y^* \rangle = 0$ for every $y \in Y$. This means $y^* = 0$, and hence A^* must be injective. The rest is similar.

Problems

1. $A(0) = A(0 + 0) = A(0) + A(0)$. Adding $-A(0)$ to both sides yields $A(0) = 0$. Similarly for $A(-x)$.

2. Follow the hint. Omitted.

3. Take a basis $\{e_1, \ldots, e_r\}$ for $M_1 \cap M_2$, and following Lemma 1.1.22, augment it with $\{f_1, \ldots, f_p\}$ and $\{g_1, \ldots, g_s\}$ to form bases for M_1 and M_2, respectively. It suffices to prove that $\{e_1, \ldots, e_r, f_1, \ldots, f_p, g_1, \ldots, g_s\}$ constitute a basis for $M_1 + M_2$. Clearly every element of $M_1 + M_2$ is expressible as a linear combination of these elements. Hence we need only show their linear independence. Suppose

$$\sum_{i=1}^{r} \alpha_i e_i + \sum_{j=1}^{p} \beta_j f_j + \sum_{k=1}^{s} \gamma_k g_k = 0.$$

This implies

$$\sum_{i=1}^{r} \alpha_i e_i + \sum_{j=1}^{p} \beta_j f_j = -\sum_{k=1}^{s} \gamma_k g_k. \tag{C.1}$$

The left-hand side belongs to M_1 while the right-hand side to M_2. Hence they both belong to $M_1 \cap M_2$. Then using the basis $\{e_1, \ldots, e_r\}$ of $M_1 \cap M_2$, we can write as

$$-\sum_{k=1}^{s} \gamma_k g_k = \sum_{i=1}^{r} \delta_i e_i.$$

Moving the left-hand side to the right, we see that $\delta_i = 0$, $i = 1, \ldots, r$, $\gamma_k = 0$, $k = 1, \ldots, s$ by the linear independence of $\{e_1, \ldots, e_r, f_1, \ldots, f_p\}$ (as a basis of M_1). Substituting this into (C.1), we obtain

$$\sum_{i=1}^{r} \alpha_i e_i + \sum_{j=1}^{p} \beta_j f_j = 0.$$

Then by the linear independence of $\{e_1, \ldots, e_r, g_1, \ldots, g_s\}$, we obtain $\alpha_i = 0$, $i = 1, \ldots, r$, $\beta_j = 0$, $j = 1, \ldots, p$. Hence $\{e_1, \ldots, e_r, f_1, \ldots, f_p, g_1, \ldots, g_s\}$ is linearly independent and forms a basis for $M_1 + M_2$.

C.2   Chapter 2

Exercise 2.1.14.   Let us show the triangle inequality only.

$$\begin{aligned}
\|f + g\| &= \sup_{a \leq t \leq b} |f(t) + g(t)| \\
&\leq \sup_{a \leq t \leq b} \{|f(t)| + |g(t)|\} \\
&\leq \sup_{a \leq t \leq b} |f(t)| + \sup_{a \leq t \leq b} |g(t)| \\
&= \|f\| + \|g\|.
\end{aligned}$$

Exercise 2.1.16.   Easily follows from $|f(t) + g(t)| \leq |f(t)| + |g(t)|$.

Exercise 2.1.24.   That every point of X is simultaneously an interior and closure point is obvious from the definitions. Hence X is an open as well as closed set. Since the empty set $\emptyset$ is the complement of X, it is also open and closed. Or else, one can directly show that the conditions are trivially satisfied since the premises are vacuous.

Exercise 2.1.26.   Let us prove that the finite intersection of open sets is open. Let $O_1, \ldots, O_n$ be open sets, and take any $x \in \cap_{i=1}^{n} O_i$. Since each O_i is open, there exists by definition $\epsilon_i > 0$ such that $B(x, \epsilon_i) \subset O_i$, $i = 1, \ldots, n$. Set $\epsilon := \min\{\epsilon_1, \ldots, \epsilon_n\}$. Then clearly $B(x, \epsilon) \subset \cap_{i=1}^{n} O_i$, and x is an interior point of $\cap_{i=1}^{n} O_i$. Since x is arbitrary, $\cap_{i=1}^{n} O_i$ is an open set. The other cases are similar.

Exercise 2.3.1.   It is obvious that M is a normed linear space. We show that M is a Banach space when X is a Banach space. Let $\{x_n\}$ be a Cauchy sequence in M. Then $\{x_n\}$ is clearly a Cauchy sequence in X, too. Since X is a Banach space, $\{x_n\}$ possesses a limit x in X, i.e., $x_n \to x$. Since M is a closed subset of X, this limit x should belong to M. That is, every Cauchy sequence of M possesses a limit in M, and hence M is a Banach space.

Problems

1.   The triangle inequality implies

$$\|x - y\| = \|x - x_n + x_n - y\| \leq \|x_n - x\| + \|x_n - y\|.$$

Since the right-hand side approaches 0, $\|x - y\| = 0$ follows. Hence by (2.2, p. 39), $x = y$.

2. Suppose $x_n \to x$. There exists N such that $\|x_n - x\| \leq 1$ for all $n \geq N$. It follows that

$$\|x_n\| = \|x_n - x + x\| \leq \|x_n - x\| + \|x\| \leq \|x\| + 1 \quad \forall n \geq N.$$

Putting $M := \max\{\|x_1\|, \dots, \|x_{N-1}\|, \|x\| + 1\}$ yields the conclusion.
3. Follows directly from the hint; draw the graph of the function $f_n(t)$.
4. Define $f(t)$ as being identically 1 on $(0, 1)$, and $1/t$ on $[1, \infty)$. Then $f \in L^2(0, \infty)$, but $f \notin L^1(0, \infty)$.

C.3   Chapter 3

Exercise 3.2.10.   Note $V - x = M$. Then $\|x + m\|$ attains its minimum when $x + m$ gives the perpendicular from x to M. Then $x + m = p_{M^\perp}(x)$, and such an m is unique (projection theorem 3.2.7, p. 67). (Hence $m = -p_M(x)$.) Thus $x_0 = x + m \perp M$ follows.
Problems
1. As indicated by the hint. Note $(z_n, e_j) = 0$, $j = 1, \dots, n - 1$.
2. Apply Theorem 3.3.1 (p. 75) for $L = A$, $y_0 = b$. Following the hint, the minimum norm solution under the definition $(x, y)_Q := y^T Q x$ is what is desired. Since the notion of adjoint operators depends on the inner product, we must first compute A^* according to this inner product. Let us denote the adjoint in this sense by A_Q^*. According to the definition (5.10) (p. 95) or (3.26), we have

$$\left(x, A_Q^* y\right)_Q = (Ax, y)_Q = y^T Q A x.$$

Put $B := (Q^T)^{-1} A^T Q^T$. This yields $(By)^T Q x = y^T Q A x$. Hence $A_Q^* = B = (Q^T)^{-1} A^T Q^T$. Applying 3.3.1 to $L = A$, $L^* = A_Q^*$, we obtain $x = (Q^T)^{-1} A^T Q^T (A(Q^T)^{-1} A^T Q^T)^{-1} b$. (Observe that the inverse on the right-hand side exists since rank $A = m$.)
3. Take any m in M. Directly by the definition of $M^\perp$, $m \perp M^\perp$, and hence $M \subset (M^\perp)^\perp$. Conversely, take any $x \in (M^\perp)^\perp$. According to Theorem 3.2.8 (p. 68), we have a unique decomposition

$$x = m + n, \quad m \in M, n \in M^\perp.$$

Since x and m are both orthogonal to $M^\perp$, $n = x - m$ should also be orthogonal to $M^\perp$. On the other hand, n is an element of $M^\perp$, so that $(n, n) = 0$, i.e., $n = 0$. Hence $x = m \in M$, and $(M^\perp)^\perp \subset M$ follows.

C.4   Chapter 4

Exercise 4.4.1.   Let $f, g \in M^\perp$, $\alpha, \beta \in \mathbb{K}$. For every m in M,

$$\langle \alpha f + \beta g, m \rangle = \alpha \langle f, m \rangle + \beta \langle g, m \rangle = 0,$$

and hence $\alpha f + \beta g$ also belongs to $M^\perp$.

C.5   Chapter 5

Exercise 5.1.2.   Omitted.

Exercise 5.1.4. Inequality (5.4) follows from

$$\|ABx\| \leq \|A\| \|Bx\| \leq \|A\| \|B\| \|x\|.$$

As an example for the strict inequality $\|AB\| < \|A\| \cdot \|B\|$, take, for example, $A = B = \begin{bmatrix} 0 & 1 \\ 0 & 0 \end{bmatrix}$. Then $AB = 0$. On the other hand, $\|A\| = \|B\| = 1$, and hence $0 = \|AB\| < 1 = \|A\| \cdot \|B\|$.

Exercise 5.1.6. Let us check the continuity of the mapping given by (1.20) (p. 15).

$$\sup_{0 \leq t \leq 1} \left| \int_0^t K(t,s)\phi(s)ds \right| \leq \sup_{0 \leq t \leq 1} \int_0^t |K(t,s)| |\phi(s)| \, ds$$

$$\leq \left\{ \int_0^1 \sup_{0 \leq t,s \leq 1} |K(t,s)| \, ds \right\} \cdot \sup_{0 \leq t \leq 1} |\phi(t)|.$$

This exhibits the continuity of (1.20). The other cases are omitted.

Exercise 5.4.2. $(A^*y, x) = \overline{(x, A^*y)} = \overline{(Ax, y)} = (y, Ax)$.

Problems

1. Recall the definition (2.26) of the standard norm of a quotient space. Since $\|\pi(x)\| = \|[x]\| = \inf_{m \in M} \|x + m\| \leq \|x + 0\| = \|x\|$, $\|\pi\| \leq 1$ follows. On the other hand, by the definition of inf (cf. Lemma A.2.1, p. 237), for every $\epsilon > 0$ there exists $m \in M$ such that $\|x + m\| - \epsilon < \|[x]\|$. Since $[x + m] = [x]$, we have $\|x + m\| - \epsilon < \|[x]\| = \|\pi(x + m)\|$. That is, there exists $x + m$ having norm $\|x + m\|$ sufficiently close (within ϵ) to $\|\pi(x + m)\|$. Since ϵ is arbitrary, $\|\pi\| \geq 1$, and hence $\|\pi\| = 1$.

2.

$$(I + B(I - AB)^{-1}A)(I - BA) = I - BA + B(I - AB)^{-1}A(I - BA)$$
$$= I - BA + B(I - AB)^{-1}(I - AB)A$$
$$= I - BA + BA = I.$$

$(I - BA)(I + B(I - AB)^{-1}A) = I$ follows similarly.

3. As indicated by the hint, and hence omitted.

4. Since $(AB)^* = B^*A^*$, $I = I^* = (AA^{-1})^* = (A^{-1})^*A^*$. This implies $(A^{-1})^* = (A^*)^{-1}$.

5. Let $\lambda \in \rho(A)$. Then $(\lambda I - A)^{-1}$ exists. Clearly $(\lambda I)^* = \bar{\lambda}I$, and then $((\lambda I - A)^{-1})^* = ((\lambda I - A)^*)^{-1} = (\bar{\lambda}I - A^*)^{-1}$ by the problem above. This implies $\bar{\lambda} \in \rho(A^*)$, and hence $\sigma(A^*) \subset \overline{\sigma(A)}$. (Here $\overline{\sigma(A)}$ denotes $\{\bar{\lambda} : \lambda \in \sigma(A)\}$.) By symmetry (note $(A^*)^* = A$), the reverse inclusion holds, and hence $\sigma(A^*) = \overline{\sigma(A)}$.

6. Take any $\lambda \neq 0$. For each $x \in \ell^2$, consider the equation $(\lambda I - S)y = x$. Writing this down componentwise, we obtain $\lambda y_1 = x_1$, $\lambda y_n - y_{n-1} = x_n$, $n = 2, 3, \ldots$. $\lambda \neq 0$, and this can be inductively solved, i.e., $y = (\lambda I - S)^{-1}x$. Note however that y does not necessarily belong to ℓ^2, nor is $(\lambda I - S)^{-1}$ guaranteed to be bounded. Observe that the relationship between y and x can be written as

$$\begin{bmatrix} y_1 \\ y_2 \\ y_3 \\ \vdots \end{bmatrix} = \begin{bmatrix} 1/\lambda & 0 & 0 & \cdots \\ 1/\lambda^2 & 1/\lambda & 0 & \cdots \\ 1/\lambda^3 & 1/\lambda^2 & 1/\lambda & 0 \\ \vdots & \vdots & \ddots & \ddots \end{bmatrix} \begin{bmatrix} x_1 \\ x_2 \\ x_3 \\ \vdots \end{bmatrix}.$$

Denote the right-hand matrix by Λ. When $|\lambda| > 1$, we have

$$\|\Lambda x\| \leq \|x/\lambda\| + \|x/\lambda^2\| + \cdots = \frac{1}{|\lambda|} \cdot \frac{1}{1 - 1/|\lambda|} \|x\|.$$

Hence $\|\Lambda\| < \infty$. On the other hand, when $|\lambda| \leq 1$, setting $x_1 = 1, x_2 = x_3 = \cdots = 0$ yields $\Lambda x = [1/\lambda, 1/\lambda^2, \ldots, 1/\lambda^n, \ldots]^T \notin \ell^2$, and Λ turns out to be unbounded. Hence the (continuous) resolvent $(\lambda I - S)^{-1}$ exists when and only when $|\lambda| > 1$, so we must have $\sigma(S) = \{\lambda : |\lambda| \leq 1\}$. Also, as already noted in Example 5.3.5, this operator S does not possess an eigenvalue.

7. A direct calculation shows that every λ with $|\lambda| < 1$ is an eigenvalue of S^*. Since a spectrum is always a closed set, we can conclude that $\sigma(S^*) = \{\lambda : |\lambda| \leq 1\}$ by taking the closure, and hence $\sigma(S^*) = \overline{\sigma(S)}$. Note, however, that if λ is an eigenvalue of S^*, $\bar{\lambda}$ is not necessarily an *eigenvalue* of S (although it is an element of the spectrum).

8. Suppose that $\lambda \neq 0$ is an element of $\rho(AB)$, i.e., $(\lambda I - AB)^{-1}$ exists and is bounded. By problem 2, $(I - BA/\lambda)^{-1}$ also exists. Hence $(\lambda I - BA)^{-1} = \lambda^{-1}(I - BA/\lambda)^{-1}$ exists, too. Hence $\rho(AB) \setminus \{0\} \subset \rho(BA) \setminus \{0\}$. The reverse inclusion holds also by symmetry, and $\rho(AB) \setminus \{0\} = \rho(BA) \setminus \{0\}$. Since $\sigma(T) = \mathbb{C} \setminus \rho(T)$ for every operator T, we have $\sigma(AB) \setminus \{0\} = \sigma(BA) \setminus \{0\}$.

9. Suppose S is compact, and let $\{x_n\}$ be a sequence weakly convergent to 0. By Theorem 5.7.3, Sx_n strongly converges to 0. By the continuity of T, TSx_n also converges strongly to 0, and hence again by Theorem 5.7.3, T is compact. A similar argument holds for the case when assuming T to be compact.

10. By continuity, T maps the unit ball of X to a bounded set. Since TX is finite-dimensional, the closure of this bounded set must be compact.

11. See the solution to the next problem 12.

12. (a) From the solution to problem 9, T^*T is compact. It is also clearly Hermitian. (b) $(T^*Tx, x) = (Tx, Tx) \geq 0$. (c) $\|Tx\|^2 = (Tx, Tx) = (T^*Tx, x) \leq \|T^*T\| \|x\|^2$. On the other hand, by Theorem 5.7.5, the norm of T^*T is attained as its maximal eigenvalue (note $T^*T \geq 0$). That is, there exists $v \neq 0$ such that $T^*Tv = \|T^*T\|v$. Hence $\|Tx\| / \|x\| \leq \sqrt{\lambda_{\max}(T^*T)}$ for every $x \neq 0$, and the equality is attained at $x = v$, where $\lambda_{\max}(A)$ denotes the maximal eigenvalue of operator A. This yields $\|T\| = \sqrt{\lambda_{\max}(T^*T)} = \sigma_{\max}(T)$.

13. (a) Solving (5.17) under the initial condition $x(0) = 0$, we obtain $Tu = x(h) = \int_0^h e^{A(h-t)} Bu(t)dt$. This corresponds to the operator in Example 5.5.2 with $K(t) = e^{A(h-t)} B$. Its adjoint operator T^* is given by $p \mapsto B^T e^{A^T(h-t)} p$. Differentiating this with an end point condition into consideration, we obtain adjoint differential equation (5.18).

(b) In view of the previous problem, the norm of T is given by the square root of the maximal eigenvalue of T^*T. On the other hand, according to problem 8, this is equal to the square root of the maximal eigenvalue of TT^*. Writing down TT^* explicitly, and changing the variable of integration, we obtain the matrix $\int_0^h e^{At} BB^T e^{A^T t} dt$.

C.6 Chapter 6

Exercise 6.6.1. For (6.31),

$$|h(t)| \leq \int_{-\infty}^{\infty} |f(\tau)||g(t - \tau)|d\tau \leq \left\{ \sup_{-\infty < t < \infty} |g(t)| \right\} \int_{-\infty}^{\infty} |f(t)|dt.$$

To see (6.32), change the order of integration by Fubini's theorem to obtain

$$\int |h(t)|dt = \int \left| \int f(t-\tau)g(\tau)d\tau \right| dt$$
$$\leq \int \int |f(t-\tau)g(\tau)|d\tau dt$$
$$= \int |g(\tau)|d\tau \int |f(t-\tau)|dt$$
$$= \|f\|_{L^1}\|g\|_{L^1}.$$

Exercise 6.6.2. First note that $\psi(\xi)$ is itself a continuous function. This follows from i) $\phi(\xi+\eta) \to \phi(a+\eta)$ in $\mathcal{D}$ as $\xi \to a$, and ii) continuity of distributions.

Directly applying the definition of a derivative, we have

$$\lim_{h\to 0}\frac{1}{h}\big(\psi(\xi+h)-\psi(\xi)\big) = \lim_{h\to 0}\langle S_{(\eta)},\frac{1}{h}\big(\phi(\xi+h+\eta)-\phi(\xi+\eta)\big)\rangle.$$

Since $(\phi(\xi+h+\eta)-\phi(\xi+\eta))/h \to \frac{\partial}{\partial\xi}\phi(\xi+\eta)$ as $h \to 0$ in $\mathcal{D}$, and the action of distributions is continuous, the right-hand side becomes $\langle S_{(\eta)}, \frac{\partial}{\partial\xi}\phi(\xi+\eta)\rangle$. Since $\partial\phi(\xi+\eta)/\partial\xi$ satisfies the same hypotheses as on ψ, we obtain the conclusion by repeating the same argument.

Problems

1.
$$\left\langle \frac{1}{h}\{\delta_{-h}-\delta\},\phi\right\rangle = \frac{1}{h}(\phi(-h)-\phi(0)) \to -\phi'(0) = \langle\delta',\phi\rangle.$$

2. First,

$$\sum_{k=-n}^{n} e^{ikt} = 1+2\cos t + \cdots + 2\cos nt$$

follows readily from $e^{ikt} + e^{-ikt} = 2\cos kt$. On the other hand, since

$$\sum_{k=-n}^{n} e^{ikt} = 1 + \frac{e^{it}(1-e^{int})}{1-e^{it}} + \frac{e^{-it}(1-e^{-int})}{1-e^{-it}}$$
$$= \frac{e^{it/2}-e^{-it/2}-e^{it/2}+e^{i(n+1/2)t}+e^{-it/2}-e^{-i(n+1/2)t}}{e^{it/2}-e^{-it/2}}$$
$$= \frac{\sin(n+1/2)t}{\sin(t/2)},$$

this agrees with (6.24).

3. For simplicity, we prove only the case $n = 1$. In this case, $e_0 = x_0$, and (6.59) becomes $x(t) = \int_0^t E(t-\tau)f(\tau)d\tau + x_0 E(t)$. Clearly $x(0) = x_0$, and the initial condition is satisfied. Since $x = HE*Hf + x(0)E$, $Dx = (D\delta*HE)*Hf + x(0)DE = \delta*Hf = Hf$, and hence $Dx = f$ is satisfied for $t > 0$.

4. a) Solve the differential equation $(d/dt - \lambda)x = 0$ under the initial condition $x(0) = 1$, and take the product of the solution $E(t)$ and the Heaviside function $H(t)$. Then in view of

section 6.8 (p. 132) this product HE gives the inverse of $\delta' - \lambda\delta$ with respect to convolution (p. 133). Thus $H(t)e^{\lambda t}$ is the desired inverse. b) Similarly, $(1/\omega)H(t)\sin\omega t$ gives the inverse of $\delta'' + \omega^2\delta$.

5. $(A_1 * A_2) * (A_1^{-1} * A_2^{-1}) = (A_1 * A_2) * (A_2^{-1} * A_1^{-1}) = A_1 * \delta * A_1^{-1} = \delta.$

C.7 Chapter 7

Exercise 7.2.2. If T is an L^2 function, its Fourier series becomes $\sum(1/2\pi)(T, e^{int})e^{int}$, and hence agrees with the usual definition.

Exercise 7.4.3. For simplicity, we consider this only for $t > 0$. (To extend the argument to $(-\infty, \infty)$, regularize this distribution following subsection 6.6.3.) It is easy to see that the function $e^{-t/2}$ is rapidly decreasing as $t \to \infty$. If e^t belonged to $\mathcal{S}'$, its action $\int e^t e^{-t/2} dt$ on $e^{-t/2}$ must be finite, but this is clearly a contradiction.

Problems

1. Compute the Fourier coefficients c_n to obtain $c_0 = 0$, $c_n = -i/n$, $n \neq 0$. Hence we have $f(t) = \sum_{n \neq 0}(-i/n)e^{int} = 2 \cdot \sum_{n=1}^{\infty}(1/n)\sin nt$.

2. Differentiate termwise the Fourier expansion obtained above. This yields $\sum_{n \neq 0} e^{int}$. Let us differentiate the original function $f(t)$ (its periodic extension to be precise) in the sense of distributions. Its derivative in the ordinary sense is -1 except at $\{n\pi : n = 0, \pm 1, \pm 2, \ldots\}$; at these points f assumes a jump of size 2π. Hence by Proposition 6.3.3 on page 118, its derivative in the sense of distribution becomes $(\sum_{n=-\infty}^{\infty} 2\pi\delta_{2\pi n}) - 1$. Hence

$$\left(\sum_{n=-\infty}^{\infty} 2\pi\delta_{2\pi n}\right) - 1 = \sum_{n \neq 0} e^{int}.$$

Moving -1 to the right, we obtain

$$\sum_{n=-\infty}^{\infty} 2\pi\delta_{2\pi n} = \sum_{n=-\infty}^{\infty} e^{int}.$$

This is nothing but Poisson's summation formula (7.20) (p. 149).

C.8 Chapter 8

Problems

1. By Proposition 6.3.3, we have

$$\frac{d}{dt}(H(t)\cos\omega t) = -H(t)\omega\sin\omega t + (H(0)\cos(0))\delta = -H(t)\omega\sin\omega t + \delta.$$

Then Corollary 8.1.3 (p. 168) yields

$$\mathcal{L}[H(t)\sin\omega t] = -\frac{1}{\omega}\left(s\frac{s}{s^2 + \omega^2} - 1\right) = \frac{\omega}{s^2 + \omega^2}.$$

2. The solution is given by $e^{At}x_0$. To calculate e^{At}, it is convenient to compute $(sI - A)^{-1}$ and then execute the inverse Laplace transformation. An easy calculation yields

$$(sI - A)^{-1} = \begin{bmatrix} \dfrac{s+3}{(s+1)(s+2)} & \dfrac{1}{(s+1)(s+2)} \\ \dfrac{-2}{(s+1)(s+2)} & \dfrac{s}{(s+1)(s+2)} \end{bmatrix},$$

and hence

$$e^{At} = \begin{bmatrix} 2e^{-t} - e^{-2t} & e^{-t} - e^{-2t} \\ -2e^{-t} + 2e^{-2t} & -e^{-t} + 2e^{-2t} \end{bmatrix}.$$

C.9 Appendix

A.5.2. If M were unbounded, then for every n there would exist $x_n \in M$ such that $\|x_n\| > n$. Suppose that there exists a convergent subsequence of $\{x_n\}$. Denote it also by $\{x_n\}$, and suppose $x_n \to x$. Then

$$\|x_n\| \le \|x_n - x\| + \|x\|.$$

Since the first term on the right approaches 0 as $n \to \infty$, the right-hand side is clearly bounded. On the other hand, $\{x_n\}$ can never be bounded by its choice. This is a contradiction.

Appendix D

Bibliographical Notes

There are many good textbooks on linear algebra. I did not consult a particular textbook, however. I attempted to draw readers' attention to some fundamental facts, concepts, and structures which often fail to be covered in standard (first) courses in linear algebra, particularly for engineers.

For the treatments of normed linear spaces and Hilbert spaces, I have benefited from the classical books by Luenberger [31], Young [67], and Mizohata [38] (available only in Japanese, however). Young [67] is compact, but nicely covers many materials, ranging from basics in Hilbert space to Hardy spaces and some introduction to H^∞ control theory. Luenberger [31] gives an extensive treatment on calculus of variations and optimization methods that this book does not cover as much; it also gives many concrete examples which are very useful when studied carefully. However, the contents are rather abundant and may be somewhat difficult to study closely in every detail. Rudin [48] covers many materials in functional analysis, ranging from Schwartz distributions to normed rings. Yosida [66] gives an extensive account on almost all aspects of functional analysis and serves as a valuable reference.

For Schwartz distributions, the basic references are [52] and [53], although the latter is more advanced and intended for mathematicians. The former gives a fairly accessible account on the subject; I benefited very much from its treatment on convolution and Fourier analysis. I have also occasionally referred to Bremermann [2], Donoghue [7], Treves [58], and Zemanian [72].

Relevant references for Fourier analysis are Hoffman [21], Katznelson [28], Papoulis [43], Walter [61], Yoshida–Kato [65], and Young [67]. Walter [61] gives a further treatment of wavelets; for wavelets, the reader is of course referred to [5, 32].

We could give only an elementary treatment of distributions, particularly in connection with topological vector spaces. But they will become indispensable for a more advanced study. The reader is referred to Köthe [29], Schaefer [51], Treves [58], and Yamanaka [64].

For real analysis, including Lebesgue integration, I have benefited from Hewitt and Stromberg [20], Ito [22], and Mizohata [37].

For complex analysis, there are many good books; from time to time, I have consulted Conway [3] and Rudin [47], but the reader can choose one depending on his or her taste.

For Hardy spaces, Hoffman [21] is a classic; but Nehari's theorem and Sarason's theorem are not treated. For the former I have consulted Young [67, 68], and for the latter the original paper [49] by Sarason and Nikol'skiĭ [42]. The proof given here relies largely on Nikol'skiĭ. I have also benefited from Tannenbaum[57], as well as [49] for the Nevanlinna–Pick interpolation theorem. See also [11] and [17] for other topics in Hardy spaces.

General topology theory can be found in Dugundji [10] and Matsusaka [33]. I have also referred to [29, 51, 58, 64] for the Hahn–Banach theorem and other topics in topological vector spaces.

A very deep and fruitful application of Hardy spaces and the generalized interpolation is found in H^∞ control theory. Chapter 10 gave just a short introduction for this modern control theory. The reader can further study H^∞ control theory through more advanced textbooks such as Francis [14], Doyle, Francis, and Tannenbaum [8], and Zhou, Doyle, and Glover [73]. The recently reprinted book [60] covers many advanced topics not treated in this book.

I have retained in the list the original references in the Japanese version. When necessary, I have supplemented them with suitable English references.

Bibliography

[1] R. M. BRACEWELL, *The Fourier Transform and Its Applications*, McGraw-Hill, New York, 1965.

[2] H. BREMERMANN, *Distributions, Complex Variables, and Fourier Transforms*, Addison-Wesley, Reading, MA, 1965.

[3] J. B. CONWAY, *Functions of One Complex Variable, I*, Springer, New York, 1975.

[4] R. F. CURTAIN AND H. J. ZWART, *An Introduction to Infinite-Dimensional Linear Systems Theory*, Springer, New York, 1995.

[5] I. DAUBECHIES, *Ten Lectures on Wavelets*, SIAM, Philadelphia, 1992.

[6] J. W. R. DEDEKIND, *Essays on the Theory of Numbers*, Dover, New York, 1901.

[7] W. F. DONOGHUE, *Distributions and Fourier Transforms*, Academic Press, New York, 1969.

[8] J. C. DOYLE, B. A. FRANCIS, AND A. R. TANNENBAUM, *Feedback Control Theory*, MacMillan, New York, 1992.

[9] J. C. DOYLE, K. GLOVER, P. P. KHARGONEKAR, AND B. A. FRANCIS, State-space solutions to standard $\mathcal{H}_\infty$ and $\mathcal{H}_2$ control problems, *IEEE Trans. Autom. Contr.*, AC-34 (1989), pp. 831–847.

[10] J. DUGUNDJI, *Topology*, Allyn and Bacon, Boston, 1973.

[11] W. L. DUREN, *Theory of H^p Spaces*, Academic Press, New York, 1970.

[12] N. J. FLIEGE, *Multirate Digital Signal Processing*, John Wiley, New York, 1994.

[13] C. FOIAS, H. ÖZBAY, AND A. R. TANNENBAUM, *Robust Control of Infinite Dimensional Systems, Lecture Notes in Control and Information Sciences 209*, Springer, New York, 1996.

[14] B. A. FRANCIS, *A Course in H_∞ Control Theory, Lecture Notes in Control and Information Science 88*, Springer, New York, 1987.

[15] B. A. FRANCIS, J. W. HELTON, AND G. ZAMES, H_∞-optimal feedback controllers for linear multivariable, *IEEE Trans. Autom. Contr.*, AC-29 (1984), pp. 888–900.

[16] P. A. FUHRMANN, On the Hamiltonian structure in the computation of singular values for a class of Hankel operators, in H^∞-Control Theory, Lecture Notes in Mathematics 1496, Springer, New York, 1991, pp. 250–276.

[17] J. B. GARNETT, Bounded Analytic Functions, Academic Press, New York, 1981.

[18] G. H. HARDY, A Mathematician's Apology, Cambridge University Press, Cambridge, UK, 1940.

[19] J. HESPANHA, Linear Systems Theory, Princeton University Press, Princeton, NJ, 2009.

[20] E. HEWITT AND K. STROMBERG, Real and Abstract Analysis, Springer, New York, 1975.

[21] K. HOFFMAN, Banach Spaces of Analytic Functions, Prentice–Hall, Englewood Cliffs, NJ, 1962.

[22] S. ITO, Introduction to Lebesgue Integration, ShoKa-Bo, 1968 (in Japanese).

[23] T. KAILATH, Linear Systems, Prentice–Hall, Englewood Cliffs, NJ, 1980.

[24] R. E. KALMAN, A new approach to recursive filtering and prediction problems, Trans. ASME, 82D (1960), pp. 35–45.

[25] R. E. KALMAN, On the general theory of control systems, in Proc. 1st IFAC World Congress, 1960, pp. 481–492.

[26] A. KANEKO, Introduction to Hyperfunctions, Kluwer Academic Publishers, Dordrecht, The Netherlands, 1988.

[27] K. KASHIMA AND Y. YAMAMOTO, Finite rank criteria for H^∞ control of infinite-dimensional systems, IEEE Trans. Autom. Contr., AC-53 (2008), pp. 881–893.

[28] Y. KATZNELSON, An Introduction to Harmonic Analysis, Dover, New York, 1976.

[29] G. KÖTHE, Topological Vector Spaces I, Springer, New York, 1969.

[30] T. A. LYPCHUK, M. C. SMITH, AND A. TANNENBAUM, Weighted sensitivity minimization: General plants in H^∞ and rational weights, Linear Algebra Appl., 8 (1988), pp. 71–90.

[31] D. G. LUENBERGER, Optimization by Vector Space Method, John Wiley, New York, 1969.

[32] S. MALLAT, A Wavelet Tour of Signal Processing, 3rd ed., Academic Press, New York, 2008.

[33] K. MATSUSAKA, Introduction to Sets and Topology, Iwanami, 1966 (in Japanese).

[34] J. C. MAXWELL, On governors, Proc. Royal Soc. London, 16 (1868), pp. 270–283.

[35] R. E. MEGGINSON, An Introduction to Banach Space Theory, Springer, New York, 2008.

[36] J. MIKUSIŃSKI, *Operational Calculus*, Pergamon, Oxford, UK, 1959.

[37] S. MIZOHATA, *Lebesgue Integration*, Iwanami, 1966 (in Japanese).

[38] S. MIZOHATA, *Introduction to Integral Equations*, Asakura, 1968 (in Japanese).

[39] S. MIZOHATA, *Partial Differential Equations*, Iwanami, 1965 (in Japanese).

[40] B. SZ. NAGY AND C. FOIAŞ, *Harmonic Analysis of Operators on Hilbert Space*, North–Holland, Amsterdam, 1970.

[41] Z. NEHARI, On bounded bilinear forms, *Ann. Math.*, 65 (1957), pp. 153–162.

[42] N. K. NIKOL'SKIĬ, *Treatise on the Shift Operator*, Springer, New York, 1986.

[43] A. PAPOULIS, *The Fourier Integral and Its Applications*, McGraw–Hill, New York, 1962.

[44] J. W. POLDERMAN AND J. C. WILLEMS, *Introduction to Mathematical Systems Theory*, Springer, New York, 1991.

[45] L. S. PONTRYAGIN, *Ordinary Differential Equations*, Kyoritsu, 1965 (in Japanese).

[46] A. ROBINSON, *Non-Standard Analysis*, North–Holland, Amsterdam, 1965.

[47] W. RUDIN, *Real and Complex Analysis*, McGraw–Hill, New York, 1966.

[48] W. RUDIN, *Functional Analysis*, McGraw–Hill, New York, 1973.

[49] D. SARASON, Generalized interpolation in H^∞, *Trans. Amer. Math. Soc.*, 127 (1967), pp. 179–203.

[50] M. SATO, Theory of hyperfunctions I, II, *J. Fac. Sci. Univ. Tokyo, Sect.* I, 8, (1959–1960), pp. 139–193, 387–437.

[51] H. H. SCHAEFER, *Topological Vector Spaces*, Springer, New York, 1971.

[52] L. SCHWARTZ, *Méthodes Mathématiques pour les Sciences Physiques*, Hermann, Paris, 1961.

[53] L. SCHWARTZ, *Théorie des Distributions*, Hermann, Paris, 1966.

[54] C. E. SHANNON, Communication in the presence of noise, *Proc. IRE*, 37 (1949), pp. 10–21.

[55] M. C. SMITH, Singular values and vectors of a class of Hankel operators, *Syst. Control Lett.*, 12 (1989), pp. 301–308.

[56] E. D. SONTAG, *Mathematical Control Theory*, Springer, New York, 1998.

[57] A. TANNENBAUM, Frequency domain methods for the H^∞-optimization of distributed systems, in *Proc. 10th Int. Conf. on Analysis and Optimization of Systems: State and Frequency Domain Approach for Infinite-Dimensional Systems*, Springer Lecture Notes in Control and Information Sciences 185, 1993, pp. 242–278.

[58] F. TREVES, *Topological Vector Spaces, Distributions and Kernels*, Academic Press, New York, 1967.

[59] P. P. VAIDYANATHAN, *Multirate Systems and Filter Banks*, Prentice–Hall, Englewood Cliffs, NJ, 1993.

[60] M. VIDYASAGAR, *Control System Synthesis—A Factorization Approach*, MIT Press, Cambridge, MA, 1985; reprinted by Morgan & Claypool Publications in 2011.

[61] G. G. WALTER, *Wavelets and Other Orthogonal Systems with Applications*, CRC Press, Boca Raton, FL, 1994.

[62] Y. YAMAMOTO, K. HIRATA, AND A. TANNENBAUM, Some remarks on Hamiltonians and the infinite-dimensional one block H^∞ problem, *Syst. Control Lett.*, 29 (1996), pp. 111–117.

[63] Y. YAMAMOTO, M. NAGAHARA, AND P. P. KHARGONEKAR, Signal reconstruction via H^∞ sampled-data control theory—beyond the Shannon paradigm, *IEEE Trans. Signal Processing*, 60 (2012), pp. 613–625.

[64] K. YAMANAKA, *Topological Vector Spaces and Generalized Functions*, Kyoritsu, 1966 (in Japanese).

[65] K. YOSIDA AND T. KATO, *Applied Mathematics–I*, ShoKa-Bo, 1961 (in Japanese).

[66] K. YOSIDA, *Functional Analysis*, 5th ed., Springer, New York, 1978.

[67] N. YOUNG, *An Introduction to Hilbert Space*, Cambridge University Press, Cambridge, UK, 1988.

[68] N. YOUNG, The Nehari problem and optimal Hankel norm approximation, *Proc. 10th Int. Conf. on Analysis and Optimization of Systems: State and Frequency Domain Approach for Infinite-Dimensional Systems*, Springer Lecture Notes in Control and Information Sciences 185, Springer, New York, 1993, pp. 199–221.

[69] G. ZAMES, Feedback and optimal sensitivity: model reference transformations, multiplicative seminorms and approximate inverses, *IEEE Trans. Autom. Control*, AC-26 (1981), pp. 301–320.

[70] G. ZAMES AND B. A. FRANCIS, Feedback, minimax sensitivity and optimal robustness, *IEEE Trans. Autom. Control*, AC-28 (1983), pp. 585–601.

[71] A. I. ZAYED, *Advances in Shannon's Sampling Theory*, CRC Press, Boca Raton, FL, 1996.

[72] A. H. ZEMANIAN, *Distribution Theory and Transform Analysis*, Dover, New York, 1987.

[73] K. ZHOU, J. C. DOYLE, AND K. GLOVER, *Robust and Optimal Control*, Prentice–Hall, Englewood Cliffs, NJ, 1996.

[74] K. ZHOU AND P. P. KHARGONEKAR, On the weighted sensitivity minimization problem for delay systems, *Syst. Control Lett.*, 8 (1987), pp. 307–312.

Index